资助项目：
山东省重点研发计划项目(节水农业和水肥一体化关键技术2018GNC2306)
山东省高等学校生化化学与分子生物学重点实验室
潍坊市科学技术发展计划项目（2017GX004）
潍坊学院科研创新团队项目（2019-10）
山东省本科教改项目（M2018X303）
山东省企业培训与职工教育重点课题（2019-302）

现代农业水肥一体化新型实用技术研究

（汉英对照）

◎ 曹慧 赵静 著　　◎ 陈志章 译

中国农业科学技术出版社

图书在版编目（CIP）数据

现代农业水肥一体化新型实用技术研究：汉英对照／曹慧，赵静著；陈志章译.—北京：中国农业科学技术出版社，2020.1

ISBN 978-7-5116-4589-0

Ⅰ.①现… Ⅱ.①曹…②赵…③陈… Ⅲ.①肥水管理-研究-汉、英 Ⅳ.①S365

中国版本图书馆 CIP 数据核字（2020）第 021817 号

责任编辑　李冠桥
责任校对　李向荣

出 版 者	中国农业科学技术出版社
	北京市中关村南大街 12 号　邮编：100081
电　　话	（010）82109705（编辑室）　（010）82109702（发行部）
	（010）82109709（读者服务部）
传　　真	（010）82106625
网　　址	http：//www.castp.cn
经 销 者	各地新华书店
印 刷 者	北京建宏印刷有限公司
开　　本	710mm×1 000mm　1/16
印　　张	24.125
字　　数	430 千字
版　　次	2020 年 1 月第 1 版　2020 年 1 月第 1 次印刷
定　　价	98.00 元

━━━━━◆ 版权所有·翻印必究 ◆━━━━━

前　言

　　受人类活动的巨大影响，全球正经历着人类历史上前所未有的环境、气候变化，淡水资源匮乏、土壤盐碱化等问题日益突出。我国淡水资源人均占有率低，尤其是农田灌溉用水量已经成为制约我国农业持续健康发展的重要因素。化肥、农药的大量使用在造成资源浪费的同时，也引起了一系列的环境问题，再加上近些年劳动力紧缺，劳动力成本大幅度提高，使得节水、节肥、省工的水肥一体化技术在我国迅速推广开来。

　　水肥一体化技术也称灌溉施肥技术，通俗地讲，就是将肥料溶解于灌溉水中，通过管道在浇水的同时施肥，将水和肥均匀、准确地输送到作物根部土壤。该技术是发展循环农业的一项关键实用技术，近些年在设施园艺作物生产，山地果树、茶树等栽培上应用十分广泛。国家农业农村部对这项节水、节肥灌溉技术也十分支持，灌溉施肥设备的推广面积越来越大。但是由于基层技术人员和广大农民缺乏对水肥一体化技术的认识，不能有效掌握使用灌溉施肥设备的核心技术，导致大量设备只被简单用来灌溉甚至被废弃，没能充分发挥节水、节肥、省工，以及提高作物产量和品质的作用，造成了资源和资金浪费。因此，我们联合长期从事设施园艺作物生产指导专家、水肥一体化设备生产研发技术人员共同出版以"生态循环农业实用技术系列丛书"为主题的著作。

　　本书共分六章：第一章主要讲了水肥一体化技术的概念、研究背景和应用前景；第二章主要讲水肥一体化技术需要的设备和系统；第三章叙述了如何选择肥料；第四章主要讲了该技术在农业中的应用；第五、第六章主要讲了水肥一体化技术在水果、蔬菜领域的应用。本书既可作为水肥一体化技术推广人员的指导用书，也可作为农民田间地头使用水肥一体化设备的参考用书，同时也可作为高等院校相关专业的教材。

　　由于水肥一体化技术比较复杂，我们掌握的理论知识和实际应用经验还不足，书中难免有疏漏和不当之处，恳请读者批评指正。同时，本书参考了相关领域专家的论著、文献等资料，在此表示感谢。

作　者

2019 年 12 月

Preface

Under the great influence of human activities, the world is experiencing unprecedented change of environment and climate, and problems are more and more serious such as lack of fresh water resources and soil salinization and so on. The low share of fresh water resources per capita in China, especially the irrigation water consumption of farmland, has become an important factor restricting the sustainable and healthy development of agriculture in China. The extensive use of chemical fertilizers and pesticides has not only caused a series of resources waste, but also caused a series of environmental problems. In addition, in recent years, the shortage of labor force has greatly increased the labor cost, so that the integrated water − fertilizer technology of saving water, saving fertilizer and saving labor has been rapidly popularized in our country.

The integrated water−fertilizer technology, also known as irrigation and fertilizer technology, generally speaking, that is, the fertilizer is dissolved in irrigation water, while ingratiating the water and fertilizer are evenly and accurately transported to the root soil of crops through the pipelines at the same time. This technology is a key practical technology for the development of circular agriculture. In recent years, it has been widely used in the horticultural crop production, and the cultivation of mountain fruit trees, tea trees and so on. The Ministry of Agriculture and Rural Affairs also supports this water − saving and fertilizer − saving irrigation technology, and the popularization area of irrigation and fertilizer equipment is increasing. However, due to the lack of understanding of the integrated technology among the grass−roots technical personnel and the majority of farmers, it is impossible to effectively master the use of irrigation and fertilizer. The core technology, resulting in a large number of equipment is simply used for irrigation or even abandoned, failed to give full play to the role of water saving, fertilizer saving, labor saving, as well as improving crop yield and quality, resulting in a waste of resources and funds. Therefore, we cooperate jointly for a long time with the experts guiding the production of horticulture crops of facilities, and the production and R&D technical personnel of integrated water-fertilizer equipment to publish the books with "*Ecological Circulating Agriculture Utilities Series*" as the theme.

This book is divided into 6 chapters: the first chapter mainly talks about the concept, research background and application prospect of the integrated water-fertilizer technology; the second chapter mainly talks about the equipment and system needed by the integrated water-fertilizer technology; the third chapter describes how to select fertilizer; the fourth chapter mainly talks about the application of the integrated water-fertilizer technology in agriculture; the fifth and sixth chapters mainly talk about the integrated water-fertilizer technology in the field of fruits and vegetables. This book can not only be used as a guide for the extension personnel of the integrated water-fertilizer technology, but also as a reference book for farmers to use the integrated water-fertilizer equipment in the field, and it can also be used as a teaching material for related specialties in colleges and universities.

Due to the complexity ofthe integrated water-fertilizer technology, our theoretical knowledge and practical application experience are still insufficient, and there are inevitably omissions and inaccuracies in the book. Readers are welcomed to criticize and correct. At the same time, this book refers to the relevant field experts' works, literature and other materials, and here we sincerely express our gratitude.

Author
December 2019

目 录

第一章 绪论 …………………………………………………………… (1)
  第一节 水肥一体化技术概述 ……………………………………… (1)
  第二节 水肥一体化技术研究进展 ………………………………… (4)
  第三节 水肥一体化技术的应用前景 ……………………………… (8)

第二章 水肥一体化技术的主要设备和系统 ………………………… (10)
  第一节 水肥一体化技术的首部枢纽 ……………………………… (10)
  第二节 水肥一体化技术系统中的施肥设备 ……………………… (28)
  第三节 水肥一体化技术的输水系统——喷灌 …………………… (38)
  第四节 水肥一体化技术的输水系统——微灌 …………………… (49)

第三章 适合水肥一体化技术的肥料的选择 ………………………… (57)
  第一节 作物吸收养分原理 ………………………………………… (57)
  第二节 肥料的选择 ………………………………………………… (63)
  第三节 养分管理 …………………………………………………… (66)

第四章 水肥一体化技术在农业生产中的应用 ……………………… (73)
  第一节 水肥一体化的规划设计 …………………………………… (73)
  第二节 水肥一体化技术的设备安装与调试 ……………………… (78)
  第三节 水肥一体化系统操作与维护 ……………………………… (85)

第五章 水果水肥一体化技术的应用 ………………………………… (95)
  第一节 苹果水肥一体化技术应用 ………………………………… (95)
  第二节 梨树水肥一体化技术应用 ………………………………… (106)
  第三节 桃树水肥一体化技术应用 ………………………………… (115)
  第四节 葡萄水肥一体化技术应用 ………………………………… (119)

第六章 蔬菜水肥一体化技术的应用 ………………………………… (124)
  第一节 番茄水肥一体化技术应用 ………………………………… (124)
  第二节 辣椒水肥一体化技术应用 ………………………………… (127)

第三节　黄瓜水肥一体化技术应用 …………………………………（130）
第四节　茄子水肥一体化技术应用 …………………………………（140）
第五节　西葫芦水肥一体化技术应用 ………………………………（143）
第六节　西瓜水肥一体化技术应用 …………………………………（146）
第七节　白菜水肥一体化技术应用 …………………………………（148）
第八节　莴苣水肥一体化技术应用 …………………………………（151）
参考文献 ……………………………………………………………（154）

Contents

Chapter Ⅰ Introduction ………………………………………… (157)
  Section Ⅰ Overview of the Integrated Water-Fertilizer Technology ……… (157)
  Section Ⅱ Research Progress of the Integrated Water-Fertilizer
            Technology ………………………………………………… (161)
  Section Ⅲ The Applicable Prospect of the Integrated Water-Fertilizer
            Technology ………………………………………………… (168)

**Chapter Ⅱ The Main Equipment and System of the Integrated
          Water-Fertilizer Technology** ……………………………… (170)
  Section Ⅰ The Head Hub of the Integrated Water-Fertilizer
            Technology ………………………………………………… (170)
  Section Ⅱ The Fertilization Equipment in the Integrated Water-Fertilizer
            Technology System ………………………………………… (193)
  Section Ⅲ Water Transfer System of the Integrated Water-Fertilizer
            Technology-Sprinkler Irrigation ………………………… (205)
  Section Ⅳ Water Transfer System of The Integrated Water-Fertilizer
            Technology-Micro-Irrigation …………………………… (220)

**Chapter Ⅲ Selection of Fertilizer Suitable for the Integrated
          Water-Fertilizer Technology** ……………………………… (230)
  Section Ⅰ The Principle of Nutrient Absorption by Crops ……………… (230)
  Section Ⅱ Selection of Fertilizers ……………………………………… (238)
  Section Ⅲ Nutrient Management ………………………………………… (243)

**Chapter Ⅳ Application of the Integrated Water-Fertilizer Technology
          to Agricultural Production** ………………………………… (252)
  Section Ⅰ Planning and Design of the Integrated Water-Fertilizer ……… (252)
  Section Ⅱ Equipment Installation and Adjustment of the Integrated
            Water-Fertilizer Technology ……………………………… (259)

Section Ⅲ	Operation and Maintenance of the Integrated Water-Fertilizer system	(270)
Chapter Ⅴ	**Application of the Integrated Water-Fertilizer Technology to Fruits**	**(284)**
Section Ⅰ	Application of the Integrated Water-Fertilizer Technology to Apples	(284)
Section Ⅱ	Application of the Integrated Water-Fertilizer Technology to Pear Trees	(301)
Section Ⅲ	Application of the Integrated Water-Fertilizer Technology to Peach Tree	(316)
Section Ⅳ	Application of the Integrated Water-Fertilizer Technology to Grapes	(322)
Chapter Ⅵ	**Application of the Integrated Water-Fertilizer Technology to Vegetables**	**(329)**
Section Ⅰ	Application of the Integrated Water-Fertilizer Technology to Tomato	(329)
Section Ⅱ	Application of the Integrated Water-Fertilizer Technology to Pepper	(334)
Section Ⅲ	Application of the Integrated Water-Fertilizer Technology to Cucumber	(339)
Section Ⅳ	Application of the Integrated Water-Fertilizer Technology to Eggplant	(355)
Section Ⅴ	Application of the Integrated Water-Fertilizer Technology to Summer Squash (Zucchini)	(360)
Section Ⅵ	Application of the Integrated Water-Fertilizer Technology to Watermelon	(364)
Section Ⅶ	Application of the Integrated Water-Fertilizer Technology to Chinese Cabbage	(368)
Section Ⅷ	Application of the Integrated Water-Fertilizer Technology to Lettuce	(372)

第一章 绪 论

水肥一体化技术是将灌溉与施肥融为一体的农业新技术。它是利用微灌系统，根据作物的需水、需肥规律和土壤水分、养分状况，将肥料和灌溉水一起适时、适量、准确地输送到作物根部土壤，供给作物吸收。水肥一体化的灌溉施肥体系投资（包括管路、施肥池、动力设备等）约为 1 000 元/亩，① 可使用 5 年左右，比常规施肥节省肥料 50%～70%，增产幅度可达 30%以上。同时，大大降低了因过量施肥而造成的水体污染问题。下面将详细描述水肥一体化技术的发展和应用。

第一节 水肥一体化技术概述

一、概念

水肥一体化是将灌溉与施肥融为一体、实现水肥同步控制的农业新技术，又称为"水肥耦""随水施肥""灌溉施肥"等。狭义地讲，就是把肥料溶解在灌溉水中，由灌溉管道带到田间每一株作物。广义地讲，就是水肥同时供应作物需要。水肥一体化是借助压力系统（或地形自然落差），根据不同土壤环境和养分含量状况、不同作物需肥特点和不同生长期需水需肥规律进行的不同需求设计，将可溶性固体或液体肥料，与灌溉水一起，通过可控管道系统供水、供肥。水肥相融后，通过管道和滴头形成滴灌，均匀、定时、定量地浸润作物根系生长发育区域，使主要根系生长的土壤始终保持疏松和适宜的水肥量。在中国 20 世纪 90 年代的田间种植浇地用水主要就是水渠灌溉，水渠灌溉最为简单，对肥料要求不高，但这种水渠灌溉不利于节水要求。然而，滴灌是根据作物需水、需肥量和根系分布进行最精确的供水、供肥，不受风力等外部条件限制；喷灌相对来说没有

---

① 亩为非法定计量单位，1 亩 = 1/15hm²。

滴灌施肥适应性广。故狭义的水肥一体化技术也称滴灌施肥。

近年来，随着全球气候变暖，干旱加剧，水资源不足已成为严重制约我国国民经济可持续发展的瓶颈。目前，我国水资源总量年约2.81万亿m^3，仅占世界水资源总量的6%，人均水资源占有量仅为2 800m^3，是世界人均水资源占有量的1/4。同时，我国水资源还存在分布不均的问题，淮河以北的广大北方地区拥有全国60%的耕地，却只有15%的水资源。农业是我国用水最多的产业，用水总量为4 000亿m^3，占总用水量的70%以上。其中，农田灌溉用水量为3 600亿~3 800亿m^3，占农业用水量的90%~95%。农业灌溉每年平均缺水300多亿m^3，平均每年受旱农田面积达214.67万hm^2，占全国耕地的1/5，旱灾农田面积860万hm^2，每年因旱灾造成粮食减产100亿~150亿kg；而且，我国在水资源短缺的情况下，水资源浪费问题却十分严重。我国传统灌溉方式是以渠道灌溉为主，传统的土渠输水渗漏损失极大，一般占输水量的50%~60%，一些土质较差的渠道输水损失高达70%以上。与传统灌溉方式相比，采用管道输水和渠道防渗输水可节水20%~30%，喷灌可节水50%，微灌可节水60%~70%，膜下滴灌可节水80%以上，地下滴灌可节水95%以上。同时，我国灌溉水利用率低，仅为43%左右，每立方米水粮食生产率为1kg左右；发达国家灌溉水利用率为70%~80%，每立方米水粮食生产率2kg以上。由于灌溉方式不合理和灌溉水利用率低，更加剧了水资源短缺的现状。根据国家水利部、中国工程院等部门的预测，我国农业用水必须维持零增长或负增长，才能保证我国用水安全和生态安全，而缓解水资源供需矛盾的重要途径之一就是发展节水灌溉。在节约灌溉用水的同时，还可以提高土地利用率，减轻劳动强度，提高农作物产量和品质。我国是肥料生产大国，同时也是消费大国，根据国际肥料工业协会数据和我国统计数据分析，2007年我国化肥使用量已占全球用量的35%左右，而且目前使用量仍以每年3.5%的速度递增。由于施肥技术不科学、肥料产品不合格等多方面原因导致肥料当季利用率低，我国肥料当季利用率氮肥为15%~35%、磷肥为10%~20%、钾肥为35%~50%，均低于日本、美国、英国、以色列等发达国家。肥料的大量和不合理施用导致部分土壤结构改变、肥力下降、重金属污染加剧、盐化碱化严重，同时也加剧了地表径流的水质污染，导致水体富营养化、地下水污染、农产品品质差等一系列危害。因此，减少化肥使用量、合理施肥、提高化肥利用率，已成为我国农业可持续发展和保障我国农产品安全的重要问题。

二、特点

传统灌溉方式有漫灌、沟灌、畦灌等，可将肥料混溶在灌溉水中，随灌溉水一同施入。漫灌是在田间不做任何沟埂，灌水时任其在地面漫流，借重力渗入土壤，是一种比较粗放的灌溉方式。该灌溉方法灌水均匀性差，水量浪费极大，肥料淋洗严重污染地下水。畦灌是用田埂将土地分割成一系列长方形小畦，灌水时，水进入畦田后沿畦长方向移动，借重力作用逐渐湿润土壤。畦灌操作简单，不需要外加灌水设施，但灌溉用水量大，水分浪费严重，容易形成漫灌，造成土壤板结，导致作物病害蔓延。沟灌是在作物行间开挖灌水沟，水从输水沟进入灌水沟借毛细管作用湿润土壤。与畦灌比较，沟灌不会破坏作物根部附近的土壤结构，对土壤团粒结构的破坏作用较轻，不会导致田面板结，可保持沟背土壤疏松，能减少土壤蒸发损失。

（一）提高灌溉水利用率

滴灌将水一滴一滴地滴进土壤，灌水时地面不出现径流，从浇地转向浇作物，减少了水分在作物株间的蒸发。同时，通过控制灌水量，土壤水深层渗漏很少，减少了无效的田间水量损失。另外，滴灌输水系统从水源引水开始，灌溉水就进入一个全程封闭的输水系统，经多级管道传输，将水送到作物根系附近，用水效率高，从而节省灌水量。

（二）提高肥料利用率

水肥被直接输送到作物根系最发达部位，可充分保证养分被作物根系快速吸收。对滴灌而言，由于湿润范围仅限于根系集中的区域，肥料利用率高，从而节省肥料。

（三）节省劳动力

传统灌溉施肥方法是每次施肥要挖穴或开浅沟，施肥后再灌水。利用水肥一体化技术，实现水肥同步管理，可节省大量劳动力。

（四）可方便、灵活、准确地控制施肥数量

根据作物需肥规律进行针对性施肥，做到缺什么补什么，缺多少补多少，实现精确施肥。

（五）有利于保护环境

不合理施肥，造成肥料浪费，大量施肥不但作物不能吸收利用，造成肥料的极大浪费，同时还会导致环境污染。水肥一体化技术通过控制灌溉深度，可避免将化肥淋洗至深层土壤，从而避免造成土壤和地下水的污染。而且，利用水肥一体化技术还能够有效地利用开发丘陵地、山地、沙石、轻度盐碱地等边缘土地。

（六）有利于应用微量元素

金属微量元素通常应用螯合态，价格较贵，通过滴灌系统可以做到精确供应，提高肥料利用率。

（七）水肥一体化技术的局限性

由于该项技术是设施施肥，前期一次性投资较大，同时对肥料的溶解度要求较高，所以快速大面积推广有一定的难度。

第二节　水肥一体化技术研究进展

一、水肥一体化技术的发展历史

水肥一体化技术是人类智慧的结晶，是生产力不断发展进步的产物，它的发展经历了很长的历史，水肥一体化技术的灵感起源是无土栽培技术。早在18世纪，英国科学家John Woodward利用土壤提取液配制了第一份水培营养液。后来水肥一体化技术经过了3个阶段的发展。

第一阶段：18世纪末到20世纪初是营养液栽培、无土栽培阶段，是水肥一体化技术的最前期的构想阶段。营养液栽培最初是指没有任何固定根系基质的水培，19世纪中期Wigmen和Pollsloff做了一项重要的实验，首次使用蒸馏水和无机盐成功培养了植物，这项实验直接验证了植物在生长过程中需要盐类物质。在这一时期最具有代表性的人物就是Van Liebig，因为他发现了植物生长中需要的C是来自于空气中的CO_2，而H和O则来源于NH_3、NO_3^-，还有一些植物必需的矿物质是由土壤环境为其提供的。Van Liebig的发现直接推翻了当时流行的腐殖质营养理论，并且初步建立了一种矿物质营养理论框架，他的学术理论至今仍应用于"现代营养耕种"，为现代农业的发展奠定了稳固的基础。

第二阶段：19世纪中期到20世纪中期无土栽培商业化生产，水肥一体化技术初步形成。第二次世界大战加速了无土栽培的发展，为了给美军提供大量的新鲜蔬菜，美国在各个军事基地建立了大型的无土栽培农场。无土栽培技术日臻成熟，并逐渐商业化。无土栽培的商业化生产开始于荷兰、意大利、英国、德国、法国、西班牙、以色列等国家。之后，墨西哥、科威特及中美洲、南美洲、撒哈拉沙漠等土地贫瘠、水资源稀少的地区也开始推广无土栽培技术。

1950年，发明了肥料与水混合后浇灌的方法，将肥料溶解在灌溉水中用于地面灌溉、漫灌和沟灌，这是水肥一体化技术的雏形。当时早期使用最多的就是氮肥，氨水和硝酸铵是氮肥中最常见的两种，由于氮肥在土壤中极易挥发的性

第一章 绪论

质,加上满地灌溉导致水的利用率很低,使得农作物生长中对氮肥的吸收率也很低。塑料管件和塑料容器的发展促进了水肥一体化技术的进一步发展,将具有合理配方的营养液通过塑料管道输送给植物。这时利用管道灌溉系统进行施肥,可以极大地提高肥料的利用率,但是通过管道进行浇灌会增加肥料的使用量,进一步促进了水泵和用于养分精细供应的肥料混合灌的设计与研发。

第三阶段:20世纪中期至今是水肥一体化技术快速发展的阶段。20世纪50年代,以色列内盖夫沙漠中哈特泽里姆基布兹的农民偶然发现水管渗漏处的农作物的生长状态要比其他地方的更好,后期做实验得到结论,滴渗灌溉可以有效减少水分的蒸发,提高了灌溉水的利用率,并且是控制水肥和农药最有效的方法。随着技术的成熟,以色列政府也加大了对滴渗灌溉技术的推广,并且在1964年成立了著名的耐特菲姆滴灌公司。以色列从落后农业国实现向现代工业国的迈进,主要得益于滴灌技术。与喷灌和沟灌相比,应用滴灌的番茄产量增加1倍,黄瓜产量增加2倍。以色列应用滴灌技术以来,全国农业用水量没有增加,农业产出却较之前翻了5倍。

耐特菲姆公司生产的第一代滴灌系统设备是用一个流量计量仪控制塑料管子中的单向水流,第二代产品是引入了高压设备控制水流,第三、第四代产品开始配合计算机使用。自20世纪60年代以来,以色列开始普及水肥一体化技术,全国43万hm^2耕地中大约有20万hm^2应用加压灌溉系统。由于管道和滴灌技术的成功,全国灌溉面积从16.5亿m^2增加到22亿~25亿m^2,耕地从16.5亿m^2增加到44亿m^2。据称以色列的滴灌技术已经发展到第六代。在温室大棚中种植的果树、花卉和温室作物,采用的都是水肥一体化施肥滴灌技术,而露天种植的蔬菜和农作物只有一小部分利用了水肥一体化技术,覆盖率并不是很高,能否利用水肥一体化技术种植要看土壤的性质和基肥施用。随着技术的发展,在微灌和微喷灌系统中也用到了水肥一体化技术,并且对农作物的生长也起到了明显的作用。随着喷灌系统由移动式改为固定式,水肥一体化技术也成功用到了喷灌系统中。20世纪80年代初期,自动推进机械灌溉系统中也开始推行水肥一体化技术,有效地改善了农作物的生长状况,促进了农业技术的发展。

水肥一体化技术不仅能够合理分配水分,还能保证农作物养分的充足,水肥一体化技术当中湿润的土壤容积只是耕种层的一小部分,特别是碰到沙质土壤更加稀少。在水肥一体化的初始阶段,利用灌溉系统给农作物施肥有两种方法:第一种是利用喷雾泵把肥料注入灌溉系统中;第二种是直接将水倒入到装有水和固体肥料的容器中,然后再引入到灌溉系统当中进行灌溉施肥。以上两种方法简单方便但其精确性不够，并且水肥灌溉也不均匀。20世纪70年代，化学技术的创新与发展生产了新形态的肥料——液体肥料，液体肥料的出现加快了水利驱动泵的研发。膜式泵是早期研发出的第一种水力驱动泵，先将肥料倒入一个敞口的大型容器中，搅拌均匀后利用膜式泵抽出将其注入灌溉系统中，这种泵产生的压力较大，大约是灌溉系统当中压力的2倍；第二种水力驱动泵就是活塞泵，利用活塞特定的物理运动将肥料吸取和注入灌溉系统中。这些肥料泵的应用实现了水肥同时供应。同时，低流量的文丘里施肥器也开始应用，利用这种施肥器肥料可均匀地溶解在灌溉水中，养分分布比较均匀，主要应用在苗圃和盆栽温室中。它的应用有效地解决了早期肥料泵在低流量下精确性差的问题。随着计算机技术的发展，对肥料的用量和流量逐渐实现了机械化设备自动控制，对肥料用量和流量的控制也越来越精确。

二、我国水肥一体化技术的发展状况

我国农业灌溉有着悠久的历史，但是大多采用大水漫灌和串畦淹灌的传统灌溉方法，水资源的利用率低，不仅浪费了大量的水资源，同时农作物的产量提高的也不明显。20世纪70年代，我国才将水肥一体化技术引入到种植领域。30多年的发展，随着微灌技术的推广和应用，水肥一体化技术也在不断提高，其发展大致分为以下3个阶段。

1974—1980年为第一阶段：在此期间，我国开始引进了灌溉设施，并且逐渐实现了国产设备的研发与制作，与此同时进行了大量的微灌实验，对相关的实验数据进行了统计。随着我国科技的发展，国内第一代成套滴灌设备于1980年成功研发生产。

1981—1996年为第二阶段：成功引进了国外先进的工艺生产技术，帮助国产设备逐步实现规模化生产。灌溉设备的规模化生产促进了微观技术在种植领域的推广，从最初的小模块试用到大面积推广，微灌试验收获了满意的答卷，并且在部分微灌试验中正在进行灌溉施肥内容的研究。

1996—2019年为第三阶段：灌溉技术方面的研究在农业领域得到了更多的重视和认可，在国内开展了大量的技术研发和培训，培养了众多农业技术人才，为我国水肥一体化技术的推广奠定了基础，早日实现大面积农用地水肥一体化技术的实行。

自20世纪90年代中期以来，我国微灌技术和水肥一体化技术迅速推广。随着农业技术的发展，我国水肥一体化技术逐渐成熟，已经从过去的小面积试行演

第一章 绪论

变为现在的整块区域内大面积种植使用，我国西北干旱地区也开始应用水肥一体化技术，并且在我国东北严寒地区和东南亚热带地区也成功使用了该项技术，技术覆盖范围广，其中包含了设施栽培和无土栽培模式，越来越多的农作物种植已经应用到水肥一体化技术，如蔬菜、苗木、大田经济作物等等。在我国经济发达的地区，科技技术先进、经济水平较高，为水肥一体化技术的推广提供了良好的条件，促进了水肥一体化技术的提高，水肥一体化技术设备配置精良，逐渐完善了大型水利设施，并且设计了一套成熟的智能自动化控制系统，一项大型示范工程正在涌现。部分地区因地制宜实施山区重力滴灌施肥，西北半干旱和干旱区协调配置日光温室集雨灌溉系统、窖水滴灌、瓜类栽培吊瓶滴灌施肥，华南地区利用灌溉注入有机肥液等，所有这些技术形式使灌溉施肥技术日趋丰富和完善。我国西北地区降水量少，土壤环境较为干旱，水肥一体化技术的引进极大地提高了当地的种植效益，改变了传统的种植模式。在大田作物种植中，新疆的棉花使用的膜下滴灌是西北地区最成功的施肥灌溉示范。新疆地处高原，风沙较大，水资源非常稀少，气候环境非常不适合大田作物种植。水肥一体化技术的出现，改善了新疆地区的种植条件，提高当地居民的经济效益，新的种植模式带来了更多的种植选择。1996年，为了提高当地的种植条件，新疆地区成功引进了滴灌技术，通过3年的研究与试验，滴灌技术日趋成熟，成功地研究出了一套低成本，适合大面积农田种植的滴灌带。农作物种植仅有灌溉技术是远远不够的，1998年又相继开展了施肥技术和栽培管理技术的研究，形成配套的技术才能够提高种植效益。作物种植中使用大功率拖拉机可以一次性完成所有的工作，如开沟、播种、施肥、覆膜等。在棉花后期的生长时期，要充分利用好灌溉设施，合理进行施肥和灌溉。

　　随着对灌溉施肥应用与理论的深入研究，我国的水肥一体化技术已经从1980年的起步发展到中等阶段。由于国家经济的提高，科学技术的进步，我国在微灌技术方面的研究已经达到了国际先进水平，研发生产的部分设施和系统已经处于国际领先行列，这些先进的技术设备是伟大的技术人员辛苦奋斗才换来的。1982年我国加入国际灌排委员会，并成为世界微灌组织成员之一，我国加强国际技术交流，重视微灌技术管理、微灌工程规划设计等的培训，培养了一大批水肥一体化技术推广管理及工程设计骨干和高学位人才。

　　水肥一体化是一项节水节肥的现代农业技术。欧洲很多地区并不缺水，但仍然采用水肥一体化技术，考虑的是该项技术的其他优点，特别是对环境的保护效应。我国水资源严重短缺，发展节水农业是形势所迫。同时，我国又是世界化肥消耗大国，单位面积施肥量居世界前列，化肥生产消耗国家大量能源，所有节省肥料的措施就是节省能源的措施。有关专家认为，水肥一体化技术的大面积推广应用，其意义绝不仅仅在于节水节肥本身，随着这项技术在更大范围的推进，它所引发的必将是我国农业由传统迈向现代化的一次具有深远意义的革命。

但是，从技术应用的角度分析，我国水肥一体化技术推广缓慢。首先，只关注了节水灌溉设备，水肥结合理论与应用研究成果较少；其次，我国灌溉施肥系统管理水平较低，培训宣传不到位，基层农技人员和农民对水肥一体化技术的应用不精通；再次，应用水肥一体化技术面积所占比例小，深度不够；最后，某些微灌设备产品，特别是首部配套设备的质量与国外同类先进产品相比仍存在着较大差距。

第三节 水肥一体化技术的应用前景

水肥一体化技术是将施肥与滴灌结合起来并使水、肥得到同步控制的一项技术，它利用灌溉设施将作物所需的养分、水分以最精确的用量供给，以此更好地节约水资源。因而，水肥一体化技术必将得到国家的大力扶植和推广，发展前景十分广阔。

一、水肥一体化技术向着科学化方向发展

水肥一体化技术向着精准农业、配方施肥的方向发展。我国国土面积很庞大，造成了各地区之间的气候环境有很大的差异性。例如，我国西北地区年降雨量很少，主要是以沙质土壤为主；我国西南地区处于亚热带气候，常年降雨，土壤湿润肥沃。所以，我国未来的水肥一体化技术发展方向，要根据不同地区、不同作物物种，还有土壤的差别综合决定。使用水肥一体化技术配料时，要综合考虑地区之间的差异性，不同土质选取不同的肥料，根据检测得出土壤的肥力特性和需肥规律，从而做出有针对性的设计配方，最终选取适合的肥料进行灌溉施肥。

另外，未来水肥一体化技术肥料的选取方向也将向科学化发展。水肥一体化技术肥料将根据水肥一体化灌溉系统的特点选取滴灌专用肥和液体化肥料。

二、水肥一体化技术向规模化、产业化方向发展

今后很长一段时间我国水肥一体化技术的市场潜力主要表现在以下几个方面：建立现代农业示范区，由政府出资引进先进的水肥一体化技术与设备作为生产示范，让农民效仿；休闲农业、观光果园等一批都市农业的兴起，将会进一步带动水肥一体化技术的应用和发展；商贸集团投资农业，进行规模化生产，建立特种农产品基地，发展出口贸易、农产品加工或服务于城市的餐饮业等；改善城镇环境，公园、运动场、居民小区内草坪绿地的发展也是水肥一体化设备潜在的市场；农民收入的增加和技术培训的到位，使农民有能力也愿意使用灌溉施肥技术和设备，以节约水、土和劳动力资源，获取最大的农业经济效益。

第二章 水肥一体化技术的主要设备和系统

一套水肥一体化技术设备包括首部枢纽、输配水管网和灌水器3部分组成。水肥一体化技术是借助于灌溉系统实现的。要合理地控制施肥的数量和浓度，必须选择合适的灌溉设备和施肥器械。常用的设施灌溉有喷灌、微喷灌和滴灌，微喷灌和滴灌简称微灌。下面会对各个设备和系统进行简单介绍。

第一节 水肥一体化技术的首部枢纽

一、加压设备

（一）泵房

加压设备一般安装在泵房内，根据灌溉设计要求确定型号类别，除深井供水外，多需要建造一座相应面积的水泵用房，并能提供一定的操作空间。水泵用房一般是砖混结构，也存在活动房形式，主要功能是避雨防盗，方便灌溉施肥器材的摆放。

取水点需要建造一个取水池，并预留一个进水口能够与灌溉水源联通，进水口需要安装一个拦污闸（材料最好用热镀锌或不锈钢），防止漂浮物进入池内。取水池的底部稍微挖深，较池外深0.5m左右，池底铺钢筋网，用混凝土铺平硬化。取水池的四周砌砖，顶上加盖，预留水泵吸水口和维修口（加小盖），并要定期清理池底，进水口不能堵塞，否则将会影响整个系统的运行。

（二）水泵

1. 水泵的选取

水泵的选取对整个灌溉系统的正常运行起着至关重要的作用。水泵选型原则是：在设计扬程下，流量满足灌溉设计流量要求；在长期运行过程中，水泵的工作效率要高，而且经常在最高效率点的右侧运行为最好；便于运行管理。

第二章 水肥一体化技术的主要设备和系统

2. 离心自吸泵

自吸泵具有一定的自吸能力,能够使水泵在吸不上水的情况下方便启动,并维持正常运行。我国目前生产的自吸泵基本上是自吸离心泵。自吸泵根据其自吸方式的不同,可分为外混式自吸泵、内混式自吸泵及带有由泵本身提供动力的真空辅助自吸泵;按输送液体和材质可分为污水自吸泵、清水自吸泵、耐腐蚀自吸泵、不锈钢自吸泵等多种自吸泵的结构形式。其主要零件有泵体、泵盖、叶轮、轴、轴承等。自吸泵结构上有独具一格的科学性,泵内设有吸液室、储液室、回液室回阀、气液分离室,管路中不需安装低阀,工作前只需保留泵体储有定量引液即可,因此简化了管路系统,又改善了劳动条件。泵体内具有涡形流道,流道外层周围有容积较大的气水分离腔,泵体下部铸有座角作为固定泵用。

ZW 型卧式离心自吸泵,是一种低扬程、大流量的污水型水泵,其电机与泵体采用轴联的方式,能够泵机分离,保养容易,在不用水的季节或有台风洪水的季节可以拆下电机,使用时再连接电机即可,同心度好、噪声小。主要应用于 50~100 亩以上、种植作物单一的农场,这样施肥时间能够统一。一个轮灌区的流量在 50~70m^3,水泵的效率能够得到较高的发挥。该水泵具有强力自吸功能、全扬程、无过载,一次加水后不需要再加水,如图 2-1 所示。

图 2-1 自吸式排污泵

其工作原理为:水泵通过电机驱动叶轮,把灌溉水从进水池内吸上来,通过出口的过滤器把水送到田间各处,在水泵的进水管 20cm 左右处,安装一个三通出口,连接阀门和小过滤器,再连接钢丝软管,作为吸肥管。

水泵在吸水的时候,进水管内部处于负压状态,这时候把吸肥管放入肥料桶的肥液中(可以是饱和肥液),打开吸肥管的阀门,肥液就顺着吸肥管被吸到水泵及管道中与清水混合,输送到田间作物根部进行施肥。与电动注射泵相比,不需要电源，不会压力过高或过低造成不供肥或过量供肥。

3. 潜水泵

潜水泵与普通的抽水机不同的是其工作在水下，而抽水机大多工作在地面上。用潜水泵作为首部动力系统，优点是简单实用，不需要加引水，也不会发生漏气和泵体气蚀余量超出等故障，水泵型号多，选择余地大。缺点是泵体电机和电线都浸在水中，在使用的过程中要防止发生漏电。

灌溉用潜水泵按出水径，通常在 DN32～DN150（指管径 32～150mm，下同），如果一台不够，可以安装两台并联使用，这样能节约用电。这里以 QS 系列潜水泵为例，介绍首部的组成，如图 2-2 所示。水泵参数：QS 潜水泵，额定流量 $65m^3$，额定扬程 18m，功率 7.5kW，电压 380V，出水口内径 DN80。QS 水泵的进水口是在水泵中上部，工作时，不会把底部淤泥吸上。冷却效果好，输出功率大，出水口在泵体顶部，方便安装。

图 2-2　QS 系列潜水泵

（三）负压变频供水设备

通常温室和大田灌溉都是用水泵将水直接从水源中抽取加压使用，无论用水量大小，水泵都是满负荷运转，所以当用水量较小时，所耗的电量与用水量大时

第二章 水肥一体化技术的主要设备和系统

一样，容易造成极大的浪费。

负压变频供水设备能根据供水管网中瞬时变化的压力和流量参数，自动改变水泵的台数和电机运行转速，实现恒压变量供水的目的，如图 2-3 所示。水泵的功率随用水量变化而变化，用水量大，水泵功率自动增大；用水量小，水泵功率自动减小，能节电 50%，从而达到高效节能的目的。

图 2-3 变频供水系统

负压变频供水设备的应用，优化了作物供水的方式。如 LFBP-天 L 系列变频供水设备，如图 2-4 所示。在原有基础上进行了优化升级，目前第三代设备已经具有自动加水、自动开机、自动关机、故障自动检索的功能，打开阀门，管道压力感应通过 PLC 执行水泵启动，出水阀全部关闭后，水泵停止工作，达到节能的效果。

负压供水设备由变频控制柜、离心水泵（DL 或 ZW 系列）、真空引水罐、远传压力表、引水筒、底阀等部件组成。电机功率一般为 5.5kW、7.5kW、11kW，一台变频控制水泵数量从一控二到一控四，可以根据现场实际用水量确定。其中 11kW 组合一定要注意电源电压。

变频恒（变）压供水设备控制柜，是对供水系统中的泵组进行闭环控制的机电一体化成套设备，如图 2-5 所示。该设备采用工业微机可变程序控制器和数字变频调整技术，根据供水系统中瞬时变化的流量和相应的压力，自动调节水泵的转速和运行台数，从而改变水泵出口压力和流量，使供水管网系统中的压力按设定压力保持恒定，达到提高供水品质和高效节能的目的。

图 2-4　LFBP-天 L 系列变频供水系统

图 2-5　变频恒压供水设备控制柜

控制柜适用于各种无高层水塔的封闭式供水场合的自动控制，具有压力恒定、结构简单、操作简便、使用寿命长、高效节能、运行可靠、使用功能齐全及完善的保护功能等特点。

控制柜具有手动、变频和工频自动三种工作形式，并可根据用户要求，追加如下各种附加功能：小流量切换或停泵，水池无水停泵，定时启停泵，双电源、双变频、双路供水系统切换，自动巡检，改变供水压力，供水压力数字显示及用户在供水自动化方面要求的其他功能。

二、过滤设备

过滤设备的作用是将灌溉水中的固体颗粒（沙石、肥料沉淀物及有机物）滤去，避免污物进入系统，造成系统和灌水器堵塞。

(一) 筛网式过滤器

筛网式过滤器是微灌系统中应用最为广泛的一种简单而有效的过滤设备，它的过滤介质有塑料、尼龙筛网或不锈钢筛网。

(1) 适用条件。筛网式过滤器主要作为末级过滤器，当灌溉水质不良时连接在主过滤器（沙砾或水力回旋过滤器）之后，作为控制过滤器使用。主要用于过滤灌溉水中的粉粒、沙和水垢等污物。当有机物含量较高时，这种类型的过滤器的过滤效果很差，尤其是当压力较大时，有机物会从网眼中挤过去进入管道，造成系统与灌水器的堵塞。筛网式过滤器一般用于二级或三级过滤（即与沙石分离器或沙石过滤器配套使用）。

(2) 分类。筛网过滤器的种类很多，按安装方式分有立式和卧式两种；按清洗方式分有人工清洗和自动清洗两种；按制造材料分为塑料和金属两种；按封闭与否分为封闭式和开敞式两种。

(3) 结构。筛网过滤器主要由筛网、壳体、顶盖等部分组成，如图2-6。筛网的孔径大小（即网目数）决定了过滤器的过滤能力，由于通过过滤器筛网的

图2-6 筛网过滤器外观及滤芯

污物颗粒会在灌水器的孔口或流道内相互挤在一起而堵塞灌水器,因而一般要求所选用的过滤器的滤网孔径大小应为所使用的灌水器孔径的 1/7~1/10。筛网的规格与孔口大小的关系见表 2-1。

表 2-1 筛网规格与孔口大小的对应关系

滤网规格/目	孔口大小		土粒类别	粒径(mm)
	mm	μm		
20	0.710	710	粗沙	0.50~0.75
40	0.420	420	中沙	0.25~0.41
50	0.181	181	细沙	0.15~0.20
100	0.151	151	细沙	0.15~0.20
120	0.126	126	细沙	0.10~0.15
150	0.105	105	极细沙	0.10~0.15
200	0.075	75	极细沙	<0.10
250	0.052	52	极细沙	<0.10
300	0.045	45	粉沙	<0.10

过滤器孔径大小的选择要根据所用灌水器的类型及流道断面大小而定。同时由于过滤器减小了过流断面,存在一定的水头损失,在进行系统设计压力的推算时一定要考虑过滤器的压力损失范围,否则当过滤器发生一定程度的堵塞时会影响系统的灌水质量。在我国现代的种植灌溉系统中,通常情况下,喷灌系统要求使用 45~85 目进行过滤,微灌系统要求使用 85~105 目进行过滤,滴灌系统要求使用 105~155 目进行过滤。但过滤数目越大,压力损失越大,能耗越大。

(二)叠片式过滤器

从其名字就能简单了解这种过滤器的原理,叠片式过滤器通过重叠的圆形过滤片进行过滤,每片过滤片的两面都会有过滤槽,当水流经过这些叠片时,水中的杂质和污物就会被盘壁和滤槽阻隔下来,这样就可以达到净化水质的效果。从原理上来看,叠片式过滤器要比筛网式过滤器的净化效果好,其过滤能力在 40~400 目,可用于初级和终级过滤,但当灌溉水中污物杂质较多时,叠片式过滤器就不适合作为第一过滤装置,这样会使得叠片式过滤器多次清洗,反而影响了灌溉系统的工作效率,如图 2-7、图 2-8 所示。

第二章 水肥一体化技术的主要设备和系统

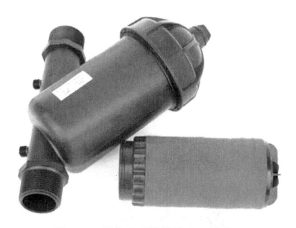

图 2-7 叠片式过滤器外观及叠片

图 2-8 自动反冲洗叠片式过滤器

(三) 离心式过滤器

离心式过滤器又称为旋流水沙分离过滤器或涡流式水沙分离器，是由高速旋转水流产生的离心力，将沙粒和其他较重的杂质从水体中分离出来，它内部没有滤网，也没有可拆卸的部件，保养维护很方便。这类过滤器主要应用于高含沙量水源的过滤，当水中含沙量较大时，应选择离心过滤器为主过滤器。它由进水口、出水口、旋涡室、分离室、储污室和排污口等部分组成，如图 2-9 所示。

· 17 ·

图 2-9 离心式过滤器

离心式过滤器的工作原理是当压力水流从进水口以切线方向进入旋涡室后做旋转运动，水流在做旋转运动的同时也在重力作用下向下运动，在旋流室内呈螺旋状运动，水中的泥沙颗粒和其他固体物质在离心力的作用下被抛到分离室壳壁上，在重力作用下沿壁面渐渐向下移动，向储污室中汇集。在储污室内横断面增大，水流速度下降，泥沙颗粒受离心力作用减小，受重力作用加大，最后沉淀下来，再通过排污管排出过滤器。而在旋涡中心的净水速度比较低，位能较高，于是作螺旋运行上升经分离器顶部的出水口进入灌溉管道系统。

离心式过滤器因其是利用旋转水流和离心作用使水沙分离而进行过滤的，因而对高含沙水流有较理想的过滤效果，但是较难除去与水密度相近和密度比水小的杂质，因而有时也称为沙石分离器。另外，在水泵启动和停机时由于系统中水流流速较小，过滤器内所产生的离心力小，其过滤效果较差，会有较多的沙粒进入系统，因而离心式过滤器一般不能单独承担微灌系统的过滤任务，必须与筛网式或叠片式过滤器结合运用，以水沙分离器作为初级过滤器，这样会起到较好的过滤效果，延长冲洗周期。离心式过滤器底部的储污室必须频繁冲洗，以防沉积的泥沙再次被带入系统。离心式过滤器有较大的水头损失，在选用和设计时一定要将这部分水头损失考虑在内，如图 2-10 所示。

第二章 水肥一体化技术的主要设备和系统

图 2-10 离心式过滤器与网式过滤器组合使用

（四）沙石过滤器

沙石过滤器又称介质过滤器。它是利用沙石作为过滤介质进行过滤的，一般选用玄武岩沙床或石英沙床，沙砾的粒径大小根据水质状况、过滤要求及系统流量确定。沙石过滤器对水中的有机杂质和无机杂质的滤出和存留能力很强，并可不间断供水。当水中有机物含量较高时，无论无机物含量有多少，均应选用沙石过滤器。沙石过滤器的优点是过滤能力强，适用范围很广，不足之处在于占的空间比较大、造价比较高。它一般用于地表水源的过滤，使用时根据出水量和过滤要求可选择单一过滤器或两个以上的过滤器组进行过滤。

沙石过滤器主要由进水口、出水口、过滤器壳体、过滤介质沙砾和排污孔等部分组成，如图 2-11 所示。其工作原理是当水由进水口进入过滤器并经过沙石过滤床时，因过滤介质间的孔隙曲折且小，水流受阻流速减小，水源中所含杂质就会被阻挡而沉淀或附着到过滤介质表面，从而起到过滤作用，经过滤后的干净水从出水口进入灌溉管道系统。当过滤器两端压力差超过 30~50kPa 时，说明过滤介质被污物堵塞严重，需要进行反冲洗，反冲洗是通过过滤器控制阀门，使水流产生逆向流动，将以前过滤阻挡下来的污物通过排污口排出。为了使灌溉系统在反冲洗过程中也能同时向系统供水，常在首部枢纽安装两个以上过滤器，其工作过程如图 2-12 所示。

图 2-11 沙石过滤器

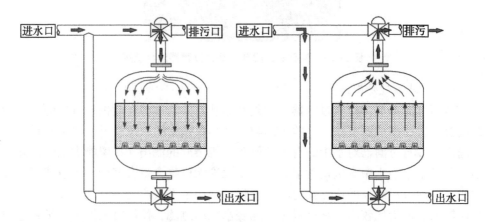

图 2-12 沙石过滤器工作状态

沙石过滤器的过滤能力主要决定于所选用的沙石的性质及粒径级配，不同粒径级配的沙石其过滤能力不同，同时由于沙石与灌溉水充分接触，且在反冲洗时会产生摩擦，因此，沙石过滤器用沙应满足以下要求：具有足够的机械强度，以防反冲洗时沙粒产生磨损和破碎现象；沙具有足够的化学稳定性，以免沙粒与酸、碱等化学物品发生化学反应，产生引起微灌堵塞的物质，更不能产生对动、植物有毒害作用的物质；具有一定颗粒级配和适当孔隙率；尽量就地取材，且价格便宜。

第二章　水肥一体化技术的主要设备和系统

（五）拦污栅（网）

很多灌溉系统以地表水作为水源，如河流、塘库等，这些水体中常含有较大体积的杂物，如枯枝残叶、藻类、杂草和其他较大的漂浮物等。为防止这些杂物进入深沉池或蓄水池中，增加过滤器的负担，常在蓄水池水泵进口处安装一种网式拦污栅，如图2-13所示，作为灌溉水源的初级净化处理设施。拦污栅构造简单，可以根据水源实际情况自行设计和制作。

图2-13　拦污栅

（六）过滤器的选型

过滤器在微灌系统中起着非常重要的作用，不同类型的过滤器对不同杂质的过滤能力不同，在设计选型时一定要根据水源的水质情况、系统流量及灌水器要求选择既能满足系统要求，又操作方便的过滤器类型及组合。过滤器选型一般有以下步骤。

第一步，根据灌溉水杂质种类及各类杂质的含量选择过滤器类型。地面水（江河、湖泊塘库等）一般含有较多的沙石和有机物，宜选用沙石过滤器进行一级过滤，如果杂质体积比较大，还需要用拦污栅作初级拦污过滤；如果含沙量大，还需要设置沉沙池作初级拦污过滤。地下水（井水）杂质一般以沙石为主，宜选用离心式过滤器作一级过滤。无论是沙石过滤器还是离心式过滤器，都可以根据需要选用筛网式过滤器或叠片式过滤器作二级过滤。对于水质较好的水源，可直接选用筛网式或叠片式过滤器。表2-2总结了不同类型过滤器对去除浇灌水中不同污物的有效性。

表 2-2 过滤器的类型选择

污物类型	污染程度	定量标准（mg/L）	离心式过滤器	沙石过滤器	叠片式过滤器	自动冲洗筛网过滤器	控制过滤器的选择
土壤颗粒	低	≤50	A	B	—	C	筛网
	高	>50	A	B	—	C	筛网
悬浮固定物	低	≤80	—	A	B	C	叠片
	高	>80	—	A	B	—	叠片
藻类	低	—	—	B	A	C	叠片
	高	—	—	A	B	C	叠片
氧化铁和锰	低	≤50	—	B	A	A	叠片
	高	>50	—	A	B	B	叠片

注：控制过滤器指二级过滤器。A 为第一选择方案；B 为第二选择方案；C 为第三选择方案。

第二步，根据灌溉系统所选灌水器对过滤器的能力要求确定过滤器的目数大小。一般来说，喷灌要求 40~80 目过滤，微喷要求 80~100 目过滤，滴灌要求 100~150 目过滤。

第三步，根据系统流量确定过滤器的过滤容量。

第四步，确定冲洗类型，在有条件的情况下，建议采用自动反冲洗类型，以减少维护和工作量。特别是劳力短缺及灌溉面积大时，自动反冲洗过滤器应优先考虑。

第五步，考虑价格因素，对于具有相同过滤效果的不同过滤器来说，选择过滤器时主要考虑价格高低，一般砂介质过滤器是最贵的，而叠片或筛网过滤器则是相对便宜的。

三、控制和量测设备

为了确保灌溉施肥系统正常运行，首部枢纽中还必须安装控制装置、保护装置、量测装置，如进排气阀、逆止阀、压力表和水表等。

（一）控制部件

控制部件的作用是控制水流的流向、流量和总供水量，它是根据系统设计灌水方案，有计划地按要求的流量将水流分配输送至系统的各部分，主要有各种阀门和专用给水部件。

1. 给水栓

给水栓是指将地下管道系统的水引出地面进行灌溉的放水口，根据阀体结构形式可分为移动式给水栓、半固定式给水栓和固定式给水栓，如图 2-14 所示。

第二章 水肥一体化技术的主要设备和系统

图 2-14 给水栓

2. 阀门

阀门是喷灌系统必用的部件，主要有闸阀、蝶阀、球阀、截止阀、止回阀、安全阀、减压阀等，如图 2-15、图 2-16、图 2-17 所示。在同一灌溉系统中，不同的阀门起着不同的作用，使用时可根据实际情况选用不同类型的阀门。

图 2-15 蝶阀

(二) 安全保护装置

灌溉系统运行中不可避免地会遇到压力突然变化、管道进气、突然停泵等一些异常情况，威胁到系统，因此在灌溉系统相关部位必须安装安全保护装置，防

图 2-16　PVC 球阀

图 2-17　止回阀

止系统内因压力变化或水倒流对灌溉设备产生破坏，保证系统正常运行。常用的设备有进（排）气阀、安全阀、调压装置、逆止阀、泄水阀等。

1. 进（排）气阀

进（排）气阀是能够自动排气和进气，且当压力水来时能够自动关闭的一种安全保护设备，主要作用是排除管内空气，破坏管道真空，有些产品还具有止回水功能。进（排）气阀是管路安全的重要设备，不可缺少。一些非专业的设计不安装进（排）气阀造成爆管及管道吸扁，使系统无法正常工作，如图2-18所示。

2. 安全阀

安全阀是一种压力释放装置，当管道的水压超过设定压力时自动打开泄压，

第二章 水肥一体化技术的主要设备和系统

图 2-18 进（排）气阀

防止水锤事故，一般安装在管路的较低处。在不产生水柱分离的情况下，安全阀安装在系统首部（水泵出水端），可对整个喷灌系统起保护作用。如果管道内产生水柱分离，则必须在管道沿程一处或几处安装安全阀才能达到防止水锤的目的，如图 2-19 所示。

图 2-19 安全阀

（三）流量与压力调节装置

当灌溉系统中某些区域实际流量和压力与设计工作压力相差较大时，就需要安装流量与压力调节装置来调节管道中的压力和流量。特别是在利用自然高差进行自压喷灌时，往往存在灌溉区管道内压力的分布不均匀，或实际压力大于喷头工作压力，导致流量与压力分布不均匀，或导致流量与压力很难满足要求，也给喷头选型带来困难。此时除进行压力分区外，在管道系统中安装流量与压力调节装置是极为必要的。流量与压力调节装置都是通过自动改变过水断面来调节流量与压力的，实际上是通过限制流量的方法达到减小流量或压力，并不会增加系统流量或压力。根据此工作原理，在生产实践中，考虑到投资问题，也有用球阀、闸阀、蝶阀等作为调节装置的，但这样一方面会影响到阀门的使用寿命，另一方面也很难进行流量与压力的精确调节。

（四）量测装置

灌溉系统的量测装置主要有压力表、流量计和水表，其作用是系统工作时实时监测管道中的工作压力和流量，正确判断系统工作状态，及时发现并排除系统故障。

1. 压力表

压力表是所有设施灌溉系统必需的量测装置，它是测量系统管道内水压的仪器，它能够实时反映系统是否处于正常工作状态，当系统出现故障时，可根据压力表读数变化的大小初步判断可能出现的故障类型。压力表常安装于首部枢纽、轮灌区入口处、支管入口处等控制节点处，实际数量及具体位置要根据喷灌区面积、地形复杂程度等确定。在过滤器前后一般各需安装1个压力表，通过两端压力差大小判断过滤器堵塞程度，以便及时清洗，防止过滤器堵塞减小过水断面，造成田间工作压力及流量过小而影响灌溉质量。喷灌用压力表要选择灵敏度高、工作压力处于压力表主要量程范围内、表盘较大、易于观看的优质产品。喷灌系统工作状态除田间观察外，主要由压力表反映，因此，必须保证压力表处于正常工作状态，出现故障要及时更换，如图2-20所示。

2. 流量计和水表

流量计和水表都是量测水流流量的仪器，两者不同之处是：流量计能够直接反映管道内的流量变化，不记录总过水量，如图2-21所示；而水表反映的是通过管道的累积水量，不能记录实时流量，要获得系统流量时需要观测计算，一般安装于首部枢纽或干管上。在配备自动施肥机的喷灌系统，由于施肥机需要按系统流量确定施肥量的大小，因而需安装一个自动量测水表。

（五）自动化控制设备

节水灌溉系统的优点之一是容易实现自动化控制。自动化控制技术能够在很大程度上提高灌溉系统的工作效率。采用自动化控制灌溉系统具有以下优点：能

第二章 水肥一体化技术的主要设备和系统

图 2-20 压力表

图 2-21 流量表

够做到适时适量地控制灌水量、灌水时间和灌水周期，提高水分利用效率；大大节约劳动力，提高工作效率，减少运行费用；可灵活方便地安排灌水计划，管理人员不必直接到田间进行操作；可增加系统每天的工作时间，提高设备利用率。节水灌溉的自动化控制系统主要由中央控制器、自动阀、传感器等设备组成，其自动化程度可根据用户要求、经济实力、种植作物的经济效益等多方面综合考虑确定。

1. 中央控制器

中央控制器是自动化灌溉系统的控制中心，管理人员可以通过输入相应的灌溉程序（灌水开始时间、延续时间、灌水周期）进行对整个灌溉系统的控制。由于控制器价格比较昂贵，控制器类型的选择应根据实际的容量要求和要实现的功能多少而定，如图 2-22 所示。

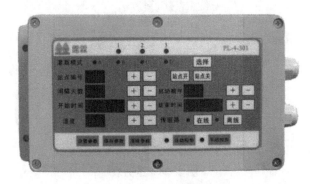

图 2-22　中央控制器

2. 自动阀

自动阀的种类很多，其中电磁阀是在自动化灌溉系统中应用最多的一种，电磁阀是通过中央控制器传送的电信号来打开或关闭阀门的，其原理是电磁阀在接收到电信号后，电磁头提升金属塞，打开阀门上游与下游之间的通道，使电磁阀内橡胶隔膜上面与下面形成压差，阀门开启，如图 2-23 所示。

图 2-23　电磁阀

第二节　水肥一体化技术系统中的施肥设备

水肥一体化技术中常用到的施肥设备主要有压差施肥罐、文丘里施肥器、泵

第二章 水肥一体化技术的主要设备和系统

吸肥法、泵注肥法、自压重力施肥法、施肥机等。

一、压差施肥罐

（一）基本原理

压差式施肥罐，由两根细管（旁通管）与主管道相接，在主管道上两条细管接点之间设置一个节制阀（球阀或闸阀）以产生一个较小的压力差，使一部分水流流入施肥罐，进水管直达罐底，水溶解罐中肥料后，肥料溶液由另一根细管进入主管道，将肥料输送到作物根区，如图2-24、图2-25、图2-26所示。

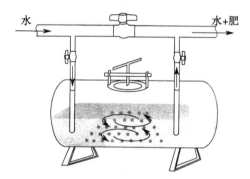

图2-24 压差施肥罐示意图

图2-25 立式金属施肥罐

肥料罐是用抗腐蚀的陶瓷衬底或镀锌铸铁、不锈钢或纤维玻璃做成，以确保

图 2-26 立式塑料施肥罐

经得住系统的工作压力和抗肥料腐蚀。在低压滴灌系统中，由于压力低，也可用塑料罐。固体可溶肥料在肥料罐里逐渐溶解，液体肥料则与水快速混合。随灌溉进行，肥料不断被带走，肥料溶液不断被稀释，养分越来越低，最后肥料罐里的固体肥料都流走了。该系统较简单、便宜，不需要用外部动力就可以达到较高的稀释倍数。然而，该系统也存在一些缺陷，如无法精确控制灌溉水中的肥料注入速率和养分浓度，每次灌溉之前都得重新将肥料装入施肥罐内。节流阀增加了压力的损失，而且该系统不能用于自动化操作。肥料罐常做成 10~300L 的规格。一般温室大棚小面积地块用体积小的施肥罐，大田轮灌区面积较大的地块用体积大的施肥罐。

（二）优缺点

压差施肥罐的优点：设备成本低，操作简单，维护方便；适合施用液体肥料和水溶性固体肥料；施肥时不需要外加动力；设备体积小，占地少。压差施肥罐的缺点：为定量化施肥方式，施肥过程中的肥液浓度不均一；易受水压变化的影响；存在一定的水头损失，移动性差，不适宜用于自动化作业；锈蚀严重，耐用性差；由于罐口小，加入肥料不方便，特别是轮灌区面积大时，每次的肥料用量大，而罐的体积有限，需要多次倒肥，降低了工作效率。

（三）适用范围

压差施肥罐适用于包括温室大棚、大田种植等多种形式的水肥一体化灌溉施肥系统。对于不同压力范围的系统，应选用不同材质的施肥罐。因不同材质的施肥罐其耐压能力不同。

第二章 水肥一体化技术的主要设备和系统

二、文丘里施肥器

(一) 基本原理

水流通过一个由大渐小然后由小渐大的管道时（文丘里管喉部），水流经狭窄部分时流速加大，压力下降，使前后形成压力差，当喉部有一更小管径的入口时，形成负压，将肥料溶液从一敞口肥料罐通过小管径细管吸取上来。文丘里施肥器即根据这一原理制成，如图2-27、图2-28所示。

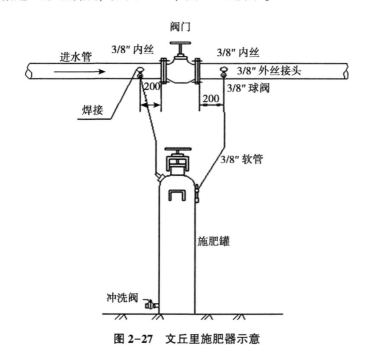

图 2-27 文丘里施肥器示意

由于文丘里施肥器会造成较大的压力损耗，通常安装时加装一个小型增压泵。一般厂家均会告知产品的压力损耗，设计时根据相关参数配置加压泵或不加泵。

文丘里施肥器的操作需要有过量的压力来保证必要的压力损耗；施肥器入口稳定的压力是养分浓度均匀的保证。压力损耗量用占入口处压力的百分数来表示，吸力产生需要损耗入口压力的20%以上，但是两级文丘里施肥器只需损耗10%的压力。吸肥量受入口压力、压力损耗和吸管直径影响，可通过控制阀和调节器来调整。文丘里施肥器可安装于主管路上（串联安装，如图2-29所示）或者作为管路的旁通件安装（并联安装，如图2-30所示）。在温室里，作为旁通件安装的施肥器其水流由一个辅助水泵加压。

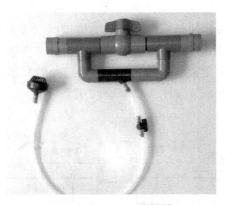

图2-28 文丘里施肥器

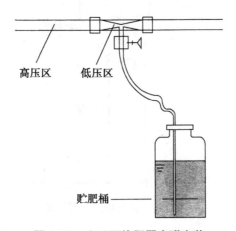

图2-29 文丘里施肥器串联安装

文丘里施肥器的主要工作参数有：一是进口处工作压力（$P_{进}$）。二是压差，压差（$P_{进}-P_{出}$）常被表达成进口压力的百分比，只有当此值降到一定值时，才开始抽吸。如前所述，这一值约为1/3的进口压力，某些类型高达50%，较先进的可小于15%。表2-3列出了压力差与吸肥量的关系。三是抽吸量。指单位时间里抽吸液体肥料的体积，单位为L/h。抽吸量可通过一些部件调整。四是流量，是指流过施肥器本身的水流量。进口压力和喉部尺寸影响着施肥器的流量。

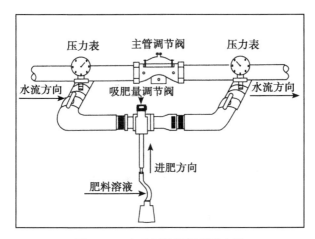

图 2-30 文丘里施肥器并联安装

流量范围由制造厂家给定。每种类型只有在给定的范围内才能准确地运行。

表 2-3 文丘里施肥器压力差与吸肥量的关系

入口压力 P_1 (kPa)	出口压力 P_2 (kPa)	压力差 ΔP (kPa)	吸肥流量 Q_1 (L/h)	主管流量 Q_2 (L/h)	总流量 Q_1+Q_2 (L/h)
150	70	80	0	1 300	1 300
150	40	110	320	2 200	2 520
150	0	150	472	2 008	2 480
100	30	70	0	950	950
100	0	100	350	2 290	2 640

文丘里施肥器具有显著的优点：不需要外部能源，直接从敞口肥料罐吸取肥料，吸肥量范围大，操作简单，磨损率低，安装简易，方便移动，适于自动化，养分浓度均匀且抗腐蚀性强。不足之处为压力损失大，吸肥量受压力波动的影响。

(二) 主要类型

1. 简单型

这种类型结构简单，只有射流收缩段，无附件，因水头损失过大，一般不宜采用。

2. 改进型

灌溉管网内的压力变化可能会干扰施肥过程的正常进行或引起事故。为防止这些情况发生，在单段射流管的基础上，增设单向阀和真空破坏阀。当产生抽吸作用的压力过小或进口压力过低时，水会从主管道流进储肥罐以至产生溢流。在抽吸管前安装一个单向阀，或在管道上装一个球阀均可解决这一问题。当文丘里施肥器的吸入室为负压时，单向阀的阀芯在吸力作用下关闭，防止水从吸入口流出，如图 2-31 所示。

图 2-31　带单向阀的文丘里施肥器

当敞口肥料桶安放在田块首部时，罐内肥液可能在灌溉结束时因出现负压被吸入主管，再流至田间最低处，既浪费肥料而且可能烧伤作物。在管路中安装真空破坏阀，无论系统何处出现局部真空都能及时补进空气。

有些制造厂提供各种规格的文丘里施肥器喉部，可按所需肥料溶液的数量进行调换，以使肥料溶液吸入速率稳定在要求的水平上。

3. 两段式

国外研制改进的两段式结构，使得吸肥时的水头损失只有入口处压力的 12%~15%，因而克服了文丘里施肥器的基本缺陷，并使之获得了广泛的应用。不足之处是流量相应降低了，如图 2-32 所示。

(三) 优缺点

文丘里施肥器的优点：设备成本低，维护费用低；施肥过程可维持肥液浓度均匀灌溉，施肥过程无需外部动力；设备重量轻，便于移动和用于自动化系统；施肥时肥料罐为敞开环境，便于观察施肥进程。文丘里施肥器的缺点：施肥时系统水头压力损失大；为补偿水头损失，系统中要求较高的压力；施肥过程中的压

第二章 水肥一体化技术的主要设备和系统

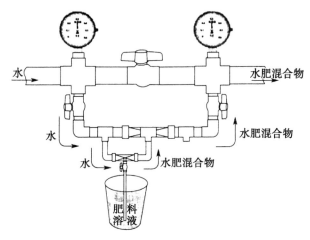

图 2-32　两段式文丘里施肥器

力波动变化大；为使系统获得稳压，需配备增压泵；不能直接使用固体肥料，需把固体肥料溶解后施用。

（四）适用范围

文丘里施肥器因其出流量较小，主要适用于小面积种植场所，如温室大棚种植或小规模农田。

（五）安装方法

在大多数情况下，文丘里施肥器安装在旁通管上（并联安装），这样只需部分流量经过射流段。当然，主管道内必须产生与射流管内相等的压力降。这种旁通运行可使用较小（较便宜）的文丘里施肥器，而且更便于移动。当不加肥时，系统也工作正常。当施肥面积很小且不考虑压力损耗时，也可用串联安装。

在旁通管上安装的文丘里施肥器，常采用旁通调压阀产生压差。调压阀的水头损失足以分配压力。如果肥液在主管过滤器之后流入主管，抽吸的肥水要单独过滤。常在吸肥口包一块100~120目的尼龙网或不锈钢网，或在肥液输送管的末端安装一个耐腐蚀的过滤器，筛网规格为120目，如图2-33所示。有的厂家产品出厂时已在管末端连接好不锈钢网。输送管末端结构应便于检查，必要时可进行清洗。肥液罐（或桶）应低于射流管，以防止肥液在不需要时自压流入系统。并联安装方法可保持出口端的恒压，适合于水流稳定的情况。当进口处压力较高时，在旁通管入口端可安装一个小的调压阀，这样在两端都有安全措施。

因文丘里施肥器对运行时的压力波动很敏感，应安装压力表进行监控。一般

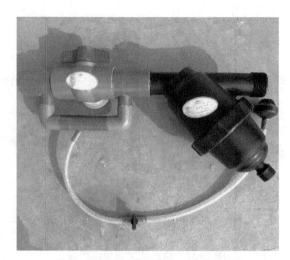

图2-33 带过滤器的文丘里施肥器

在首部系统都会安装多个压力表。节制阀两端的压力表可测定节制阀两端的压力差。一些更高级的施肥器本身即配有压力表供监测运行压力。

三、重力自压式施肥法

(一) 基本原理

在应用重力滴灌或微喷灌的场合，可以采用重力自压式施肥法。在南方丘陵山地果园或茶园，通常引用高处的山泉水或将山脚水源泵至高处的蓄水池。通常在水池旁边高于水池液面处建立一个敞口式混肥池，池大小在 $0.5 \sim 5.0 m^3$，可以是方形或圆形，方便搅拌溶解肥料即可。池底安装肥液流出的管道，出口处安装PVC球阀，此管道与蓄水池出水管连接。池内用20~30cm长的大管径。管入口用100~120目尼龙网包扎。如图2-34所示。

(二) 应用范围

我国华南、西南、中南等地区有大面积的丘陵山地果园、茶园、经济林地及作物大田，非常适合采用重力自压灌溉。在很多山地果园在山顶最高处建有蓄水池，果园一般采用拖管淋灌或滴灌。此时采用重力自压施肥非常方便做到水肥结合。在华南地区的柑橘园、荔枝园、龙眼园有相当数量的果农采用重力自压施肥。重力自压施肥简单方便，施肥浓度均匀中，农户易于接受。不足之处是必须把肥料运送到山顶。

第二章 水肥一体化技术的主要设备和系统

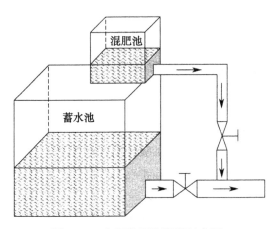

图 2-34 自压灌溉施肥器示意图

四、泵吸肥法

(一) 基本原理

泵吸肥法是利用离心泵直接将肥料溶液吸入灌溉系统，适合于几十公顷以内面积的施肥。为防止肥料溶液倒流入水池而污染水源，可在吸水管上安装逆止阀。通常在吸肥管的入口包上 100~120 目滤网（不锈钢或尼龙），防止杂质进入管道，如图 2-35 所示。

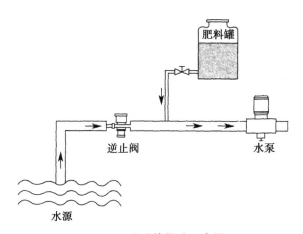

图 2-35 泵吸施肥法示意图

(二) 优缺点

该法的优点是不需外加动力,结构简单,操作方便,可用闭口容器盛肥料溶液。施肥时通过调节肥液管上阀门,可以控制施肥速度,精确调节施肥浓度。缺点是施肥时要有人照看,当肥液快完时应立即关闭吸肥管上的阀门,否则会吸入空气,影响泵的运行。

五、泵注肥法

泵注肥法是利用加压泵将肥料溶液注入有压管道,通常泵产生的压力必须要大于输水管的水压,否则肥料注不进去。对用深井泵或潜水泵抽水直接灌溉的地区,泵注肥法是最佳选择。泵施肥法施肥速度可以调节,施肥浓度均匀,操作方便,不消耗系统压力。不足是要单独配置施肥泵。对施肥不频繁地区,普通清水泵可以使用,施完肥后用清水清洗,一般不生锈。但对于频繁施肥的地区,建议用耐腐蚀的化工泵。

第三节 水肥一体化技术的输水系统——喷灌

一、喷灌的概念

喷灌是利用水泵加压或自然落差将水通过压力管道送到田间,经喷头喷射到空中,形成细小的水滴,均匀喷洒在农田上,为作物正常生长提供必要水分条件的一种先进灌水方法。与传统的地面灌水方法相比,喷灌具有明显的特点。

二、喷灌的技术特点

(一) 喷灌的优点

(1) 节约用水。由于喷灌可以对不同土壤各种作物进行适时、适量、均匀和有计划的小定额灌溉,所以不产生地面径流和深层渗透,因而提高了水的有效利用率,达到了节水的目的,而且灌水均匀。若与地面灌溉相比,喷灌较沟、畦灌省水30%~50%。若透水性强、保水能力差的沙质土壤或山坡岗地,节水在70%以上。

(2) 适应性强。喷灌适用于各种地形和土壤条件,不一定要求地面平整,对于不适合地面灌溉的山地、丘陵、坡地等地形较复杂的地区和局部有高丘、坑洼的地区，都可以应用喷灌；除此以外，喷灌可应用于多种作物，对于所有密植浅根系作物，如叶菜类蔬菜、块根类蔬菜、马铃薯等都可以采用喷灌。同时对透水性强或沉陷性土壤及耕作表层土薄且底土透水性强的沙质土壤而言，最适合运用喷灌技术。

（3）节省劳力和土地。喷灌的机械化程度高，又便于采用小型电子控制装置实现自动化，可以节省大量劳动力，如果采用喷灌施肥技术，其节省劳动力的效果更为显著，此外，采用喷灌还可以减少修田间渠道、灌水沟畦等用工。同时，喷灌利用管道输水，固定管道可以埋于地下，减少田间沟、渠、畦、埂等的占地，比地面灌溉节省土地7%~15%。

（4）增加产量，改善品质。首先，喷灌能适时适量地控制灌水量，采用少灌勤灌的方法，使土壤水分保持在作物正常生长的适宜范围内，同时喷灌像下雨一样灌溉作物，对耕层土壤不会产生机械破坏作用，保持了土壤团粒结构，有效地调节了土壤水、肥、气、热和微生物状况。其次，喷灌可以调节田间小气候，增加近地层空气湿度，调节温度和昼夜温差，又避免干热风、高温及霜冻对作物的危害，具有明显的增产效果，一般蔬菜增产1~2倍。再次，喷灌能够根据作物需水状况灵活调节灌水时间与灌水量，整体灌水均匀，且可以根据作物生长需求适时调整施肥方案，有效提高农产品的产量和品质。

（二）喷灌的缺点

（1）喷洒作业受风影响较大。由喷头喷出来的水滴在落洒地面的过程中受风的影响很大。在其影响下，喷头在各方向的射程、水量分布发生明显变化，从而影响均匀性，甚至漏喷。灌溉季节多风的地区应在设备选型和规划设计上充分考虑风的不利影响，如难以解决，则应考虑采用其他灌溉方法。

（2）蒸发损失大。水滴在落到地面前会产生蒸发损失，尤其在干旱、多风及高温季节，蒸发损失更大，其损失量与风速、气温、空气湿度有关。

（3）设备投资高。喷灌系统工作压力较高，对设备的耐压要求也高，因而设备投资一般较高。这也是当前制约喷灌发展的主要因素。与此相关的另一个问题是，目前喷灌设备质量不高，加上管理不善，造成设备损坏、丢失，甚至系统提前报废，投入得不到相应的回报。因此建设喷灌工程必须切实把好设备和施工质量关，必须提高管理。

（4）耗能和运行费用高。喷灌系统需要加压设备提供一定的压力，才能保证喷头的正常工作，达到均匀灌水的要求，在没有自然水压的情况下只有通过水泵进行加压，这需要消耗一部分能源（电、柴油或汽油），增加了运行费用。为解决这类问题，目前喷灌正向低压化方向发展。另外，在有条件的地方要充分利用自然水压，可大大减少运行费用。

（5）表面湿润较多，深层湿润不足。与滴灌相比，喷灌的灌水强度要大得多，因而存在表层湿润较多，而深层湿润不足的缺点，这种情况对深根作物不利，但是如在设计中恰当地选用较低的喷灌强度，或用延长喷灌时间的办法使水分充分渗入下层，则会大大缓解此类问题。

此外，对于尚处于小苗时期的作物，由于没有封行，在使用喷灌系统灌溉，尤其是将灌溉与施肥结合进行时，一方面很容易滋生杂草，从而影响作物的正常生长，另一方面，又加大了水肥资源的浪费；而在高温季节，特别是在南方，在使用喷灌系统灌溉时，在作物生长期间容易形成高温、高湿环境，引发病害的发生传播等。

半固定管道式喷灌系统中干管为固定设置，但支管移动使用，大大提高了支管的利用率，减少支管用量，使亩投资低于固定管道式喷灌系统。这种形式在我国北方小麦产区具有很大的发展潜力。为便于移动支管，管材应为轻型管材，如薄壁铝管、薄壁镀锌钢管，并且配有各类快速接头和轻便的连接件、给水栓。

软管牵引卷盘式喷灌机属于行喷式喷灌机，规格以中型为主，同时也有小型的产品。国外还应用钢索牵引卷盘式喷灌机，但仅适用于牧草的灌溉。软管牵引卷盘式喷灌机结构紧凑，机动性好，生产效率高，规格多，单机控制面积可达150~300亩，喷洒均匀度较高，喷灌水量可在几毫米至几十毫米的范围内调节。这种机型适合我国目前的经济条件和管理水平，只要形成农业的适度规模经营或统一种植，即可在一定范围内推广应用。

轻小型喷灌机组指以柴油机或电动机配套的喷灌机组，有手抬式和手推式两种，均属定喷式喷灌机。轻小型喷灌机组是适应20世纪70年代我国农村的动力情况发展起来的，经过20年的不懈努力，目前已形成动力2~12kW，配套完整、规格齐全、批量生产的喷灌主导产品之一。轻小型喷灌机组适应水源小而分散的山丘区和平原缺水区，具有一次性投资少，操作简单，保管维修方便，喷灌面积可大可小，适用于抗旱等优点。

三、设备及选型

喷灌设备又称喷灌机具，主要包括喷头、喷灌用水泵、喷灌管材及附件、喷灌机等。下面主要讲述喷头、喷灌机的使用和类型。

(一) 喷头的分类、性能及选型

1. 喷头类型

喷头的种类很多，通常按喷头工作压力或结构形式进行分类。

(1) 按工作压力分类。按工作压力分类，可以把喷头分为低压喷头、中压喷头和高压喷头，其中低压喷头的工作压力小于200kPa，射程小于15.5m，流量小于2.5m³/h；中压喷头的工作压力为200~500kPa，射程为15.5~42m，流量为2.5~32m³/h；高压喷头的工作压力大于500kPa，射程大于42m，流量大于32m³/h。

(2) 按结构形式分类。按结构形式，可以把喷头分为旋转式喷头、固定式喷头和喷洒孔管三类。

一是旋转式喷头。旋转式喷头又称为射流式喷头，其特点是边喷洒边旋转，水从喷嘴喷出时成集中射流状，故射程较远，且流量范围大，喷灌强度较低，是目前我国农田灌溉中应用最普遍的一种喷头形式。旋转式喷头的共同缺点是竖管不垂直时，喷头转速不均匀，因而会影响喷灌的均匀性。

a. 垂直摇臂式喷头。这是一种反应式喷头。垂直摇臂式喷头是一种中、高压型的喷头，除幼嫩作物外，可适应于各种作物。特别是应用在行走喷洒的系统中稳定性好。另外，它还可以喷洒污水或粪液等混合液体。

b. 全射流喷头。它的最大优点是无撞击部件，结构较简单，喷洒性能好。

二是固定式喷头。固定式喷头又称漫射式喷头或散水式喷头。此类喷头的特点是水流向全圆周或部分圆周（扇形）同时喷洒，射程短，湿润圆半径一般只有3~9m，喷灌强度较高，一般为15~20mm/h以上，多数喷头的水量分布是近处喷灌强度比平均喷灌强度大得多，通常雾化程度较高。

三是喷洒孔管。喷洒孔管是由一根或几根较小直径的管子组成，在管子的顶部分布有一些小的喷水孔，喷水孔直径仅1~2mm。根据喷水孔分布的形式，其可分为单列孔管和多列孔管两种。

a. 单列孔管。喷水孔按照一条直线等距排列，喷水孔间距为60~150cm，两根孔管之间距离通常为16m，孔管支架架在田间并借助自动摆动器的作用可在90°范围内绕管轴旋转，使得孔管两侧均可以喷到。自动摆动器可以是水轮带动的也可以是活塞带动的齿轮或涡轮组。单列孔管多为固定式的。主要用于苗圃及蔬菜地的喷灌。这样操作方便、生产率高，但基建投资相当高而且支架对耕作及其他田间作业有一定影响。可移动式的单列孔管应用较少，它通常支撑在较矮的支架或者其他便于移动的支撑物上，这样设备投资较低，但移动式孔管的劳动量则较大。

b. 多列孔管。由可移动的轻便管子构成，在管子的顶部钻有许多小孔，孔的排列可以保证两侧 6~15m 宽的田地能均匀地受到喷灌。多列孔管的工作压力较低，所以较适用于利用自然压力进行喷灌，而且它不需要自动摆动器，结构上比单列孔管简单得多。

2. 喷头的选型

喷头的选择包括喷头型号、喷嘴直径和工作压力的选择。在选定喷头之后，喷头的流量、射程等性能参数也就确定了。

（1）喷头的选型原则。按照国家标准《喷灌工程技术规范》的规定，喷头选择原则如下：一是组合后的喷灌强度不超过土壤的允许喷灌强度值；二是组合后的喷灌均匀系数不低于规范规定的数值；三是雾化指标值应符合作物要求的数值；四是有利于减少喷灌工程的年费用。

（2）喷头选型分析。小喷头要求的工作压力较低，能量消耗少，意味着运行成本较低，但由于其射程小，要求管道布置得较密，管道用量增大。大喷头射程远，管道间距大，要求的工作压力大，能量消耗较大，运行成本较高。所以在初选喷头时应根据具体条件经过技术经济分析多方面加以考虑。

（二）喷灌机的分类及选择

喷灌机是将喷头、管道、水泵、动力机等按一定配套方式组合，并在机械、水力、运行操作等方面符合要求的一种灌水机械。喷灌机自成体系，能独立在田间移动作业。喷灌机进行大面积喷灌时，应当在田间布置供水系统或水源，供水系统可以是明渠，可以是有压管道。如果有压管道的水压力能满足喷灌机的入机压力要求，喷灌机可以不配装动力机和水泵。

1. 喷灌机的分类

为了适应不同地形和作物的要求，喷灌机的形式多种多样。根据运行中喷头的喷洒方式可将喷灌机分成定喷式和行喷式两大类。定喷式是指喷灌机停在一个位置上进行喷洒，一个位置喷完后，喷灌机按设计要求移动到下一个新位置后再进行喷灌作业，直至全部控制面积上都喷灌完毕。定喷式喷灌机包括：手提式喷灌机、手抬式喷灌机、手推式喷灌机、拖拉机悬挂式喷灌机、滚移式喷灌机等。

行喷式喷灌机是指一边移动一边进行喷洒作业的喷灌机。行喷式喷灌机包括卷盘式、中心支轴式喷灌机、平移式喷灌机等。

2. 喷灌机的结构特点及其使用范围

（1）手提式喷灌机。手提式喷灌机可由一个人搬移，它的动力使用 0.37~

1.5kW 的微型电动机或 1.5 马力①的风冷内燃机；水泵采用配手压泵的微型高速离心泵或微型喷灌离心泵，水泵与动力机多直联；管道可用锦塑软管或低密度聚乙烯管；喷头采用工作压力为 150~250kPa 的低压喷头。手提式喷灌机结构简单，重量轻，安装操作容易，工作压力低，耗能少，价格便宜，能用 220V 的单相电源，还能充分利用小水源。它适合我国经济体制改革后农村一家一户的经营方式，可用于喷灌小面积的大田作物、蔬菜和经济作物，也可用于喷灌庭院作物和绿地。

（2）手抬式喷灌机。它的动力机一般为 3~5kW 或 3~6 马力的风冷柴油机；水泵采用自吸离心泵以适应经常移动的特点，水泵与动力机多用联轴器直联，管道常用锦（维）塑软管或薄壁铝，喷头多采用工作压力合金管，管径一般为 50mm 或 65mm，喷头多采用工作压力为 300~350kPa 的中压摇臂式喷头。图 2-36 所示为手抬式喷灌机结构示意图。

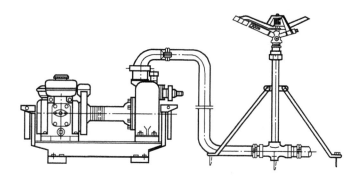

图 2-36 手抬式喷灌机结构示意图

手抬式喷灌机结构紧凑，重量较轻，操作、保养简单易行，无需留机行道，适应性强，价格较便宜，但当喷头用竖管直接装在水泵出口处时，由于机组振动大，会影响喷灌质量。

手抬式喷灌机适用于分散的小地块，特别是山区丘陵的复杂地形，可喷灌粮食作物、蔬菜、经济作物、苗圃及果树等。

（3）手推式喷灌机。当动力机和水泵稍大些时，采用两人手抬来移动就比较费力，因此，就将水泵、动力机及传动机构固定在装有胶轮的车架上，喷头安装在引出的支管上，可进行多喷头组合喷洒，手推式喷灌机的动力机一般采用

---

① 1 马力约为 735W，全书同。

7.5kW 的电动机或 10~12 马力的柴油机；水泵配用自吸离心泵或有自吸装置的普通离心泵，水泵与动力机可以直联，也可以用 V 带传动；管道多采用由快速接头连接的铝合金管，按一条或两条支管配置；喷头采用中、低压摇臂式喷头，一般配置 8~12 个喷头。图 2-37 所示为手推式喷灌机示意图。

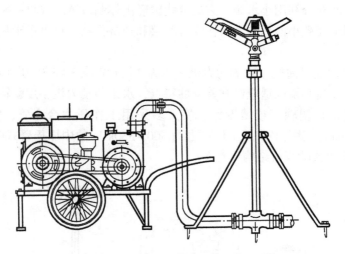

图 2-37 手推式喷灌机示意图

手推式喷灌机结构简单，投资及运行费用较低，使用维修机动灵活，特别是当选用柴油机作动力时，由于 10~12 马力的柴油机是目前我国农村保有量最大的动力机，可实现一机多用，节省投资。它的主要缺点是：当在黏重土壤上使用时，若喷湿了机组周围，会造成道路泥泞、移动困难。这种机型适用于各种作物，特别是平原地区的小块地。

（4）滚移式喷灌机。滚移式喷灌机是一种较成熟的机械移动式定喷喷灌机。在国外，如美国、德国等有一定使用，我国也生产这种机型。

滚移式喷灌机实际上就是利用机械来整体移动装有喷头并连接好的支管，喷灌机采用单元组装多支点的结构，由轻便高强度的铝合金管、大直径钢圈式高强度的铝合金管、中央驱动车及带快速接头的进水软管等部分组成。图 2-38 所示为滚移式喷灌机工作示意图。

这种机型的优点是结构简单，维修容易，控制面积大，生产效率较高，配备机手少，日移动次数少，操作方便；田间工程量少，占地少。它的主要缺点是受滚轮直径的影响，不能灌溉高秆作物；适应地形坡度和土壤的能力较差。

第二章 水肥一体化技术的主要设备和系统

1—水源；2—抽水机组；3—输水干管；4—给水栓；5—连接软管；6—钢圈式轮；7—喷头；8—喷洒支管；9—驱动车。

图 2-38 滚移式喷灌机工作图

这种机型适应在坡度小于10%的较平整的非黏性土的地块中使用。常用于喷灌小麦、谷物、豆类、蔬菜、瓜类等矮秆作物，特别适合于牧草喷灌。

（5）卷盘式喷灌机。卷盘式喷灌机是指用软管输水，在喷洒作业时利用喷灌压力水驱动卷盘旋转，卷盘上卷绕软管（或钢索），牵引装有一个高压喷头的小车或装有若干个固定式喷头的悬臂桁架，使其沿作业线自行移动并喷洒的喷灌机。图 2-39 所示为软管牵引卷盘式喷灌机示意图。

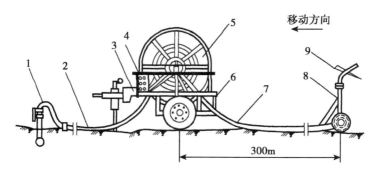

1—给水栓；2—供水管；3—水动力机；4—调速器；5—卷盘车；6—机架（底盘）；7—PE半软管；8—喷头车；9—远射程喷头。

图 2-39 软管牵引卷盘式喷灌机示意图

这种喷灌机一般由喷头车、卷盘车两大部分组成，利用压力干管或移动式抽水装置供给压力水。卷盘车包括卷盘、半软管、机架、行走轮、动力机、调速装置及安全机构等。喷头车较简单,包括喷头和车架。运输时,喷头车多数可以装载在卷盘车上。这种喷灌机由于机体较大,故在田间作业和运输时需由拖拉机牵引。卷盘式喷灌机与其他大中型喷灌机相比,有如下优缺点。

优点为:①结构简单,亩投资较低。②规格多,机动性好,适用范围广。③操作技术简单,可自动控制,生产率较高,一台机器可由一个人管理(另请一拖拉机手临时辅助工作),每天移动 1~2 处,喷头车按 20~40m/h 的速度自行移动,每天可喷灌 20~60 亩。大型卷盘式喷灌机的控制面积为 200~300 亩。

缺点为:①受机型限制管径小、长度大,而且要逐层缠绕,因而水头损失大、耗能多,运行费高。如管径 110mm、长 270m 时,供水管加上水涡轮的损失,达到喷头工作压力的 76%~98%。一般管路损失占入机压力的 43%~49%,即一半左右的能量在管道中损失掉了。如喷灌次数频繁,应对其适用性慎加考虑。②要求较宽的机行道(运输道、喷头车工作道),占地较多。为此,应统筹考虑喷灌机的规格和田间规划,以尽可能减少占地。

卷盘式喷灌机能够适应各种大小形状和地形起伏的地块,适应灌溉各种高秆和矮秆作物(如玉米、大豆、土豆、牧草等),以及某些果树和经济作物(如甘蔗、茶叶、香蕉等),但要求土质不要太黏重。

四、管道式喷灌系统设计

(一)喷灌的技术要求

(1)适时适量地给作物提供水分。要做到这一点,必须制订合理的灌溉体系,保证在干旱年或半干旱年作物正常生长对水分的要求。也就是说,喷灌工程的设计标准必须满足灌溉保证率不低于 85%,必须按这个标准为喷灌工程配套容量合理、工程结构可靠、运行安全方便、各部分的规格尺寸能保证喷灌灌溉制度实施的水源工程。

(2)有较高的喷灌均匀度。这里的喷灌均匀度指的是组合均匀度,它与单个喷头的水量分布情况、喷头的工作压力、喷头的布置形式与间距、喷头转速的均匀性、竖管安装的倾斜度、地面坡度、风速风向等因素有关。要求在设计风速下,定喷式喷灌系统的组合均匀系数不低于 75%,行喷式喷灌机的组合均匀系数不低于 85%。

(二)田间管道系统的布置

田间管道系统的布置取决于田块的形状、地面坡度、耕作与种植方向、灌溉季节的风速与风向、喷头的组合间距等情况,需进行多方案比较,择优选用,主要有以下两种形式。

（1）丰字形布置。如图 2-40 所示。

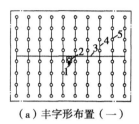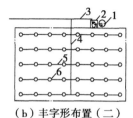

（a）丰字形布置（一）　　（b）丰字形布置（二）

（a）1—井；2—泵站；3—干管；4—支管；5—喷头。（b）1—蓄水池；
2—泵站；3—干管；4—分干管；5—支管；6—喷头。

图 2-40　丰字形布置

（2）梳齿形布置。如图 2-41 所示。

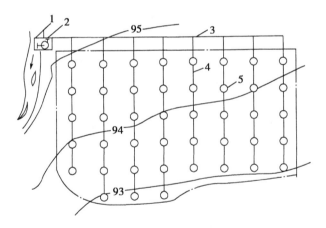

1—河渠；2—泵站；3—干管；4—支管；5—喷头。

图 2-41　梳齿形布置

1. 田间管道系统布置的原则

（1）应符合喷灌工程规划的要求。

（2）喷洒支管应尽量与耕作和作物种植方向一致。

（3）喷洒支管最好平行等高线布置，如果条件限制，至少也应尽量避免逆坡布置。

（4）在风向比较恒定的喷灌区，支管最好垂直于主风向布置，应尽量避免平行主风向布置。

(5) 喷洒支管与上一级管道的连接,应避免锐角相交,支管铺设应力求平顺、减少折点。

在贯彻以上原则时,有时会出现矛盾,这时应根据具体情况进行分析比较,分清主次,因地制宜地确定布置方案。

2. 影响田间管道系统布置的主要因素

影响田间管道系统布置的因素很多,经常会遇到各因素之间相互制约的现象,造成在相同的条件下常常能作出多种可能的布置方案来。为了从中选出技术上、经济上最有利的方案,下面把影响管道布置的主要因素介绍如下。

(1) 地形条件。在地形起伏的喷灌区,喷洒支管常常无法全部沿等高线布置。这时支管应顺坡垂直等高线或与等高线斜交铺设,以下降的地形高度来弥补支管沿程的水头损失。如果地形坡降正好等于或接近支管的水力坡降则最为理想;如果地形坡降比支管的水力坡降大得多,则应于适当位置布置减压阀或采用减小管径的办法来解决。对于上一级管道只能布置在低处的情况,逆坡铺设的支管不能太长。

(2) 地块形状。地块形状不规则会给田间管道布置带来困难。一般对半固定式和移动式管道系统来说,支管在地块中的走向应一致,应尽量使多数支管的长度相同。

(3) 耕作与种植方向。有的喷灌区处于慢坡地带,传统的耕作、种植方向是顺坡。若喷洒支管按平行于等高线布置,则与耕作、种植方向不能保持一致。当喷洒支管移动使用时会造成很多困难,甚至损伤作物。这时应按耕种方向顺坡布置喷洒支管。有时在同一地块内存在不同的耕作种植方向,这时就应通过技术经济分析和方案比较,将耕作方向调整统一。

(4) 风向和风速。风对喷灌的灌水质量影响很大。在喷灌季节若喷灌区内风速很小,则喷洒支管的布置可不考虑风向,而以满足别的要求为主。若风速达到或超过 $2m/s$ 且有主风向时,喷洒支管应垂直主风向布置,这样在风的作用下喷头横向射程可用密支管上的喷头数来弥补。

(5) 水源位置。这里主要是指平原井灌区,一眼井控制 200 亩左右土地,形成一个小系统。

第四节 水肥一体化技术的输水系统——微灌

一、微灌的概念

微灌，即是按照作物需水要求，通过低压管道系统与安装在末级管道上的特制灌水器，将水和作物生长所需的养分以较小的流量均匀、准确地直接输送到作物根部附近的土壤表面或土层中的灌水方法。与传统的地面灌溉和全面积都湿润的喷灌相比，微灌只以少量的水湿润作物根区附近的部分土壤，因此又叫局部灌溉。

微灌灌水流量小，一次灌水延续时间较长，灌水周期短，需要的工作压力较低，能够较精确地控制灌水量，能把水和养分直接地输送到作物根部附近的土壤中去。按灌水时水流出流方式的不同，可以将微灌分为如图2-42所示四种形式。

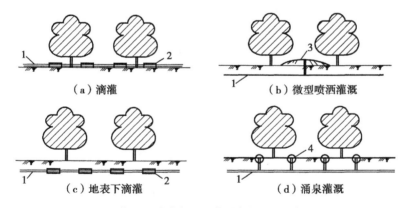

(a) 滴灌　　(b) 微型喷洒灌溉
(c) 地表下滴灌　　(d) 涌泉灌溉

1—毛管；2—微喷头；3—微型喷头；4—涌水器。

图2-42　微灌的形式

二、微灌的特点

（一）微灌的优点

（1）节省能源，减少投资。微喷头也属于低压灌溉，设计工作压力一般在150~200kPa，同时微喷灌系统流量要比喷灌小，因而对加压设施的要求要比喷灌小得多，可节省大量能源，发展自压灌溉对地势高差也比喷灌小。同时由于设计工作压力低，系统流量小，又可减少各级管道的管径，降低管材压力，使系统的总投资大大下降。

（2）调节田间小气候，易于自动化。由于微喷灌水滴雾化程度大，可有效增加近地面空气湿度，在炎热天气可有效降低田间温度，甚至还可将微喷头移至树冠上，以防止霜冻灾害等。

（3）水分利用率高，增产效果好。微喷灌也属于局部灌溉，因而实际灌溉面积要小于地面灌溉，减少了灌水量。同时微喷灌具有较大的灌水均匀度，不会造成局部的渗漏损失，且灌水量和灌水深度容易控制，可根据作物不同生长期需求规律和土壤含水量状况适时灌水，提高水分利用率，管理较好的微喷灌系统比喷灌系统用水可减少25%~35%。微喷灌还可以在灌水过程中进行喷施可溶性化肥、叶面肥和农药，具有显著的增产作用，尤其对一些对温度和湿度有特殊要求的作物增产效果更明显。

（4）灵活性大，使用方便。微喷灌的喷灌强度由单喷头控制，不受邻近喷头的影响，相邻的两微喷头间喷洒水量不相互叠加。这样可以在果树不同生长阶段通过更换喷嘴来改变喷洒直径和喷灌强度，以满足果树生长需水量。微喷头可移动性强，根据条件的变化可随时调整其工作位置，如树上、行间或株间等，有些情况下微喷灌系统还可以与滴灌系统相互转化。

（二）缺点

微灌的局限性通常会有以下几种情况：首先，田间微喷灌易受杂草、作物茎秆的阻挡而影响喷洒质量；其次，在作物未封行前，微喷灌结合喷肥会造成杂草大量生长；再次，对水质要求较高。水中的悬浮物等容易造成微喷头的堵塞，因而要求对灌溉水进行过滤；然后，灌水均匀度受风影响较大。在大于3级风的情况下，微喷水滴容易被风吹走，灌水均匀度降低。因而微喷头的安装高度在满足灌水要求的情况下要尽可能低一些，以减少风对喷洒的影响。

三、设备及选型

（一）灌水器

为了满足微灌发展的需要，国外已经研制出各式各样的灌水器，其分类如下。

1. 按灌水器与毛管的连接方式分类

（1）管间式。把灌水器安装在两段毛管的中间，使灌水器本身成为毛管的一部分。例如，把管式滴头两端带倒刺的接头分别插入两段毛管内，使绝大部分水流通过滴头体内腔流向下一段毛管，而很少的一部分水流通过滴头体内的侧孔进入滴头流道内，经过流道消能后再流出滴头。图 2-43 所示为管间式滴头。

图 2-43　管间式滴头

（2）管上式。直接插装在毛管壁上的灌水器，如傍插式滴头、微管、涌水器、孔口滴头和微喷头等，均属于管上式灌水器。图 2-44 所示为管上式滴头。

图 2-44　管上式滴头

2. 按灌水器的出水方式分类

（1）滴水式。滴水式灌水器的出流特征是毛管中的压力水流经过消能后以不连续的水滴或细流形式向土壤灌水。如管式滴头、孔口滴头、涡流滴头等均属于滴水式灌水器。图 2-45 所示为滴水式滴头。

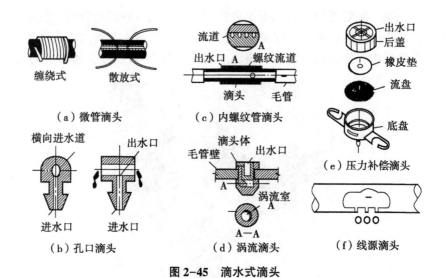

图 2-45 滴水式滴头

(2) 喷水式。压力水流通过灌水器的孔口以喷射方式向土壤灌水。根据喷洒方式不同又可分为射流旋转式和折射式两种。图 2-46 所示为喷水式喷头。

图 2-46 喷水式喷头

(3) 涌泉式。毛管中的压力水流以涌泉的方式通过灌水器向土壤灌水。涌泉灌溉的优点是工作水头低，孔口直径较大，不易堵塞。图 2-47 所示为涌水器示意图。

(4) 渗水式。毛管中的压力水通过毛管壁上的许多微孔或毛细管渗出管外而进入土壤。渗水毛管有两种形式，即多孔透水毛管和边缝式薄膜管。边缝式薄膜管是利用结合缝形成的毛细通道来渗水的。毛细通道一般宽为 0.1~0.25mm，

第二章 水肥一体化技术的主要设备和系统

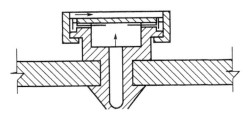

图 2-47 涌水器示意图

高为 0.7~2.5mm，长为 150~600mm。

（5）间歇式。毛管中的压力水流以间歇、脉冲的方式流出灌水器而灌入土壤。因此，又把间歇式称作脉冲灌水方式。图 2-48 所示为间歇式灌水器示意图。

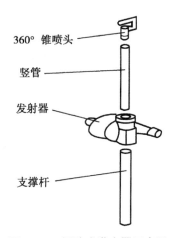

图 2-48 间歇式灌水器示意图

（二）滴头及微喷头的选择

1. 滴头

通过流道或孔口将毛管中的压力水流变成滴状或细流状的装置称为滴头，其流量一般不大于 12L/h。按滴头的消能方式可把它分为以下几种。

（1）长流道型滴头。长流道型滴头是靠水流与流道管壁之间的摩阻消能来调节出水量大小的。如微灌滴头、内螺纹管式滴头等，如图 2-49、图 2-50 所示。

（2）孔口型滴头。孔口型滴头靠孔口出流造成的局部水头损失来消能调节

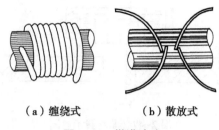

（a）缠绕式　　（b）散放式

图 2-49　微灌滴头

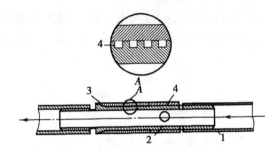

1—毛管；2—滴头；3—滴头出水；4—螺纹。

图 2-50　内螺纹管式滴头

出流量的大小，如图 2-51 所示。

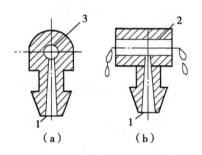

1—进口；2—出口；3—横向出水道。

图 2-51　孔口型滴头

（3）涡流型滴头。涡流型滴头靠水流进入灌水器的涡室内形成的涡流来消能调节出水量的大小。水流进入涡室内，由于水流旋转产生的离心力迫使水流趋向涡室的边缘，在涡流中心产生一个低压区，使中心的出水口处压力较低，因而调节流量，如图 2-52 所示。

第二章 水肥一体化技术的主要设备和系统

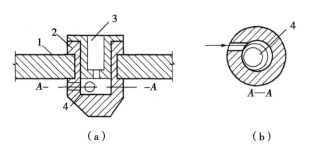

1—毛管壁；2—滴头体；3—出水口；4—涡流室。

图 2-52 涡流型滴头

（4）压力补偿型滴头。压力补偿型滴头是利用水流压力对滴头内的弹性体（片）的作用，使流道（或孔口）形状改变或过水断面面积发生变化，即当压力减小时，增大过水断面面积；压力增大时，减小过水断面面积，从而使滴头出流量自动保持稳定，同时还具有自清洗功能。滴头名称和代号表示方法如图 2-53 所示。

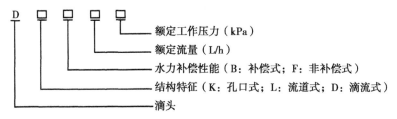

图 2-53 滴头名称和代号表示方法

2. 微喷头

微喷灌与一般喷灌相比，具有流量小、喷洒射程近、喷水孔直径小、工作压力低等特点，与一般喷灌的最主要区别在于喷头。因此，微喷灌常常作为微灌的一部分，但与滴灌相比，微喷灌的湿润面积大，抗堵塞性能好。

微喷灌除喷头以外，管网部分与滴灌系统或喷灌系统相同。微喷头种类很多，根据微喷头的喷洒方式，可分为折射式微喷头、旋转式微喷头、离心式微喷头、缝隙式微喷头等几类。

（1）折射式微喷头。折射式微喷头主要靠喷嘴顶部的折射板使喷嘴的压力水流改变方向，向四周喷射，在空气阻力的作用下，水流被粉碎成雾化的水滴降落到地面进行灌溉。根据折射板的结构和形状，折射式微喷头又可分为单向折射微喷头、双向折射微喷头、角度折射微喷头、梅花形折射微喷头等形式。折射式微喷头的喷洒形状如图2-54所示。喷洒的水流有线状和全圆两种。

图2-54 折射式微喷头的喷洒形状

（2）旋转式微喷头。旋转式微喷头依靠喷嘴的射流，使带有曲线形导流槽的旋转臂旋转，将水流喷洒在以喷嘴为中心的地面上。旋转式微喷头转动形式有旋臂式、旋轮式、旋转长臂式等多种。其中，旋转长臂式还有双臂双喷嘴、三臂三喷嘴、四臂四喷嘴等几种类型。旋转式微喷头一般为全圆喷灌。

（3）离心式微喷头。离心式微喷头依靠喷嘴中螺旋形流道产生的离心作用将水流分散，这种喷头有塑料和铜制两种，流量和射程均可调节。射程范围在3~9m。离心式微喷头与旋转式微喷头一样，一般为全圆喷灌。

（4）缝隙式微喷头。缝隙式微喷头通过喷头上多个出流小孔或缝隙喷洒，喷洒方式与折射式微喷头类似，有不同喷洒角度和不同范围的喷头。

第三章 适合水肥一体化技术的肥料的选择

本章主要讲解了作物吸收养分原理、肥料的选择和养分管理。重点在于肥料的选择，详细地介绍了在什么样的情况下适合用什么肥料，并且也讲到两种肥料之间是否有相互作用。

第一节 作物吸收养分原理

一、作物必需的营养元素

一种元素是否为植物在生长的过程中所需要的营养元素（图3-1）的标准有3个：一是缺少一种元素导致植物不能完成整个生命周期；二是缺少一种元素之后，植物在生长的过程中会出现奇怪的症状，只有补充这种元素，这些奇怪的症状才会消失；三是这种元素对植物的新陈代谢有着直接的营养作用，不会对植物的生长环境产生作用。

大量元素包括碳、氢、氧、氮、磷、钾。在植物体内一般都是百分之几的含量。碳、氢、氧也是有机物的组成元素，主要来源是水和空气，植物对这三种元素需求量比较大，然而土壤中含量比较少，只能通过施肥来满足需求，氮、磷、钾肥是植物需要量较多的肥料。

中量元素包括钙、镁、硫3种元素。在植物体内一般都是千分之几的含量，在土壤中这几种元素含量比较多，所以不需要通过施肥来补充，然而在南方地区降水量比较大，所以需要通过施肥来补充它们的含量。

微量元素包括铁、铜、锌、锰、钼、硼和氯。在植物体内一般都是万分之几的含量，虽然这些元素比较少，但是对植物的作用是很重要的，土壤中含量容易满足植物的需求，但有的微量元素在土壤中不能满足植物的需求，也需要通过施肥来进行补充。

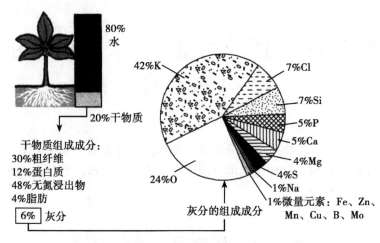

图 3-1 植物矿质元素构成

有些元素称为有益元素,是对植物的生长有一些作用,但不是必需的元素,或者是对某些元素在特定的条件下有作用。例如钠、硅、钴、钒、硒、铝、碘、铬、砷、铈等。有益元素的使用量要适中,不然对于植物的损伤是十分严重的,过少会影响植物正常生长,过多会使植物发生毒害。植物对于有益元素的需求量一般都很少,所以需要控制好有益元素的使用量。

二、适量施肥的重要规律

(一)养分的归还(补偿)学说

养分归还学说是德国农业化学家李比希(J. V. Liebig)在1840年《化学在农业和生理学上的应用》一书中提出的。李比希从自然科学的观点出发,把整个地球上的生命现象看作是一个运动的循环过程,认为农业是人类和自然界进行物质交换的基础,人类和动物通过食物摄取了植物从土壤和大气中吸收同化的营养物质,并通过植物本身和动物排泄物归还于土壤。他提出,由于人类在土地上种植作物并把这些产物拿走,这就必然会使地力逐渐下降,从而使土壤所含的养分愈来愈少。因此,要想恢复地力就必须归还从土壤中拿走的全部东西,不然就难于指望再获得过去那样高的产量,为了增加产量应该向土壤中施加肥料。

养分归还(补偿)学说的内涵包括以下几个要点:①植物每次在收获的时候会把土壤中的一些养分给带走,久而久之,土壤中的养分会越来越少;②随之土壤中的养分越来越少,植物的产量就越来越少,所以需要及时地补充土壤中的养分，保持土壤肥力；③通过施肥来保持土壤中各种元素的平衡以提高产量。

养分归还（补偿）学说也存在片面性：①李比希认为大气中的碳酸盐是作物氮素营养的唯一来源，所以，他对土壤养分消耗估计只着眼于磷、钾等矿质元素上；②李比希基于矿质营养学说，过高评价矿质营养成分，而过低评价了有机营养的作用，同时，反对布森高关于厩肥主要供给氮素的说法，错误地认为植物所需要的矿物质是以厩肥的形式供给土地的；③反对布森高关于豆科作物能丰富土壤氮素的说法，错误地认为植物仅能从土壤中摄取生活所需的营养矿物质，每次收获都从土壤中取走物质而使土壤贫化；④要想恢复地力必须归还从土壤中拿走的全部东西。而实际施肥并非需要归还作物从土壤的取走的任何营养物质，比如土壤养分含量较高的物质，或作物生长非必需的营养元素等。

（二）报酬递减律与米采利希学说

报酬递减律（图3-2）是18世纪经济学家杜尔哥（A. R. Turgot）和安德森（J. Anderson）提出的，报酬递减律表述为："从一定土地上所得到的报酬随着向该土地投入的劳动和资本量的增加而有所增加，但随着投入的单位劳动和资本的增加，报酬的增加却在逐渐减少。"米采利希（E. A. Mitscherlich）等在20世纪初期，在前人工作的基础上深入探讨了施肥量与产量的关系（燕麦磷肥的砂培试验），通过试验发现：①在其他技术条件相对稳定的前提下，随着施肥量的渐次增加，作物产量也随之增加，但作物的增产量却随施肥量的增加而呈递减趋势，即（$\Delta Y_1/X_1 > \Delta Y_2/X_2 > \Delta Y_3/X_3 \cdots$），因而与前人提出的报酬递减律相吻合；②如果一切条件都符合理想的话，作物将会产生出某种最高产量，相反，只要有任何某种主要因素缺乏时，产量便会相应地减少。

米采利希公式只反映了施肥增产的规律但未能反映当施肥量超过某一范围后作物减产的现象。费佛尔（Pfeiffer）等提出了反映施肥量与产量之间的抛物线模式，其一元二次方程为：$y = b_0 + b_1 x + b_2 x_2$，方程式中 b_0、b_1、b_2 为系数，可以根据试验数据求出。此式反映的方程效应为：当施肥量很低时，作物产量几乎成直线上升，而当施肥量超过最高产量施肥量时，增施肥料不但不能增产，相反产量会下降。这一结果已被国内外许多试验所证实。

（三）同等重要律

作物需要的元素，不管是大量元素还是微量元素，都是缺一不可的，即使作物需要这种元素不是很多，但是缺少这种元素也会影响作物的正常生长，导致某些功能衰减。同等重要律表示，微量元素、稀有元素和大量元素对于作物来说同等重要。

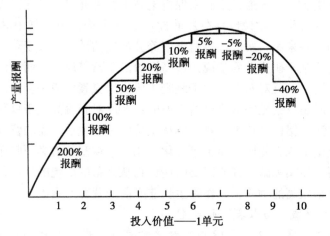

图 3-2 报酬递减律图示说明

（四）不可替代律

作物体内的各种元素都有一定的功效，不可以让其他元素替代，比如在缺少钾的情况下是不可以让氮进行替换的。作物体内缺少什么元素，就必须施用含有这种元素的化肥。

（五）因子综合作用律

影响作物正常生长发育的因子有：水分、养分、光照、空气、品种以及耕作条件等。在这其中肯定有一个是起主导作用的限制因子，产量也在一定程度受这种因子制约着。为了提高农业的经济效益，在施肥的时候必须与一些农业技术措施相互配合，并且各个养分之间也必须相互配合施用。

三、影响作物吸收营养元素的因素

作物一般都是通过根在土壤中吸收营养元素。所以，除了作物本身的遗传特性外，还有一些因素会影响作物对营养元素的吸收，比如，土壤和其他环境因子。

影响养分吸收的因素主要包括介质中的养分浓度、温度、光照强度、土壤水分、通气状况、土壤 pH 值、养分离子的理化性质、根的代谢活性、苗龄、生育时期植物体内养分状况等。

（一）介质中养分浓度

通过大量的研究发现，在浓度比较低的时候，离子的吸收率会随着养分的浓度上升而上升,但是速度不是特别的快,在浓度比较高的时候,离子的吸收选择性较低,主要是因为蒸腾速率对离子的吸收产生了很大的影响。所以,在使用化肥施肥时,一般选择分批使用,这样有利于作物的吸收。

(二) 介质中的养分种类

介质中的养分离子间具有一定的拮抗作用与协助作用。所谓的离子间的拮抗作用是指在溶液中某一离子存在能抑制另一离子吸收的现象。表现在阳离子与阳离子之间,如 K^+、Rb^+、Cs^+ 之间,Ca^{2+}、Mg^{2+}、Ba^{2+} 之间,NH_4^+ 和 H^+ 对 Ca^{2+}、K^+ 对 Fe^{2+} 之间;表现在阴离子与阴离子之间,如 Cl^-、Br^- 和 I^- 之间,$H_2PO_4^-$ 和 OH^- 之间,$H_2PO_4^-$ 和 Cl^- 之间,NO_3^- 和 Cl^- 之间,SO_4^{2-} 和 SeO_4^{2-} 之间。

所谓离子间的协助作用是指在溶液中,某一离子的存在有利于根系对另一些离子的吸收。表现在阴离子与阳离子之间,如 NO_3^-、SO_4^{2-} 等对阳离子的吸收有利;表现在二价或三价阳离子对一价阳离子之间,如 Ca^{2+}、Mg^{2+}、Al^{3+} 等能促进 K^+、Rb^+、Br^- 以及 NH_4^+ 的吸收。

(三) 根系的代谢及代谢产物的影响

由于离子和其他溶质在很多情况下是逆浓度梯度的累积,所以需要直接或间接地消耗能量。在不进行光合作用的细胞和组织中(包括根),能量的主要来源是呼吸作用。因此,所有影响呼吸作用的因子都可能影响离子的累积。

(四) 苗龄和生育阶段

(1) 作物种子的营养吸收。在种子发芽的前后,一般都是依靠贮存营养物质进行吸收,到了三叶时期,则需要通过介质来提供营养了。

(2) 作物在不同的生长阶段所吸收的养分是不同的,在生长初期,吸收的比较少,但是随着时间的推移,吸收的养分就越来越多。到了之后的成熟阶段,作物对于营养元素的吸收也就逐渐变小了。

(五) 植物营养临界期与植物营养最大效率期

植物的营养临界期和营养最大效率期是在施肥时最关键的两个时期。植物的营养临界期主要是指过多或过少的营养元素,造成植物体内营养元素不平衡,对作物的正常生长有着明显的不良影响的那段时期。多出现在作物生育前期,如磷素营养的临界期多出现在幼苗期。植物的营养最大效率期主要是指吸收的营养物质在植物体内能发挥最大功效的那段时期。多出现在作物生长最旺盛的时期,在这一时期,由于作物生长能力迅速,吸收的养分也特别的多,如果想要达到增产的目的,那么在这段时间就必须满足作物对养分的需求。在肥水管理过程中既要重视植物需肥的关键时期,又要正视植物吸肥的连续性,采用基肥、追肥、种肥相结合的方法。

四、施肥误区

在如今,有许多的施肥方法、施肥时间、施肥量运用的都不是很恰当,出现很多误区,这样施肥后会导致作物出现死亡的现象。主要的误区主要有以下几点:

(一) 偏施氮肥,忽视磷、钾肥

肥料的施用只注重氮肥会造成茎叶徒长、组织柔软,抗病能力降低,会影响作物后期的正常生长发育,影响营养物质的转化,使产量降低、品质下降等。所以,在施用氮肥时要适量、适时,并且需要与钾肥和磷肥配合起来施用,以此来保障作物的正常生长,提高产量。

(二) 忽视微量元素的施用

只注重大量元素的施用,而忽略微量元素的施用,这样不仅微量元素跟不上,而且大量元素的吸收也会受到影响,造成不必要的浪费,忽略微量元素会导致植株畸形、落花落果、作物产量和品质下降等。所以,应该根据作物的特性在施用大量元素的同时,配合着微量元素的施用,以此来保证作物的正常生长发育。

(三) 施肥越多越好

虽然有时候施肥量过大,能够增加作物的产量,但是由于成本过高,实际下来的收益却不怎么高;有的时候施肥量过大会导致作物生殖器官发育不足,导致产量下降,这样做只会适得其反。所以,应该根据作物的特性,以及土壤肥力和作物的种植密度等找出最佳的施肥方案,避免出现浪费的现象,充分发挥肥料的效果,以此来增加经济效益。

(四) 出现缺肥症状后再施肥

由于肥料在施入后,还需要一段时间作物才能吸收利用。如果在作物出现症状后再施入肥料,不但会延长作物缺少肥料的时间,还会导致作物减产,所以,肥料的施入时间应该根据作物的一般特性和光、温、水、施肥方法等因素来确定。

(五) 施肥表施

肥料由于易挥发、流失或者难以到达作物的根部,导致作物很难吸收,致使肥料利用率极低。所以,在施肥的时候一定要确定好施肥的位置,保障土壤中的根部成功吸收到,确保施肥的效果。

第三章 适合水肥一体化技术的肥料的选择

第二节 肥料的选择

一、化肥的特性及选择

肥料的种类很多，一般分为化学肥料和有机肥料两大类。化学肥料又称为矿质肥料或无机肥料，是指用化学方法合成或某些矿物质经过机械加工而生产的肥料，有些属工业副产品。化学肥料与有机肥料相比较，既有许多优点，也有不少缺点。

化学肥料的成分是无机物质，养分种类集中。如化学氮肥的主要营养成分是氮素，化学磷肥的主要营养成分是磷素，化学钾肥的主要营养成分是钾素等。复混肥料中虽含有几种营养元素，但主要营养成分还是氮、磷、钾中的二种或三种元素。化学肥料的营养元素含量比有机肥料高得多，如碳酸氢铵含氮17%，硫酸铵含氮21%，尿素含氮46%；而有机肥料养分含量较低，如人粪尿一般含氮（N）0.5%~0.8%、含磷（P_2O_5）0.2%~0.4%、含钾（K_2O）0.2%~0.3%，厩肥平均含氮（N）0.55%、磷（P_2O_5）0.22%、钾（K_2O）0.55%。1kg尿素相当于60~70kg人粪尿、100kg厩肥中的含氮量。此外，化学肥料运输和贮存方便，如管理适当，养分损失很少；而有机肥料的积制、堆腐、贮存等管理过程比较麻烦，如措施不当，很容易损失养分。所以，相比较而言，化学肥料养分含量高，施用量少，肥效显著，也便于运输、贮存和施用。

常用化学肥料多数是水溶性的，施入土壤后，很容易溶解于水，能迅速被作物根系吸收，肥效较快，如硫酸铵、尿素等，施入土壤后几天内就可见效。由于化学肥料养分含量高，肥效快而及时，对作物的增产效果十分明显。据全国化肥试验网的结果统计，每千克化肥可增产粮食5~10kg。据联合国粮农组织（FAO）的统计资料，化肥对农作物增产的贡献份额占40%~60%。而有机肥料虽养分种类多，但肥效缓慢、稳定而持久，并有较长后效，在改良和培肥土壤方面较化学肥料优越。作物生长需要适宜的养分比例，而化肥的养分单一，可以按不同比例人为地配合施用，这样一方面可以满足各种作物对不同养分的要求比例，另一方面配合施用比单独施用增产效果显著，这种现象称为肥料的交互效应。化肥的施用具有很强的针对性，可以按土、按作物施用不同种类和特性的肥料，按作物不同生育期采用不同的施肥方式，如不同时期追施不同肥料、叶面喷施技术等。特别是在现代配方施肥技术中，用化肥容易操作。

化肥也存在一些缺点。目前，我国常用的化学肥料仍以单元肥料或单质肥料（即以一种主要营养元素为主的化学肥料）为主，如氮肥、磷肥、钾肥等，这类肥料养分种类单一，含量高，若施用不当，容易造成作物吸收的养料比例失调。常用的化学肥料多属水溶性，施入土壤后容易流失、挥发和被土壤固定，肥效不持久，肥料利用率低。如铵态氮肥在碱性土壤中表施极易造成氨的挥发损失，在多雨地区或灌溉条件下施用硝态氮肥容易随水流失，水溶液磷肥施入土壤后极易发生固定作用，从而降低肥效。长期大量施用单施化肥，在一定程度上会造成土壤板结，土壤结构变差，土壤盐分含量增加，发生次生盐渍化，使土壤理化性质变坏。此外，化学肥料与有机肥料的另一个重要区别是不含有微生物，对土壤养分的转化和土壤生物性质的改善不利。因此，化学肥料在改土培肥方面不及有机肥料。化学肥料是高耗能产品，它的生产和使用成本较高，单施化肥会增加农业生产成本；而有机肥料可以就地积制，就地使用，有利于降低农业生产成本。另外，从施肥对环境的影响来看，化学肥料较有机肥料更易造成环境污染。

二、用于灌溉施肥的化肥种类

灌溉施肥技术有着严格的要求，特别是对于设备、肥料以及管理方式。滴灌灌水器的流道比较细小，所以在使用肥料时只能选择水溶性固态肥料或者液态肥，防止在施肥时把流道给堵塞。流道比较大的是喷灌的喷头，而且喷出来的就像下雨一样，所以可以喷洒叶面肥。在使用喷灌施肥时肥料的选择并没有太大的要求。

由于尿素、碳酸氢铵、氯化铵、硫酸铵、硫酸钾、氯化钾等肥料纯度比较高，杂质也相对较少，和水相溶之后不会产生沉淀，并且都符合国家的标准和行业标准，都可以用作追肥。缺少磷元素的时候就使用磷酸二氢钾可溶性肥料作为追肥。用追肥补充微量元素的时候，不能和补充磷元素的追肥一起使用，很容易形成难溶性的沉淀物质，堵塞滴头或者喷头。

在配制时要特别注意：①含磷酸根的肥料与含钙、镁、铁、锌等金属离子的肥料混合后产生沉淀；②含钙离子的肥料与含硫酸根离子的肥料混合后会产生沉淀；③最好现用现配；④对于混合后会产生沉淀的肥料应采用分别单独注入的办法来解决。

各种滴灌专用肥，如山东青岛生产的日本注商肥料，配方有 16：16：16、19：7：19、16：6：21、20：9：11 等型号，可直接使用。各种冲施液体肥或沼液经过滤后的肥液（图3-3），可直接使用。

图3-3　主要的冲施肥

有机肥用于微灌系统，在使用时必须解决两方面的问题，分别是：①有机肥料必须是液体；②必须经过多次的过滤。一般的有机肥都适合微灌施肥，特别是易沤腐、残渣少的。然而含有纤维素、木质素的有机肥则很难通过微灌系统进行施肥。有的有机肥通过沤腐之后，由于残渣太多也不适合使用微灌系统进行施肥。只要通过沤腐之后，不存在有残渣堵塞微灌系统的颗粒，就可以使用微灌系统来进行施肥。

三、肥料间各种因素的相互作用

（一）肥料混合时的反应

为了防止肥料混合后可以产生堵塞微灌系统的沉淀，应该在微灌系统中使用两个以上的储存罐，一个放置钙、镁和微量元素，另一个放置硫酸盐和磷酸盐，这样可以有效地避免肥料的混合，可安全有效地灌溉施肥。

（二）肥料溶解时的温度变化

不一样的肥料在溶解的时候都会产生热反应。比如，磷酸溶解时会放出大量的热量，使水的温度升高；尿素在溶解时会吸收周围的热量，使水的温度下降，这些反应对于田间配置营养液有很重要的指导意义。比如，在天气气温比较低的时候，为了防止盐析作用，应该把肥料按照合理的顺序使它们进行溶解，并利用它们溶解时产生的热量来防止盐析作用。不同种类的肥料在不同温度下的每升溶解度都是不一样的，见表3-1。

表 3-1　化肥在不同温度下的溶解度　　　　　　　　　　　（克）

化合物	分子式	0℃	10℃	20℃	30℃
尿素	$CO(NH_2)_2$	680	850	1 060	1 330
硝酸铵	NH_4NO_3	1 183	1 580	1 950	2 420
硫酸铵	$(NH_4)_2SO_4$	706	730	750	780
硝酸钙	$Ca(NO_3)_2$	1 020	1 240	1 294	1 620
硝酸钾	KNO_3	130	210	320	460
硫酸钾	K_2SO_4	70	90	110	130
氯化钾	KCl	280	310	340	370
硝酸氢二钾	K_2HPO_4	1 328	1 488	1 600	1 790
磷酸二氢钾	KH_2PO_4	142	178	225	274
磷酸二铵	$(NH_4)_2HPO_4$	429	628	692	748
磷酸一铵	$NH_4H_2PO_4$	227	295	374	464
氯化镁	$MgCl_2$	528	540	546	568
硫酸镁	$MgSO_4$	260	308	356	405

第三节　养分管理

一、土壤养分检测

对于生长在土壤中的植物，土壤测试是确定肥料需求的必要手段。土壤分析应阐明土壤中某种营养元素的含量对要种植的某种具体作物而言是充足还是缺乏。土壤本身含有各种养分，通过先前施用化肥或有机肥也会有养分残留。但是土壤中的养分只有一小部分能被植物吸收利用，即对植物有效。氮主要存在于有机物中，并且只有被微生物分解形成硝态氮和铵态氮才能被植物吸收利用。土壤中的磷只有一小部分是速效磷，但土壤磷库会释放磷以维持土壤溶液的磷浓度。土壤中只有交换性钾和存在于溶液中的钾才能被植物吸收利用，但是随着有效钾不断被吸收，它与固定态钾之间的动态平衡被打破，钾会被转化释放到土壤溶液。测出土壤的营养元素总含量并不能说明它们对植物的有效性。现已有只浸提出潜在有效养分的分析方法，这些方法广泛应用于土壤分析实验室，分析数据可以可靠地估测养分的有效性。

不同元素和不同土壤的浸提方法也不同，一些方法用弱酸或弱碱作浸提剂，一些则使用离子交换树脂，以模拟根对养分的吸收。阳离子有效性（如钾离子）通常都是测定浸提的可交换性部分。在用分析数据进行诊断之前，必须要用田间试验中作物对养分的反应结果来校正分析数据。

确定一种作物对养分的需要量时，必须将作物对养分的总需求量减去土壤所含的有效养分含量。另外，灌溉施肥中使用的水溶性养分，特别是磷肥，在土壤中会发生反应而使有效性降低，对于用土壤种植的作物，施肥时必须考虑到这一点。例如，磷肥的施用量通常比植物实际需要的量大，从而满足植物的吸收。

土壤和生长基质测试应包含另外两个参数：电导率（EC）和pH值。土壤或生长基质中水浸提物的电导率可反映可溶性盐分含量。施肥后没有被植物吸收的或没有淋失的那部分养分及灌溉水本身会造成盐分累积，盐分浓度增加会使根际环境的渗透压升高、根系对水分和养分的吸收减少，从而造成减产。一些离子过量会对植物有毒害作用，并会对土壤结构产生副作用。

土壤和生长基质浸提物的pH值反映了土壤和生长基质的酸碱度。大多数植物在pH值接近中性时长势最好。一些肥料具有酸化作用，如施用铵化合物会因氧化成硝态氮而使酸度增加。缓冲作用很弱的介质如粗质地沙壤土，酸化作用比细质地土壤更为明显。当灌溉水含有过量钠离子时，土壤会碱化。

二、植物养分检测

（一）植株测试的方法

1. 植株测试的方法

按测试方法分类有化学分析法、生物化学法、酶学方法、物理方法等。

化学分析用法是最常用的、最有效的植株测试方法，按分析技术的不同，又将其分为植株常规分析和组织速测，植株常规分析多采用干样品，组织测定指分析新鲜植物组织汁液或浸出液中活性离子的浓度，前一方法是评价植物营养的主要技术，后者具有简便、快速的特点宜于田间直接应用。

生物化学法是测定植株中某种生化物质来表征植株营养状况的方法，如测定水稻叶鞘或叶片中天门冬酸胺，或用淀粉—碘反比应作为氮的营养诊断法。

酶学方法：作物体内某些酶的活性与某些营养元素的多少有密切关系，根据这种酶活性的变化,即可判断某种营养元素的丰缺。

物理方法:如叶色诊断,叶片颜色 ⇔ 叶绿素 ⇔ 氮。

2. 测定部位

一般来说,植株不同部位的养分浓度与全株养分浓度间是有一定关系的,然而,同种器官不同部位的养分浓度也有很大差异,而且不同养分之间这种差异是不一致的,叶、根中氮浓度对氮素供应的变化更敏感一些。因此,可作为敏感的指标。

(二)植株测试中指标的确定

1. 诊断指标的表示方法

(1)临界值。

(2)养分比值:由于营养元素之间的相互影响,往往一种元素浓度的变化常引起其他元素的改变。因此,用养分的比值作为诊断指标,要比用一种元素的临界值能更好地反映养分的丰缺关系。

(3)相对产量。

(4)DRIS 法:诊断施肥综合法。DRIS 法基于养分平衡的原理,用叶片诊断技术,综合考虑营养元素之间的平衡情况和影响作物产量的诸因素,研究土壤、植株与环境条件的相互关系,以及它们与产量的关系。

指数法:先进行大量叶片分析,记载其产量结果和可能影响产量的各种参数,将材料分为高产组(B)和低产组(A),将叶片分析结果以 N%、P%、K%、N/P、N/K、K/P、NP、NK、PK 等多种形式表示,计算各形式的平均值,标准差(SD)、变异系数(CV)、方差(S)及两种的方差比(S_A/S_B),选择保持最高方差比的形式作为诊断的表示形式,对 N、P、K 的诊断,一般多用 N/P、N/K、K/P 的表示形式。

应用时,将测定结果按下式求出 N、P、K 指数:

N 指数 =+ [f(N/P)+f(N/K)] /2
P 指数 =- [f(N/P)+f(K/P)] /2
K 指数 =+ [f(N/P)-f(N/K)] /2

当实测 N/P>标准 N/P 时,则

F(N/P) = 100× [N/P(实测)/N/P(标准)-1] ×10/CV

当实测 N/P<标准 N/P 时,则

F(N/P) = 100× [1-N/P(标准)/N/P(实例)] ×10/CV

f(N/K),f(K/P)类推

三个指数的代数和为0。负指数值越大,养分需要强度越大,正指数越大,养分需要强度越小。

如一小麦求得 N=-13,P=-31,K=44,则需肥强度 P>N>K。

评价:该法在多种作物上有较高的准确性,不受采样时间、部位、株龄和品种的影响,优于临界值法,但它只指出作物对某种养分的需求程度,而未确定施肥数量。

2. 确定诊断指标的方法

诊断指标应通过生产试验获得大量试验数据的情况下才确定。通常采用以下方法。

(1) 大田调查诊断。在一个地区选代表性地块,在播前或生育期进行化学诊断,并结合当地经验,收集各种数据,统计整理,从中找出不同条件下产量、养分等变化幅度的规律,划分成不同等级作为诊断标准。

(2) 田间校验。即养分丰缺指标的划分研究。利用田间多点试验,找出养分测定值与相对产量之间的曲线,一般把指标划分为"高、中、低、极低"四级。田间校验的优点是能全面反映当地的自然条件,把影响养分供应量的诸因素都表现在分级指标中,所得指标准确性高,但需时间及重复。

田间试验分短期、长期两种,短期多为一年试验,一般分为施与不施某养分两处理,要求多个试验点的重复,根据相对产量划分等级,由于年限短,不能反映肥料的叠加效应。因此,结果只能决定是否需施肥,而不能确定施肥量。

长期田间试验可为确定养分用量提供数据,一个点上相同肥料及用量多年试验,可得测试值与产量相关性数据。为确定施肥量作基础。

(3) 对比法。在品种、土壤类型相同条件下,选正常、不正常的健壮植株,多点测土壤植物养分含量。二者比较确定指标。

诊断指标的确定应是营养诊断、大田生产和肥料试验三者结合,并进行多次诊断找出规律。任何诊断指标都是在一定生产条件下取得的。外地的指标,只能作参考,不可生搬硬套。引用时要经产地生产的检验,才可使用。

3. 应用诊断指标应注意的问题

(1) 作物种类与品种特性。不同作物、同一作物不同品种,同一品种不同生育期对养分的需求和临界浓度不同,应用指标时应考虑。在有些情况下,养分含量高,生长量或产量并不一定高。

(2) 营养元素间的相互关系。拮抗作用,例如,Ca^{2+} 与 Mg^{2+};协助作用,例如,Ca^{2+}、Mg^{2+}、Al^{3+} 能促进 K^+、NH_4^+ 等的吸收。

因此，应用某种元素的诊断指标时，不仅要了解该养分的相对量，也要了解有关元素的相互关系。

（3）诊断的技术条件要求一致，采样分析等应和拟定指标时一致，应有可比性，否则指标就无应用价值。指标应随生产水平和技术措施的改变而不断修正。

总之，既要应用诊断指标施肥，又不要孤立地应用指标，必须因地制宜，根据具体情况灵活运用指标，从而使诊断指标更具实用价值，使诊断技术逐步完善。

三、施肥方案制定

施肥方案必须明确施肥量、肥料种类、肥料的使用时期。施肥量的确定要受到植物产量水平、土壤供肥量、肥料利用率、当地气候、土壤条件及栽培技术等综合因素的影响。确定施肥量的方法也很多，如养分平衡法、田间试验法等。这里仅以养分平衡法为例介绍施肥量的确定方法。

（一）施肥量确定

（1）植物计划产量的养分需求总量。土壤肥力是决定产量高低的基础，某一种植物计划产量多高要依据当地的综合因素而确定，不可盲目过高或过低，确定计划产量的方法很多，常用的方法是以当地前3年植物的平均产量为基础，再增加10%~15%的产量作为计划产量。按照计划产量，按下列公式算出植物计划产量所需要氮、磷、钾的总量。

植物计划产量所需养分量（kg）=（计划产量/100）×100kg产量所需养分量

（2）土壤供肥量。土壤供肥量是指植物达到一定产量水平时从土壤中吸收的养分量（不含施用的肥料养分量）。获得这一数值的方法很多，一般来讲，土壤的供肥量多以该种土壤上无肥区全收获物中养分的总量来表示，各地应按土壤类型，对不同植物进行多点试验，取得当地的可靠数据后，按下式估算土壤供肥量：土壤供肥量=土壤养分测定值（mg/kg）×0.15×校正系数。

（3）肥料利用率。肥料利用率是指植物吸收来自所施肥料的养分占所施肥料养分总量的百分率。它是合理施肥的一个重要标志，也是计算施肥量时所需的一个重要参数，它可以通过田间试验和室内的化学分析结果按下式求得：肥料利用率（%）=［（施肥区植物地上部分该元素的吸收量−无肥区植物地上部分该元素的吸收量）/所施肥料中该元素的总量］×100。

第三章 适合水肥一体化技术的肥料的选择

知道了实现计划产量所需的养分总量、土壤供肥量和将要施用的肥料利用率及该种肥料中某一养分的含量,就可依据下面公式估算出计划施肥量:计划施肥量(kg)=(计划产量所需的养分总量-土壤供肥量)/(肥料中有效养分含量×肥料利用率)。

(二)施肥时期的确定

掌握植物的营养特性是实现合理施肥的最重要依据之一。不同的植物种类其营养特性是不同的,即便是同一种植物在不同的生育时期其营养特性也是各异的,只有了解植物在不同生育期对营养条件的需求特征,才能根据不同的植物及其不同的时期,有效地应用施肥手段调节营养条件,达到提高产量、改善品质和保护环境的目的。植物的一生要经历许多不同的生长发育阶段,在这些阶段中,除前期种子营养阶段和后期根部停止吸收养分的阶段外,其他阶段都要通过根系或叶等其他器官从土壤中或介质中吸收养分,植物从环境中吸收养分的整个时期,叫植物的营养期。植物不同生育阶段从环境中吸收营养元素的种类、数量和比例等都有不同要求的时期,叫作植物的阶段营养期。植物对养分的要求虽有其阶段性和关键时期,但决不能忘记植物吸收养分的连续性。任何一种植物,除了营养临界期和最大效率期外,在各个生育阶段中适当供给足够的养分都是必需的。

(三)施肥环节的确定

植物有营养期且有阶段营养期,在植物营养期内就要根据苗情而施肥,所以施肥的任务不是一次就能完成的。对于大多数一年生或多年生植物来说,施肥应包括基肥、种肥和追肥3个时期(或环节)。每个施肥时期(或环节)都起着不同的作用。

(1)基肥。群众也常称为底肥,它是在播种(或定植)前结合土壤耕作施入的肥料。其作用是双重的,一方面是培肥和改良土壤,另一方面是供给植物整个生长发育时期所需要的养分。通常多用有机肥料,配合一部分化学肥料作基肥。基肥的施用应按照肥土、肥苗、土肥相融的原则施用。

(2)种肥。播种(或定植)时施在种子附近或与种子混播的肥料。其作用是给种子萌发和幼苗生长创造良好的营养条件和环境条件。因此,种肥一般多用腐熟的有机肥或速效性的化学肥料以及细菌肥料等。同时,为了避免种子与肥料接近时可能产生的不良作用,应尽量选择对种子或根系腐蚀性小或毒害轻的肥料。凡是浓度过大、过酸或过碱、吸湿性强、溶解时产生高温及含有毒性成分的肥料均不宜作种肥施用。例如,碳酸氢铵、硝酸铵、氯化铵,以及土法生产的过磷酸钙等均不宜作种肥。

（3）追肥。指在植物生长发育期间施入的肥料。其作用是及时补充植物在生育过程中所需的养分，以促进植物进一步生长发育，提高产量和改善品质，一般以速效性化学肥料作追肥。

（四）施肥方法的确定

施肥方法：①撒施。撒施是施用基肥和追肥的一种方法，即把肥料均匀撒于地表，然后把肥料翻入土中。凡是施肥量大的或密植植物如小麦、水稻、蔬菜等封垄后追肥以及根系分布广的植物都可采用撒施法。②条施。条施也是基肥和追肥的一种方法，即开沟条施肥料后覆土。一般在肥料较少的情况下施用，玉米、棉花及垄栽红薯多用条施；再如小麦，在封行前可用施肥机或耧把肥料耩入土壤。③穴施。穴施是在播种前，把肥料施在播种穴中，而后覆土播种。其特点是施肥集中，用肥量少，增产效果较好，果树、林木多用穴施法。④分层施肥。将肥料按不同比例施入土壤的不同层次内。⑤随水浇施。在灌溉（尤其是喷灌）时将肥料溶于灌溉水而施入土壤的方法。这种方法多用于追肥方式。⑥根外追肥。把肥料配成一定浓度的溶液，喷洒在植物叶面，以供植物吸收。⑦环状和放射状施肥。环状施肥常用于果园施肥，是在树冠外围垂直的地面上，挖一环状沟，深、宽各30~60cm，施肥后覆土踏实。翌年再施肥时可在第一年施肥沟的外侧再挖沟施肥，以逐年扩大施肥范围。放射状施肥是在距树木一定距离处，以树干为中心，向树冠外围挖4~8条放射状直沟，沟深、宽各50cm，沟长与树冠相齐，肥料施在沟内，翌年再交错位置挖沟施肥。

第四章 水肥一体化技术在农业生产中的应用

水肥一体化技术在农业生产的作物栽培中推广使用时，要讲究客观规律，科学的方法，做到合理规划，精心布局，因地制宜，正确规范操作，使作物增产增收。

第一节 水肥一体化的规划设计

一、水肥一体化智能灌溉系统概述

基于物联网的智能化灌溉系统涉及传感器技术、自动控制技术、数据分析和处理技术、网络和无线通信技术等关键技术，是一种应用潜力广阔的现代农业设备。该系统通过土壤墒情监测站实时监测土壤含水量数据，结合示范区的实际情况（如灌溉面积、地理条件、作物种类的分布、灌溉管网的铺设等）对传感数据进行分析处理，依据传感数据设置灌溉阈值，进而通过自动、定时或手动等不同方式实现水肥一体化智能灌溉。中心站管理员可通过电脑或智能移动终端设备，登录系统监控界面，实时监测示范区内作物生长情况，并远程控制灌溉设备（如固定式喷灌机等）。

基于物联网的智能化灌溉系统，能够实现示范区的精准和智能灌溉，可以提高水资源利用率，缓解水资源日趋紧张的矛盾，增加作物产量，降低作物成本，节省人力资源，优化管理结构。

二、水肥一体化智能灌溉系统总体设计方案

（一）水肥一体化智能灌溉系统总体设计目标

智能化灌溉系统实现对土壤含水量的实时采集，并以动态图形的形式在管理界面上显示。系统依据示范区内灌溉管道的布设情况及固定式喷灌机的安装位

置，预先设置相应的灌溉模式（包含自动模式、手动模式、定时模式等），进而通过对实时采集的土壤含水量值和历史数据的分析处理，实现智能化控制。系统能够记录各个区域每次灌溉的时间、灌溉的周期和土壤含水量的变化，有历史曲线对比功能，并可向系统录入各区域内作物的配肥情况、长势、农药的喷洒情况以及作物产量等信息。系统可通过管理员系统分配使用权限，对不同的用户开放不同的功能，包括数据查询、远程查看、参数设置、设备控制和产品信息录入等功能。

（二）水肥一体化智能灌溉系统架构

系统布设土壤墒情监测站和远程设备控制系统、智能网关和摄像头等设备，实现对示范区内传感数据的采集和灌溉设备控制功能；示范区现场通过 2G/3G 网络和光纤实现与数据平台的通信；数据平台主要实现环境数据采集、阈值告警、历史数据记录、远程控制、控制设备状态显示等功能；数据平台进一步通过互联网实现与远程终端的数据传输；远程终端实现用户对示范区的远程监控（图 4-1）。

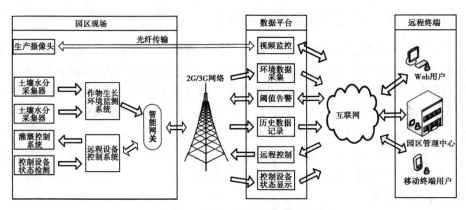

图 4-1 水肥一体化智能灌溉系统整体结构图

依据灌溉设备以及灌溉管道的布设和区域的划分，布设核心控制器节点，通过 ZigBee 网络形成一个小型的局域网，通过 GPRS 实现设备定位，然后再通过嵌入式智能网关连接到 2G/3G 网络的基站，进而将数据传输到服务器；摄像头视频通过光纤传输至服务器；服务器通过互联网实现与远程终端的数据传输（图 4-2）。

第四章 水肥一体化技术在农业生产中的应用

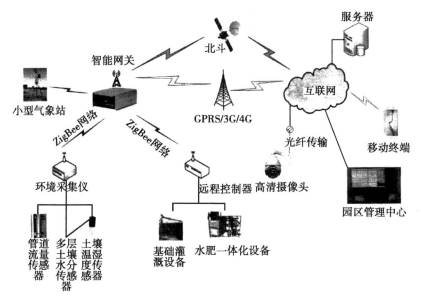

图 4-2 水肥一体化智能灌溉系统实现框

（三）水肥一体化智能灌溉系统组成

智能化灌溉系统可分为 6 个子系统：作物生长环境监测系统、远程设备控制系统、视频监测系统、通信系统、服务器、用户管理系统。

(1) 作物生长环境监测系统。作物生长环境监测系统主要为土壤墒情监测系统（土壤含水量监测系统）。土壤墒情监测系统是根据示范区的面积、地形及种植作物的种类，配备数量不等的土壤水分传感器，以采集示范区内土壤含水量，将采集到的数据进行分析处理，并通过嵌入式智能网关发送到服务器。示范区用户根据种植作物的实际需求，以采集到的土壤墒情（土壤含水量）参数为依据实现智能化灌溉。通过无线网络传输数据，在满足网络通信距离的范围内，用户可根据需要调整采集器的位置。

(2) 远程设备控制系统。远程设备控制系统实现对固定式喷灌机以及水肥一体化基础设施的远程控制。预先设置喷灌机开闭的阈值，根据实时采集到的土壤含水量数据生成自动控制指令，实现自动化灌溉功能。也可通过手动或者定时等不同的模式实现喷灌机的远程控制。此外，系统能够实时检测喷灌机的开闭状态。

(3) 视频监测系统。视频监测系统实现对示范区关键部位的可视化监测，根据示范区的布局安置高清摄像头，一般安装在作物的种植区内和固定式喷灌机的附近，视频数据通过光纤传输至监控界面，园区管理者可通过实时的视频查看作物生长状态及灌溉效果。

（4）通信系统。如果地域范围比较广阔，地形复杂，则有线通信难度较大。本系统拟采用 ZigBee 网络实现示范区内的通信。ZigBee 网络可以自主实现自组网、多跳、就近识别等功能，该网络的可靠性好，当现场的某个节点出现问题时，其余的节点会自动寻找其他的最优路径，不会影响系统的通信线路。

ZigBee 通信模块转发的数据最终汇集于中心节点进行数据的打包压缩，然后通过嵌入式智能网关发送到服务器。

（5）服务器。服务器为一个管理数据资源并为用户提供服务的计算机，具有较高的安全性、稳定性和处理能力，为智能化灌溉系统提供数据库管理服务和 Web 服务。

（6）用户管理系统。用户可通过个人计算机和手持移动设备，通过 Web 浏览器登录用户管理系统。不同的用户需要分配不同的权限，系统会对其开放不同的功能。例如，高级管理员一般为示范区相关主要负责人，具有查看信息、对比历史数据、配置系统参数、控制设备等权限；一般管理员为种植管理员、采购和销售人员等，具有查看数据信息、控制设备、记录作物配肥信息和出入库管理等权限；访问者为产品消费者和政府人员等，具有查看产品生长信息、园区作物生长状况等权限。用户管理系统安装在园区的管理中心，具体设施包括用户管理系统操作平台，可供实时查看示范区作物生长情况。

（四）水肥一体化智能灌溉系统功能

智能化灌溉系统能实现如下功能：环境数据的显示查看及分析处理、智能灌溉、作物生长记录、产品信息管理等。

（1）环境数据的显示查看及分析处理。一是环境数据的显示查看。在系统界面上能显示各个土壤墒情采集点的数据信息，可设定时间刷新数据。数据显示类型包含实时数据和历史数据，能够查看当前实时的土壤水分含量和任意时间段的土壤水分含量（如每月或当天示范区土壤的墒情数据）；数据显示方式包含列表显示和图形显示，可以根据相同作物的不同种植区域或相同区域不同时间段的数据进行对比，以曲线、柱状图等形式出现。二是环境数据的分析处理。根据采集到的土壤水分含量，结合作物实际生长过程中对土壤水分含量的具体需求，设置作物打开灌溉阀门的水分含量阈值；依据不同作物对土壤水分含量的需求，设定灌溉时间、灌溉周期等。

（2）智能灌溉。本系统可实现三种灌溉控制方式。一是按条件定时定周期灌溉。根据不同区域的作物种植情况任意分组，进行定时定周期灌溉。二是多参数设定灌溉。对不同作物设定适合其生长的多参数的上限与下限值，当实时的参数值超出设定的阈值时，系统就会自动打开相对应区域的电磁阀，对该区域进行灌溉，使参数值稳定在设定数值内。三是人工远程手动灌溉。管理员可通过管理系统，手动进行远程灌溉操作。

（3）作物生长记录。通过数据库记录各个区域的环境数据、灌溉情况、配肥信息、作物长势以及产量等信息。

（4）产品信息管理。园区管理员录入各区域内作物的配肥情况、长势、农药的喷洒情况、产品产量质量、产品出入库管理、仓库库存状况以及农作物产品的品级分类等信息。

（五）水肥一体化智能灌溉系统特点

本系统采用了扩展性的设计思路，在设计架构上注重考虑系统的稳定性和可靠性。整个系统由多组网关及 ZigBee 自组织网络单元组成，每个网关作为一个 ZigBee 局域网络的网络中心，该网络中包含多个节点，每一个节点由土壤水分采集仪或远程设备控制器组成，分别连接土壤水分传感器和固定式喷灌机。本系统可以根据用户的需求，方便快速地组建智能灌溉系统。用户只需增加各级设备的数量，即可实现整个系统的扩容，原有的系统结构无需改动。

（六）水肥一体化智能灌溉系统设计

（1）系统布局。由于本系统的通信子模块采用具有结构灵活、自组网络、就近识别等特点的 Zigbee 无线局域网络，对于土壤湿度传感器的控制器节点的布设相对灵活。根据园区种植作物种类的不同及各种作物对土壤含水量需求的不同布设土壤湿度传感器；根据园区内铺设的灌溉管道、固定式喷灌机位置及作物的分时段、分区域供水需要安装远程控制器设备（每套远程控制器设备包括核心控制器、无线通信模块、若干个控制器扩展模组及其安装配件），每套控制器设备依据就近原则安装在固定式喷灌机旁，实现示范区灌溉的远程智能控制功能；此外，通过控制设备自动检测固定式喷灌机开闭状态信号及视频信号，远程查看，实时掌握灌溉设备的开闭状态。

在项目的实施中，根据示范区的具体情况（包括地理位置、地理环境、作物分布、区域划分等）安装墒情监测站。远程控制设备后期需要安装在灌溉设备的控制柜旁，通过引线的方式实现对喷灌机包括水肥一体化基础设施的远程控制。

(2) 网络布局。土壤墒情监测设备和远程控制器设备分别内置 ZigBee 模块和 GPRS 模块，都作为通信网络的节点。嵌入式智能网关是一定区域内的 ZigBee 网络的中心节点，共同组成一个小型的局域网络，实现园区相应区域的网络通信，并通过 2G/3G 网络实现与服务器的数据传输。

该系统均采用无线传输的通信方式，包括 ZigBee 网络传输及 GPRS 模块定位。由于现场地势平坦，无高大建筑物或其他东西遮挡，因此具备无线传输的条件。

第二节　水肥一体化技术的设备安装与调试

水肥一体化技术的设备安装主要包括首部设备安装、管网设备安装和微灌设备安装等环节。

一、首部设备安装与调试

(一) 负压变频供水设备安装

负压变频供水设备安装处应符合控制柜对环境的要求，柜前后应有足够的检修通道，进入控制柜的电源线径、控制柜前的低压柜的容量应有一定的余量，各种检测控制仪表或设备应安装于系统贯通且压力较稳定处，不应对检测控制仪表或设备产生明显的不良影响。如安装于高温（高于45℃）或具有腐蚀性的地方，在签订订货单时应做具体说明。在已安装时发现安装环境不符合，应及时与原供应商取得联系进行更换。

水泵安装应注意进水管路无泄漏，地面应设置排水沟，并应设置必需的维修设施。水泵安装尺寸见各类水泵安装说明书。

(二) 离心自吸泵安装

1. 安装使用方法

第一步，建造水泵房和进水池，泵房占地 3m×5m 以上，并安装一扇防盗门，进水池 2m×3m。第二步，安装 ZW 型卧式离心自吸泵，进水口连接进水管到进水池底部，出口连接过滤器，一般两个并联。外装水表、压力表及排气阀（排气阀安装在出水管墙外位置，水泵启停时排气阀会溢水，保持泵房内不被水溢湿）。第三步，安装吸肥管，在吸水管三通处连接阀门，再接过滤器，过滤器与水流方向要保持一致，连接钢丝软管和底阀。第四步，施肥桶可以配 3 只左右,每只容量200L左右,通过吸肥管分管分别放进各肥料桶内,可以在吸肥时把不能同时混配的肥料分桶吸入,在管道中混合。第五步,施肥浓度,根据进出水管的口径,配置吸肥管的口径,保持施肥浓度在5%~7%。通常4英寸[①]进水管,3英寸出水管水泵,配1英寸吸肥管,最后施肥浓度在5%左右。肥料的吸入量始终随水泵流量大小而改变,而且保持相对稳定的浓度。田间灌溉量大,即流量大,吸肥速度也随之加快,反之,吸肥速度减慢,始终保持浓度相对稳定。

2. 注意事项

施肥时要保持吸肥过滤器和出水过滤器畅通,如遇堵塞,应及时清洗;施肥过程中,当施肥桶内肥液即将吸干时,应及时关闭吸肥阀,防止空气进入泵体产生气蚀。

(三)潜水泵安装

1. 安装方法

拆下水泵上部出水口接头,用法兰连接止回阀,止回阀箭头指向水流方向。管道垂直向上伸出池面,经弯头引入泵房,在泵房内与过滤器连接,在过滤器前开一个直径20mm的施肥口,连接施肥泵,前后安装压力表。水泵在水池底部需要垫高0.2m左右,防止淤泥堆积,影响散热。

2. 施肥方法

第一步,开启电机,使管道正常供水,压力稳定。第二步,开启施肥泵,调整压力,开始注肥,注肥时需要有操作人员照看,随时关注压力变化及肥量变化,注肥管压力要比出水管压力稍大一些,保证能让肥液注出水管,但压力不能太大,以免引起倒流,肥料注完后,再灌15min左右的清水,把管网内的剩余肥液送到作物根部。

(四)山地微蓄水肥一体化

山地微蓄水肥一体化技术是利用山区自然地势高差获得输水压力,对地势相对较低的田块进行微灌,即将"微型蓄水池"和"微型滴灌"组合成"微蓄微灌"。这种方式不需要电源和水泵等动力的配置,适合山区、半山区以及丘陵地带的作物种植园的灌溉。水池出水口位置直接安装过滤器及排气阀等设备,然后连接管网将灌溉水肥输送到植物根部。

山地水肥一体化技术的使用,每年每亩可以节约用工15个以上,作物增产15%以上。水肥一体可以使肥料全部进入土壤耕作层中,减少了肥料浪费流失及

---

① 1英寸约为2.54cm,全书同。

表面挥发，能节肥15%以上。

山地水肥一体化技术的首部设备主要由引水池、沉沙池、引水管、蓄水池、总阀门、过滤器以及排气阀等组成（图4-3）。这种首部设置简单、安全可靠，如果过滤器性能良好，在施肥过程中基本不需要护理。

引水池、沉沙池起到初步过滤水源、蓄水的作用，将水源中的泥沙、枝叶等进行拦截。引水管是将水源中的水引到蓄水池的管道，引水管埋在地下0.3~0.4m为宜，防止冻裂和人为破坏。如果管路超过1km，且途中有起伏坡地，需要在起伏高处设置排气阀，防止气阻。

蓄水池与灌溉地的落差应在10~15m，蓄水池大小根据水源大小、需灌溉面积确定，一般以50~120m³为宜。蓄水池建造质量要求较高，最好采用钢筋混凝土结构，池体应深埋地下，露出地面部分以不超过池体的1/3为宜。建池时，预装清洗阀、出水阀和溢水口，特别注意要在建造蓄水池的同时安装，使其与水池连成一体，不能在事后打孔安装，否则容易漏水。顶部加盖留维修口，以确保安全。

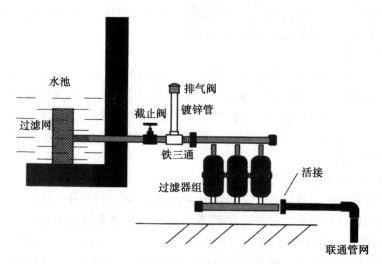

图4-3 山地水肥一体化首部结构示意图

过滤器安装在出水阀处，最好同时安装2套为一组的过滤器，方便清洗。施肥池也可以用施肥桶代替，容积为1~2m³，连接出水管。

在实践中，有两种施肥方式适合山地水肥一体化技术。一是直接施肥。在肥量和供水量确定的情况下，可以根据水池蓄水量，按照施肥浓度加入易于溶解的肥料。肥料需要事先在小桶中搅拌至充分溶解，滤除残渣，倒入蓄水池中再均匀混合，灌溉时打开各处阀门，肥液直接流到作物根部。在施肥完成后，用清水充入管道15min左右，把灌水器及管道内的剩余肥液冲洗干净，防止肥液结晶堵塞滴头。二是压差施肥。在稍高于蓄水池最高液面处放置施肥桶，用施肥管连接出水管，并将施肥口开在过滤器的前面，使肥液经过过滤器去除杂质，灌溉时打开出水管总阀，再打开施肥阀，让饱和肥液与灌溉水在管道内混合均匀后经各级管网输送到作物区，施肥完成后，再用清水冲管即可。

二、管网设备安装与调试

(一) 平地管网

在水肥一体化设施建设过程中，除了选择合适的首部设备，还需要布局合理、经济实用的供水管网。近几年，塑料管业飞速发展，品质日趋成熟，塑料管道以价廉质优的优势代替镀锌管道。目前，灌溉管网的建设大多采用塑料管道，其中应用最广的有聚氯乙烯（PVC）和聚乙烯（PE）管材管件，其中PVC管需要用专用胶水黏合，PE管需要热熔连接。

1. 开沟挖槽

铺设管网的第一步是开沟挖槽，一般沟宽0.4m，深0.6m左右，呈U形，挖沟要平直，深浅一致，转弯处以90°和135°处理。沟的坡面呈倒梯形，上宽下窄，防止泥土坍塌导致重复工作。在适合机械施工的较大场地可以用机械施工，在田间需要人工作业（图4-4）。

图4-4 开挖沟槽

2. PVC管道安装

与PVC管道配套的是PVC管件，管道和管件之间用专用胶水粘接，这种胶水能把PVC管材、管件表面溶解成胶状，在连接后物质相互渗透，72h后即可连成一体。所以，在涂胶的时候应注意胶水用量，不能太多，过多的胶水会沉积在管道底部,把管壁部分溶解变软,降低管道应力,在遇到水锤等极端压力的时候,此处最容易破裂,导致维修成本增高,还影响农业生产。

3. PE 管道安装

PE 管道采用热熔方式连接,有对接式热熔和承插式热熔两种,一般大口径管道(DN100mm 以上)都用对接热熔连接,有专用的热熔机,具体可根据机器使用说明进行操作。DN80mm 以下均可以用承插方式热熔连接,优点是热熔机轻便,可以手持移动,缺点是操作需要 2 人以上,承插后,管道热熔口容易过热缩小,影响过水。

(二)山地管网

山地灌溉管网适合选用 PE 管,常规安装方向同平地管网。铺设方法与平地有些不同,主管从蓄水池沿坡而下铺设,高差每隔 20~30m 安装减压消能池,消能池内安装浮球阀。埋地 0.3m 以下,支管垂直于坡面露出地面,安装阀门,阀门用阀门井保护。

各级支管依照设计要求铺设,关键是滴灌管的铺设需要沿等高线方向铺设,出水量更加均匀。

三、微灌设备安装与调试

(一)微喷灌的安装与调试

微喷灌系统包括水源、供水泵、控制阀门、过滤器、施肥阀、施肥罐、输水管、微喷头等。这里以温室大棚微喷灌安装为例。

材料选择与安装:吊管、支管、主管管径宜分别选用 4~5mm、8~20mm、32mm,壁厚 2mm 的 PVC 管,微喷头间距 2.8~3m,工作压力 0.18MPa 左右,单相供水泵流量 8~12L/h,要求管道抗堵塞性能好,微喷头射程直径为 3.5~4m,喷水雾化要均匀,布管时两根支管间距 2.6m,把膨胀螺栓固定在温棚长度方向距地面 2m 的位置上,将支管固定,把微喷头、吊管、弯头连接起来,倒挂式安装好微喷头即可。

1. 安装步骤

(1)工具准备。钢锯、轧带、打孔器手套等。

(2)安装方式。大棚内,微喷头一般是倒挂安装(图 4-5),这种方式不仅不占地,还可以方便田间作业。根据田间试验和实际应用效果,微喷头间距以 2.5~2.6m 为宜,下挂长度以地面以上 1.8~2m 较合适,一般选择 G 型微喷头,微喷头 G 型桥架朝向要朝一个方向,这样喷出的水滴可以互补,提高均匀度。

第四章　水肥一体化技术在农业生产中的应用

图 4-5　倒挂微喷头

（3）防滴器安装。在安装过程中，可以安装防滴器，使微喷头在停止喷水的时候阻止管内剩余的水滴落，以免影响作物生长。也可不装，其窍门是在安装微喷头的时候，调整作畦位置和支管安装位置，使喷头安装在畦沟地正上方，剩余的水滴落在畦沟里。

（4）端部加喷头。大棚的端部同时安装两个喷头，高差10cm，其中一个喷头流量40L/h。其作用是使大棚两端湿润更均匀。

（5）喷头预安装。裁剪毛管（以预定长度均匀裁剪）—装喷头—安装对夹配重—成品。

（6）固定黑管。把黑管沿大棚方向纵向铺开，调整扭曲部分，使黑管平顺地铺在地上，按预定距离打孔，再安装喷头，从大棚末端开始，预留2m，开始把装好喷头的黑管捆扎固定在棚管上，注意不宜用铁丝类金属丝捆扎，因为其在操作中容易丝勾外翘，扎破大棚膜或者生锈。

2. 安装选型

（1）喷道选择。一套大棚安装几道微喷，要根据大棚宽幅确定。

8m大棚两道安装，喷头选择70L，双流道，型号LF-GWPS6000，喷幅6m。两道黑管距离4m左右，喷头间距2.5~2.6m，交叉排列。

6m大棚单道安装，喷头流量120L，单流道，型号LFGWP8000，喷幅8m，间距2.5~2.8m。大棚两端双个安装，高差10cm，其中一个喷头70L/h。

（2）喷管选择。喷管通常选用黑色低密度（高压）聚乙烯管，简称黑管。这种管材耐老化，能适应严酷的田间气候环境，新料管材能在田间连续使用10年以上。

（3）管径选择。根据单道喷灌长度，通过计算得出管道口径，一般长度30m以内可以用外径16mm的黑管，30~50m以内可以用外径20mm的黑管，50~70m以内用外径25mm的黑管，70~90m用外径32mm的黑管。一般长度不超过100m，这样可以节约成本；长度100m以上，建议从中间开三通过水。

（二）滴灌设备安装与调试

作物的生物学特征各异，栽培的株距、行距也不一样（通常15~40cm），为了达到灌溉均匀的目的，所要求滴灌带滴孔距离、规格也一样。通常滴孔距离有15cm、20cm、30cm、40cm，常用的有20cm、30cm。这就要求滴灌设施实施过程中，需要考虑使用单条滴灌带端部首端和末端滴孔出水量均匀度相同且前后误差在10%以内的产品。在设计施工过程中，需要根据实际情况选择合适规格的滴灌带，还要根据这种滴灌带的流量等技术参数确定单条滴灌带的最佳铺设长度。

1. 滴灌设备安装

（1）灌水器选型。大棚栽培一般选用内镶式滴灌带，规格16mm×200mm或300mm，壁厚可以根据农户投资需求选择0.2mm、0.4mm、0.6mm，滴孔朝上，平整地铺在畦面的地膜下面。

（2）滴灌带数量。可以根据作物种植要求和投资意愿，决定每畦铺设的条数，通常每畦至少铺设一条，两条最好。

（3）滴灌带安装。棚头横管用25英寸，每棚一个总开关，每畦另外用旁通阀，在多雨季节，大棚中间和棚边土壤湿度不一样，可以通过旁通阀调节灌水量。

铺设滴灌带时，先从下方拉出，由一人控制，另一人拉滴灌带，当滴管带略长于畦面时，将其剪断并将末端折扎，防止异物进入。首部连接旁通或旁通阀，要求滴灌带用剪刀裁平，如果附近有滴头，则剪去不要，把螺旋螺帽往后退，把滴灌带平稳套进旁通阀的口部，适当摁住，再将螺帽往外拧紧即可。滴灌带尾部折叠并用细绳扎住，打活结，以方便冲洗（用带用堵头也可以，只是在使用过程中受水压、泥沙等影响，不容易拧开冲洗，直接用线扎住方便简单）。

把黑管连接总管，三通出口处安装球阀，配置阀门井或阀门箱保护。整体管网安装完成后，通水试压，冲出施工过程中留在管道内的杂物，调整缺陷处，然后关水，滴灌带上堵头，25英寸黑管上堵头。

2. 设备使用技术

（1）滴灌带通水检查。在滴灌受压出水时，正常滴孔的出水量是呈滴水状的，如果有其他洞孔，出水是呈喷水状的，在膜下会传出水柱冲击的响声，所以要巡查各处，检查是否有虫咬或其他机械性破洞，发现后及时修补。在滴灌带铺设前，一定要对畦面的地下害虫或越冬害虫进行一次灭杀。

（2）灌水时间。初次灌水时，由于土壤团粒疏松，水滴容易直接往下顺着土块空隙流到沟中，没能在畦面实现横向湿润。所以要短时间、多次、间歇灌水，让畦面土壤形成毛细管，促使水分横向湿润。

瓜果类作物在营养生长阶段要适当控制水量，防止枝叶生长过旺影响结果。在作物挂果后，滴灌时间要根据滴头流量、土壤湿度、施肥间隔等情况决定。一般在土壤较干时滴灌 3~4h；而当土壤湿度居中且仅以施肥为目的时，水肥同灌约 1h 较合适。

（3）清洗过滤器。每次灌溉完成后，需要清洗过滤器。每 3~4 次灌溉后，特别是水肥灌溉后，需要把滴灌带堵头打开冲水，将残留在管壁内的杂质冲洗干净。作物采收后，集中冲洗 1 次，收集备用。如果是在大棚内，只需要把滴灌带整条拆下，挂到大棚边的拱管上即可，下次使用时再铺到膜下。

第三节　水肥一体化系统操作与维护

一、水肥一体化系统操作

水肥一体化系统操作包括运行前的准备、灌溉操作、施肥操作、轮灌组更替和结束运行前的操作等工作。

（一）运行前的准备

运行前的准备工作主要是检查系统是否按设计要求安装到位，检查系统主要设备和仪表是否正常，对损坏或漏水的管段及配件进行修复。

1. 检查水泵与电机

检查水泵与电机所标示的电压、频率与电源电压是否相符，检查电机外壳接地是否可靠，检查电机是否漏油。

2. 检查过滤器

检查过滤器安装位置是否符合设计要求，是否有损坏，是否需要冲洗。介质过滤器在首次使用前，首先在罐内注满水并放入一包氯球，搁置 30min 后按正常使用方法各反冲 1 次。此次反冲也可预先搅拌介质，使之颗粒松散，接触面展开。然后充分清洗过滤器的所有部件,紧固所有螺丝。离心式过滤器冲洗时先打开压盖,将沙子取出冲净即可。网式过滤器手工清洗时,扳动手柄,放松螺杆,打开压盖,取出滤网,用软刷子刷洗筛网上的污物并用清水冲洗干净。叠片过滤器要检查和更换变形叠片。

3. 检查肥料罐或注肥泵

检查肥料罐或注肥泵的零部件和与系统的连接是否正确,清除罐体内的积存污物以防进入管道系统。

4. 检查其他部件

检查所有的末端竖管是否有折损或堵头丢失。前者取相同零件修理,后者补充堵头。检查所有阀门与压力调节器是否启闭自如,检查管网系统及其连接微管,如有缺损应及时修补。检查进排气阀是否完好,并打开。关闭主支管道上的排水底阀。

5. 检查电控柜

检查电控柜的安装位置是否得当。电控柜应防止阳光照射,并单独安装在隔离单元内,要保持电控柜房间的干燥。检查电控柜的接线和保险是否符合要求,是否有接地保护。

(二)灌溉操作

水肥一体化系统包括单元系统和组合系统。组合系统需要分组轮灌。系统的简繁不同、灌溉作物和土壤条件不同都会影响到灌溉操作。

1. 管道充水试运行

在灌溉季节首次使用时,必须进行管道充水冲洗。充水前应开启排污阀或泄水阀,关闭所有控制阀门,在水泵运行正常后缓慢开启水泵出水管道上的控制阀门,然后从上游至下游逐条冲洗管道,充水中应观察排气装置的工作是否正常。管道冲洗后应缓慢关闭泄水阀。

2. 水泵启动

要保证动力机在空载或轻载下启动。启动水泵前,首先关闭总阀门,并打开准备灌水的管道上所有排气阀排气,然后启动水泵向管道内缓慢充水。启动后观察和倾听设备运转是否有异常声音,在确认启动正常的情况下,缓慢开启过滤器及控制田间所需灌溉的轮灌组的田间控制阀门,开始灌溉。

3. 观察压力表和流量表

观察过滤器前后的压力表读数差异是否在规定的范围内,压差读数达到 7m 水柱,说明过滤器内堵塞严重,应停机冲洗。

第四章 水肥一体化技术在农业生产中的应用

4. 冲洗管道

新安装的管道（特别是滴灌管）第一次使用时，要先放开管道末端的堵头，充分放水冲洗各级管道系统，把安装过程中集聚的杂质冲洗干净后，封堵末端堵头，然后才能开始使用。

5. 田间巡查

要到田间巡回检查轮灌区的管道接头和管道是否漏水，各个灌水器是否正常。

（三）施肥操作

1. 压差式施肥罐

（1）压差施肥罐的运行。压差施肥罐的操作运行顺序如下。第一步，根据各轮灌区具体面积或作物株数（如果树）计算好当次施肥的数量。称好或量好每个轮灌区的肥料。第二步，用两根各配一个阀门的管子将旁通管与主管接通，为便于移动，每根管子上可配用快速接头。第三步，将液体肥直接倒入施肥罐，若用固体肥料则应先行单独溶解并通过滤网注入施肥罐。有些用户将固体肥直接投入施肥罐，使肥料在灌溉过程中溶解，这种情况下用较小的罐即可，但需要5倍以上的水量以确保所有肥料被用完。第四步，注完肥料溶液后，扣紧罐盖。第五步，检查旁通管的进出口阀均关闭而节制阀打开，然后打开主管道阀门。第六步，打开旁通进出口阀，然后慢慢地关闭节制阀，同时注意观察压力表，得到所需的压差（1~3m水压）。第七步，对于有条件的用户，可以用电导率仪测定施肥所需时间，或用 Amos Teitch 的经济公式估计施肥时间。施肥完毕后关闭进口阀门。第八步，要施下一罐肥时，必须排掉部分罐内的积水。在施肥罐进水口处应安装一个1/2英寸的进排气阀或1/2英寸的球阀。打开罐底的排水开关前，应先打开排气阀或球阀，否则水排不出去。

（2）压差施肥罐施肥时间监测方法。压差施肥罐是按数量施肥的方式，开始施肥时流出的肥料浓度高，随着施肥进行，罐中肥料越来越少，浓度越来越稀。阿莫斯特奇（Amos Teich）总结了罐内不断降低的溶液浓度的规律，即在相当于4倍罐容积的水流过罐体后，90%的肥料已进入灌溉系统（但肥料应在一开始就完全溶解），流入罐内的水量可用罐入口处的流量来测量。灌溉施肥的时间取决于肥料罐的容积及其流出速率：

$$T = 4V/Q$$

式中，T 为施肥时间，小时；V 为肥料罐容积，L；Q 为流出液速率，L/h；4是指流入肥料罐中需480L水才能把120L肥料溶液全部带入灌溉系统中。

因为施肥罐的容积是固定的，当需要加快施肥速度时，必须使旁通管的流量增大。此时要把节制阀关得更紧一些。Amos Teich 公式是在肥料完全溶解的情况下获得的一个近似公式。在田间情况下很多时候用固体肥料（肥料量不超过罐体的 1/3），此时肥料被缓慢溶解。有研究者比较了等量的氯化钾和磷酸钾肥料在完全溶解和固体状态两种情况下倒入施肥罐，在相同压力和流量下的施肥时间。用监测滴头处灌溉水的电导率的变化来判断施肥的时间，当水中电导率达到稳定后表明施肥完成。将 50kg 固体硝酸钾或氯化钾（或溶解后）倒入施肥罐，罐容积为 220L，每小时流入罐的水量为 1 600L，主管流量为 37.5m³/h，通过施肥罐的压力差为 0.18kg/cm²，灌溉水温度为 30℃。结果表明，在流量、压力、用量相同的情况下，不管是直接用固体肥料，还是将其溶解后放入施肥罐，施肥的时间基本一致。两种肥料大致在 40min 施完。施肥开始后约 10min，滴头处才达到最大浓度，这与测定时轮灌区的面积有关（施肥时面积约 150 亩）。面积越大，开始施肥时肥料要走的路越远，需要的时间越长。由于施肥的快慢与经过施肥罐的流量有关，当需要快速施肥时，可以增大施肥罐两端的压差；反之减小压差。在有条件的地方，可以用下列方法测定施肥时间。

①EC 法（电导率法）。肥料大部分为无机盐（尿素除外），溶解于水后使溶液的电导率增加。监测施肥时流出液电导率的变化即可知每罐肥的施肥时间。将某种单质肥料或复合肥料倒入罐内约 1/3 容积，称重，记录入水口压力（有压力表情况下）或在节制阀的旋紧位置做记号（入水口无压力表），用电导率仪测量流出液的 EC 值，记录施肥开始的时间。施肥过程中每隔 3min 测量 1 次，直到 EC 值与入水口灌溉水的 EC 值相等，此时表明罐内无肥，记录结束的时间。开始与结束的时间差即为当次的施肥时间。

②试剂法。利用钾离子与铵离子能与 2% 的四苯硼钠形成白色沉淀来判断，做法同 EC 法相似。试验肥料可用硝酸钾、氯化钾、硝酸铵等含钾或铵的肥料。记录开始施肥的时间。每次用 50mL 的烧杯取肥液 3~5mL，滴入 1 滴四苯硼钠溶液，摇匀，开始施肥时变白色沉淀，之后随浓度越来越稀而无反应。此时的时间即为施肥时间。

尿素是灌溉施肥中最常用的氮肥。但上述两种方法都无法检测尿素的施肥时间。通过测定等量氯化钾的施用时间，根据溶解度来推断尿素的施肥时间。如在常温下，氯化钾溶解度为 34.7g/100g 水，尿素为 100g/100g 水。当氯化钾的施肥时间为 30min 时，因尿素的溶解度比氯化钾更大，等重量的尿素施肥完成时间同样也应为 30min。或者将尿素与钾肥按 1∶9 的比例加入罐内，用监测电导率的办法了解尿素的施肥时间。因钾肥的溶解度比尿素小，只要监测不到电导率的增加，表明尿素已施完毕。

③流量法。根据 Amos Teich 公式 $T=4V/Q$，当施肥时所使用的是液体肥料或溶解性较好的固体肥料如尿素时，可推算出一次施肥所需要的时间。因此，可在压差施肥罐的出水口端安装一流量计，从开始施肥到流量计记录的流量约为 4 倍的压差施肥罐体积时，表明施肥罐中肥料已基本施完，此时段所经历的时间即为施肥时间。

了解施肥时间对应用压差施肥罐施肥具有重要意义。当施下一罐肥时必须要将罐内的水放掉至少 1/2~2/3，否则无法加放肥料。如果对每一罐的施肥时间不了解，可能会出现肥未施完但停止施肥，将剩余肥料溶液排走而浪费肥料，或肥料早已施完但仍在灌溉，若单纯为施肥而灌溉时会浪费水源或电力，增加施肥人工。特别在雨季或土壤不需要灌溉而只需施肥时更需要加快施肥速度。

研究者在田间调查发现施肥罐使用中存在一些问题。在田间大面积灌溉区，有些施肥罐体积太小，应该配置 300L 以上的施肥罐，以方便用户施肥。有些施肥罐上不安装进、排气阀，导致操作困难。有些施肥罐的进水和出肥管管径太小，无法调控施肥速度。一般对 200L 以上的施肥罐，应该采用 32mm 的钢丝软管。从倒肥的操作便利性来看，卧式施肥罐优于立式施肥罐。通常一包肥料 50kg，倒入齐腰高的立式罐难度较大。

2. 文丘里施肥器

虽然文丘里施肥器可以按比例施肥，在整个施肥过程中保持恒定浓度供应，但在制订施肥计划时仍然按施肥数量计算。比如一个轮灌区需要多少肥料要事先计算好。如用液体肥料，则将所需体积的液体肥料加到储肥罐（或桶）中。如用固体肥料，则先将肥料溶解配成母液，再加入储肥罐。或直接在储肥罐中配制母液。当一个轮灌区施完肥后，再安排下一个轮灌区。

当需要连续施肥时，对每一轮灌区先计算好施肥量。在确定施肥速度恒定的前提下，可以通过记录施肥时间或观察施肥桶内壁上的刻度来为每一轮灌区定量。对于有辅助加压泵的施肥器，在了解每个轮灌区施肥量（肥料母液体积）的前提下，安装一个定时器来控制加压泵的运行时间。在自动灌溉系统中，可通过控制器控制不同轮灌区的施肥时间。当整个施肥可在当天完成时，可以统一施肥后再统一冲洗管道，否则必须将施过肥的管道当日冲洗。冲洗的时间要求同旁通罐施肥法。

3. 重力自压式施肥法

利用自重力施肥由于水压很小（通常在3m以内），用常规的过滤方式（如叠片过滤器或筛网过滤器）由于过滤器的堵水作用，往往使灌溉施肥过程无法进行。张承林等在重力滴灌系统中用下面的方法解决过滤问题。在蓄水池内出水口处连接一段1~1.5m长的PVC管，管径为90mm或110mm。在管上钻直径30~40mm的圆孔，圆孔数量越多越好，将120目的尼龙网缝制成管大小的形状，一端开口，直接套在管上，开口端扎紧。用此方法大大地增加了进水面积，虽然尼龙网也照样堵水，但由于进水面积增加，总的出流量也增加。泥肥池内也用同样方法解决过滤问题。当尼龙网变脏时，更换一个新网或洗净后再用。经几年的生产应用，效果很好。由于尼龙网成本低廉，容易购买，用户容易接受和采用。

4. 泵吸肥法

根据轮灌区的面积计算施肥量，然后倒入施肥池。开动水泵，放水溶解肥料。打开出肥口处开关，肥料被吸入主管道。通常面积较大的灌区吸肥管用50~70mm的PVC管，方便调节施肥速度。一些农户选用的出肥管管径太小（25mm或32mm），当需要加速施肥时，由于管径太小无法实现。对较大面积的灌区（如500亩以上），可以在肥池或肥桶上划刻度。一次性将当次的肥料溶解好，然后通过刻度分配到每个轮灌区。假设一个轮灌区需要一个刻度单位的肥料，当肥料溶液到达一个刻度时，立即关闭施肥开关，继续灌溉冲洗管道。冲洗完后打开下一个轮灌区，打开施肥池开关，等到达第二个刻度单位时表示第二轮灌区施肥结束，依次进行操作。采用这种办法对大型灌区施肥可以提高工作效率，减轻劳动强度。

在北方一些井灌区水温较低，肥料溶解慢。一些肥料即使在较高水温中溶解也慢（如硫酸钾）。这时在肥池内安装搅拌设备可显著加快肥料的溶解，一般搅拌设备由减速机（功率1.5~3.0kW）、搅拌桨和固定支架组成。搅拌桨通常要用304不锈钢制造。

5. 泵注肥法

南方地区通常都有打药机。许多农民利用打药机作注肥泵用。具体做法如下。在泵房外侧建一个砖水泥结构的施肥池，一般3~4m³。通常高1m，长宽均2m，以不漏水为质量要求。池底最好安装一个排水阀门，方便清洗排走肥料池的杂质。施肥池内侧最好用油漆划好刻度，以0.5m为一格。安装一个吸肥泵将池中溶解好的肥料注入输水管。吸肥泵通常用旋涡自吸泵，扬程须高于灌溉系统设计的最大扬程，通常的参数为：电源220V或380V，功率0.75~1.1kW，扬程

50m，流量3~5m³/h。这种施肥方法肥料有没有施完看得见、施肥速度方便调节的特点，它适合用于时针式喷灌机、喷水带、卷盘喷灌机、滴灌等灌溉系统。它克服了压差施肥罐的所有缺点。特别是使用地下水的情况下，由于水温低（9~10℃），肥料溶解慢，可以提前放水升温，自动搅拌溶解肥料。通常减速搅拌机的电机功率为1.5kW。搅拌装置用不生锈材料做成倒T形。

6. 移动式灌溉施肥机

移动式灌溉施肥机是针对没有电力供应的种植地块而研发的，主要由汽油泵、施肥罐、过滤器和手推车组成，可直接与田间的灌溉施肥管道相连使用，移动方便、迅速。当用户需要对田间进行灌溉施肥时，可以用机车将灌溉施肥机拉到田间，与田间的管道相连，轮流对不同的田块进行灌溉施肥。移动式灌溉施肥机可以代替泵房固定式首部系统，成本低廉，便于推广，能够满足小面积田块灌溉施肥系统的要求。目前，移动式灌溉施肥机的主管道有2寸①和3寸两种规格，每台移动式灌溉施肥机可负责50~100亩的面积。

（1）移动式灌溉施肥机操作规程。第一步，移动施肥机每次使用前，都要检查机油和汽油的油位，如果油不足，要在使用前加入足量的油；加入机油时轻轻抬起汽油机的一端，不宜过于倾斜，机油也不宜过多。检查所有接头是否连接完好；检查渠道水位是否处于开机的安全水位。第二步，水泵注水室应在启动前加满预注水，否则会损坏水泵密封圈，同时也会有抽不上水的现象发生。第三步，启动施肥机时，首先要开启燃油开关，关闭阻风门拉杆，将发动机开关置于开启位置，同时将气门拉杆稍向左侧移动，然后抖动启动手柄。当移动施肥机开启后，慢慢开启阻风门，同时将节气门置于所需速度位置。第四步，移动施肥机系统平稳启动后，观察压力表读数，等水抽上来后，压力表显示为水泵工作扬程时，一定要慢慢打开系统总阀门进行灌溉，以防压力大的水一下子冲掉出水口接头。第五步，正常出水时，左边的压力表读数在0.08~0.20MPa（8~20m）之间，右边读数在0~0.05MPa（0~5m）之间。第六步，过滤器正常工作时，过滤器两边的压力表读数相差为0.01~0.03MPa，表明此时过滤是清洁的；而当过滤器两边压力表读数相差超过0.04MPa的时候，必须尽快清洗过滤器。第七步，在系统运行时，管理人员应要去相关轮灌区巡查，看运行是否正常。发现破管、断管、堵塞、灌水器损坏、漏水等现象应及时处理。不能处理时，应立即通知有关技术人员协助处理。第八步，系统停机时，将节气门拉杆向左移到底，同时关

---

① 1寸约为3.3cm，全书同。

闭发动机开关、燃油开关，然后关闭球阀开关。

（2）施肥操作。第一步，将计算好的肥料倒入肥料桶，加水搅拌溶解后方可打开施肥开关。第二步，施肥前，先打开要施肥区的开关开始灌水。等到田间所有灌水器都正常出水后，打开施肥开关开始施肥。施肥时间控制在 30~60min 为宜，越慢越好（具体情况可以根据田间的干湿状况调整）。施肥速度可以通过肥料池的开关控制。第三步，施完肥后，不能立即关闭灌溉系统，还要继续灌溉 10~30min 清水，将管道中的肥液完全排出。如果在阴雨天气施肥，此措施可以待天晴后施肥时补洗。不然的话，会在灌水器处长藻类、青苔、微生物等，造成滴头堵塞（这个措施非常重要，也是滴灌成功的关键）。

（四）轮灌组更替

根据水肥一体化灌溉施肥制度，观察水表水量确定达到要求的灌水量时，更换下一轮灌组地块，注意不要同时打开所有分灌阀。首先打开下一轮灌组的阀门，再关闭第一个轮灌组的阀门，进行下一轮灌组的灌溉。操作步骤按以上重复。

（五）结束灌溉

所有地块灌溉施肥结束后，先关闭灌溉系统水泵开关，然后关闭田间的各开关。对过滤器、施肥罐、管路等设备进行全面检查，达到下一次正常运行的标准。注意冬季灌溉结束后要把田间位于主支管道上的排水阀打开，将管道内的水尽量排净，以避免管道留有杂质。

二、水肥一体化系统的维护保养

要想保持水肥一体化技术系统的正常运行和提高其使用寿命，关键是要正确使用及良好地维护和保养。

（一）水源工程

水源工程有地下取水、河渠取水、塘库取水等多种形式，保持这些水源工程建筑物的完好、运行可靠以及确保设计用水的要求，是水源工程管理的首要任务。

对泵站、蓄水池等工程经常进行维修养护，每年非灌溉季节应进行年修，保持工程完好。对蓄水池沉积的泥沙等污物应定期排除洗刷。开敞式蓄水池的静水中藻类易于繁殖，在灌溉季节应定期向池中投放绿矾，可防止藻类滋生。

灌溉季节结束后，应排除所有管道中的存水，封堵阀门和井。

第四章　水肥一体化技术在农业生产中的应用

（二）施肥系统

在进行施肥系统维护时，关闭水泵，开启与主管道相连的注肥口和驱动注肥系统的进水口，排除压力。

1. 注肥泵

先用清水洗净注肥泵的肥料罐，打开罐盖晾干，再用清水冲净注肥泵，然后分解注肥泵，取出注肥泵驱动活塞，用随机所带的润滑油涂抹部件，进行正常的润滑保养，最后擦干各部件重新组装好。

2. 施肥罐

首先仔细清洗罐内残液并晾干，然后将罐体上的软管取下并用清水洗净，软管要置于罐体内保存。每年在施肥罐的顶盖及手柄螺纹处涂上防锈液，若罐体表面的金属镀层有损坏，立即清锈后重新喷涂。注意不要丢失各个连接部件。

3. 移动式灌溉施肥机的维护保养

对移动式灌溉施肥机的使用应尽量做到专人管理，管理人员要认真负责，所有操作严格按技术操作规程进行；严禁动力机空转，在系统开启时一定要将吸水泵浸入水中；管理人员要定期检查和维护系统，保持整洁干净，严禁淋雨；定期更换机油（半年），检查或更换火花塞（1年）；及时人工清洗过滤器滤芯，严禁在有压力的情况下打开过滤器；耕翻土地时需要移动地面管，应轻拿轻放，不要用力拽管。

（三）田间设备

1. 排水底阀

在冬季来临前，为防止冬季将管道冻坏，把田间位于主支管道上的排水底阀打开，将管道内的水尽量排净，此阀门冬季不关闭。

2. 田间阀门

将各阀门的手动开关置于打开的位置。

3. 滴灌管

在田间将各条滴灌管拉直，勿使其扭折。若冬季回收也要注意勿使其扭曲放置。

（四）预防滴灌系统堵塞

1. 灌溉水和水肥溶液先经过过滤或沉淀

在灌溉水或水肥溶液进入灌溉系统前，先经过过滤器或沉淀池，然后经过滤器后才进入输水管道。

2. 适当提高输水能力

根据试验，水的流量在 4~8L/h 范围内，堵塞减到很小，但考虑到流量越大则费用越高，故最优流量约为 4L/h。

3. 定期冲洗滴灌管

滴管系统使用 5 次后，要打开滴灌管末端堵头进行冲洗，把使用过程中积聚在管内的杂质冲洗出滴灌系统。

4. 事先测定水质

在确定使用滴灌系统前，最好先测定水质。如果水中含有较多的铁、硫化氢、丹宁，则不适合滴灌。

5. 使用完全溶于水的肥料

只有完全溶于水的肥料才能进行滴灌施肥。不要通过滴灌系统施用一般的磷肥，因为磷会在灌溉水中与钙反应形成沉淀，堵塞滴头。最好不要混合几种不同的肥料，避免发生相关的化学作用而产生沉淀。

（五）细小部件的维护

水肥一体化系统是一套精密的灌溉装置，许多部件为塑料制品，在使用过程中要注意各步操作的密切配合，不可猛力扭动各个旋钮和开关。打开各个容器时，注意一些小部件要依原样安回，不要丢失。

水肥一体化系统的使用寿命与系统保养水平有直接关系，保养得越好，使用寿命越长，效益越持久。

第五章　水果水肥一体化技术的应用

灌溉和施肥是果园的两项重要管理措施，传统上灌溉和施肥都是分开进行的，这无疑是成本巨大的。灌溉和施肥同时进行是最好的措施，果树根系一边吸水，一边吸肥，就会大大提高肥料的利用率，果树生长壮旺。水肥同时管理的技术就叫水肥一体化管理技术。特别是采用管道灌溉和施肥后（果园最适宜用滴灌或微喷灌），可以大幅度节省灌溉和施肥的人工，托普云农水肥一体化技术是一种科学、节省、高效的水肥管理技术。

第一节　苹果水肥一体化技术应用

苹果是我国的重要果树，栽培面积大，主要分布于环渤海湾和西北黄土高原等地区。目前我国的苹果种植技术正朝着机械化、标准化、简约化方向发展。苹果是多次施肥和频繁灌溉的果树，水肥管理与产量品质有密切关系。目前在苹果生产中存在劳动力短缺，劳动力价格逐年上升，施肥成本不断增加，过量施肥，不平衡施肥，土壤盐化、酸化、板结和根系微生态环境变差等问题。施肥和灌溉是两项重要的田间作业，如何做到合理和科学，广大农户急需理论和技术的指导。特别是简单易懂、图文并茂的科普读本，更受果农欢迎。

一、栽培苹果的起源

苹果是世界上落叶果树中最普遍栽培的一种果树，早在5 000多年以前就已经为人们所栽培了。但是，关于栽培苹果的起源问题，迄今仍无定论。许多国家的学者都曾进行过研究，其主要原因在于苹果栽培历史悠久、地理分布广泛、品种繁杂，种和品种、类型之间易于杂交，为解决此一问题带来了很多困难。

大多数学者认为栽培苹果起源于多中心。从苹果属植物在世界上的分布来看，栽培苹果可能起源于森林苹果、高加索苹果和塞威氏苹果3种。根据波诺马林科的意见，通常认为栽培苹果所属种，实际上根本不存在，因为迄今在自然界中一直还不曾发现此种的野生类型。

波诺马林科对栽培苹果的起源问题进行了大量的研究。他对苹果属植物的野生种进行了广泛调查、收集和对比观察研究，特别是经过对一些认为可能是栽培苹果的野生原始种的形态、生物学特性、地理分布与当今栽培品种性状的对比观察研究认为，塞威氏苹果乃是当今栽培苹果的真正祖先。

森林苹果是一个分布很广的种，西到西班牙的比斯开海湾（Bay of Biscay），东到伏尔加河流域，北到斯堪的纳维亚半岛，南到地中海与黑海都有分布，但其类型贫乏，相当单纯，其果实果肉疏松，味酸涩，间或带有苦味，色泽绝大部分为绿色或黄色，当地称其为酸苹果。在典型的森林苹果中看不到有现今栽培苹果品种具有的甜或酸甜风味的果实。

高加索苹果像森林苹果一样，类型单纯，果实酸涩，虽经大量调查，始终未发现有味甜的类型，其果实色泽也为绿色，只有在高海拔的南向山坡处偶尔看到一些带有红晕的果实。

塞威氏苹果则与上述两种不同，其形态和生物类型十分多样。仅就果实来看，其果实大小直径可为 1.5~7cm 不等，单果重可为 6~50g 或更重；形状有圆形、扁形、长形等，棱起多样；色泽有绿、淡黄直至鲜红，且有的有锈和果粉；果肉颜色有白、乳黄、粉红；果实风味也十分多样，酸而不可食的类型到酸甜味美的类型，样样具有；成熟期也有早熟、晚熟之分。据此，波诺马林科认为中心分布于中亚哈萨克斯坦山区和我国天山山区的塞威氏苹果才是迄今世界栽培苹果的祖先。

波诺马林科认为，塞威氏苹果与其他野生苹果种最重要的区别在于其在遗传上具有决定栽培苹果品种的决定性状，即具有又甜又大的果实，正是大而甜的果实性状产生了当今的栽培品种。

波诺马林科根据自己的调查研究，认为世界上的栽培苹果是由中亚西亚传至欧洲的。他认为世界上最古老的果园产生在中亚西亚地区。当时，人们还不会嫁接，是用根蘖繁殖的，而塞威氏苹果易于发生根蘖。利用根蘖栽植把优良单株的特性保存了下来。另外，塞威氏苹果还可以用种子进行繁殖。通过古代人们不断地从野果林中选优和栽培，一些优良类型不断得以改进。这些优良类型逐渐地由中亚西亚传至邻近的国家，如伊朗、阿富汗、阿尔巴尼亚、希腊和土耳其，后来经过高加索传至欧洲。

波诺马林科所叙述有关塞威氏苹果的多样性，我国的学者在调查新疆新源的野果林时，也看到了类似的情况。

二、苹果的生长习性

苹果是一种落叶乔木果树,每年入春芽萌动生长,入冬则落叶休眠,年年如此。苹果新梢的生长势随其年龄时期的不断演进也不断在发生变化。幼龄时期,植株生长旺盛,新梢粗壮,年生长量可达120cm以上。进入盛果期,生长势显著变弱,新梢年生长量一般30~50cm。盛果期末期,新梢生长的强度更弱,年生长量常不到20cm。进入衰老期后,外围新梢生长量很少,多在5cm左右。

苹果幼树除表现为生长旺盛、新梢粗壮外,树冠枝条多不开张,直立生长习性也强。由于幼树有这种特性,在幼龄时期,对于大多数品种来说,要特别注意开张枝条角度,通过开张角度可以削弱枝条生长势,同时又改善了冠内的通风透光条件,有利于幼树适龄结果。

有些苹果品种的枝条着生角度较大,结果后枝条易下垂,对于这样的品种在幼龄时期可不必开张骨干枝的角度。

大多数品种进入结果期后,由于产量的负担,枝条即逐渐开张。

苹果植株生长的另一特性是枝条的层性分布,枝条生长分布的层性特性是在其长期的系统发育过程中形成的,这是对外界条件适应的一种表现。苹果植株由于逐年不断地生长,为了解决下部早先发出枝条的光照问题,在其系统发育的过程中就形成了每年新梢都是顶端数芽抽生为长枝,以下各芽则萌发为短枝、叶丛枝或呈潜伏状态不萌发,这样先一年的枝条和当年发出的枝条自然也就成为层状分布。

苹果枝条层性分布现象的内在原因还在于芽在枝条上着生的异质性和枝条的顶端生长优势。一个枝条上的芽由于其生发的时间和当时树体营养状况的差异,一般表现为枝条顶端的芽发育都比较充实,而基部的芽则发育瘦弱,加以春季萌发生长后,水分、养分都优先流向顶端的芽,而下部的芽得到的水分和养分较少,因此下部的芽生长势弱、不能抽生为长枝。这样连年如此生长,结果也就形成了层性。

层性的明显与否和苹果品种发枝力的强弱有很大关系。通常发枝力强的品种,如红玉、金冠、秦冠等,层性表现不明显;而发枝弱的品种,如国光、青香蕉等,则层性比较明显。

苹果枝条层性分布的显著程度,随着植株年龄的增长也在发生变化。幼龄时期枝条生长健旺,层性表现比较明显;进入盛果期后,枝条生长势减弱,层性也就逐渐变得不明显了。至衰老时期,枝条几乎停止增长,就更看不出枝条的层性分布了。

苹果的大多数品种在年生长周期中都表现有新梢二次生长的现象,即当年生新梢有春梢和秋梢之分,只是在我国北方地区新梢生长长度较短而已。根据中国农业科学院果树研究所的观察,在辽宁兴城地区无论早、中、晚熟品种新梢生长高峰多在 5 月下旬,个别品种出现在 6 月初,早熟品种和大多数中熟品种只有一次生长高峰,晚熟品种有两次生长高峰,多出现在 8 月上旬。春梢生长停止期始于 7 月初,7 月末开始二次生长。

根据江苏省徐州市果园的观察,在黄河故道地区,苹果新梢 4 月上旬开始生长,5 月上旬为新梢旺盛生长期。红玉品种在 6 月初还有一次生长高峰,以后生长逐渐减缓。秋梢多在 7 月初开始生长,中旬进入高峰,8 月中旬停止生长。生长旺盛的树还能继续发育出三次梢来,直到 9 月初结束。由于其生长结束较晚,顶芽充实度一般欠佳。

在黄河故道地区,苹果幼树新梢生长量一般多为 90~120cm,春秋梢长度差别很小。有些品种,如伏红、金冠、华冠、祝光等,常在秋梢上形成大量腋花芽或顶花芽。在保证骨干枝正常生长和良好整形的原则下,修剪时应注意的对此类腋花芽利用。

三、苹果所需要的各种营养成分及其在果树生理上的作用

果树是一种绿色植物,它与其他有生命的有机体的重要差别在于:果树可以利用自然界的无生命的物质(无机物)来作为自己的营养,这也就是说,植物的营养来自于无机物。但需要指出的是:许多研究证明植物也具有直接利用氨基酸、核酸以及其他有机化合物中氮的能力。植物摄取土壤中的无机物的过程与土壤微生物的活动是密切不可分的,地球上 90% 的植物都与土壤微生物处于共生状态中。

绿色植物具有行使光合作用的器官,可以利用太阳能把矿物质元素转化为有机物形态。植物细胞中的内含物——原生质和细胞核,就是植物利用碳、氮、磷和硫组成的,植物细胞的正常活动还需要其他一些矿物质元素,如钾、钙、镁、铁、锰、锌、铜等。

氮和其他灰分虽然在植物体内所占比例很小,但它们在植物的生命活动中起着重要的作用。叶片的同化作用、叶绿素的形成,不仅需要光、温度和水分,同时还必须有各种矿物质元素。因此,要想调节植物的营养条件,必须了解各种营养元素在植物生理活动中的作用。

（一）氮

氮素对于植物来说，其重大作用在于它是植物最重要的含氮物质（氨基酸）的组成成分，而氨基酸则是构成蛋白质和核酸分子（细胞核的内含物）的物质基础，另外，在光合作用中起极其重要作用的叶绿素的形成离不开氮。因此，氮是植物进行整个生命活动不可缺少的元素。氮素如果不足则氮素有机物的合成受阻，从而导致生长停止、叶片变小、色泽变淡。果树植株在氮供应不足的情况下，只能利用根、干和枝条内贮存的含氮有机物以保证某些新梢的生长，但是由于自身氮的含量急剧减少，最后引起坐果率低和幼果大量脱落。在果树植株旺盛生长的阶段如能获得充足的氮素供应，则可促进氮素有机物质的大量合成，促进其生长过程，加速树体的发育。但若氮素供应过多则常导致新梢久久不能停止生长、枝条成熟度差而导致抗寒能力降低。

氮的供应对于嘎拉品种果实的大小有重大的影响。氮的水平提高可以增大叶片的氮水平，促使果实增大和产量提高。要使嘎拉果实增大，叶片的氮含量不应低于2.0%，但如超过2.5%时则将影响果实的贮藏能力。

与其他果树相比，苹果对氮的需求量并不高，成年果园每公顷的纯氮需要量为30~100kg。许多报告称，在肥沃的土壤上就是不施氮肥，苹果也能正常生长和丰产（Atkinson，1980）。通常推荐的苹果施氮量为每公顷施纯氮75~150kg。

氮素在生理上对果树植株影响的大小，与植株所处的生长阶段、水分与其他营养物质特别是磷、钾的供应有密切关系。

果树植株不同器官氮的含量不一，其含量最多的为叶片、果实和结果枝，营养枝含量很少，而以干和多年生枝及根内的含量最少。许多研究证明，苹果植株的不同器官氮的含量（占其干重的百分比）大体为：果实为0.40%~0.80%，叶片为2.30%，营养枝为0.54%，结果枝为0.88%，干和多年生枝为0.49%，根为0.32%。

果树植株的氮含量以及其他营养物质的含量随着植株器官年龄的增长、年周期内不同的生长阶段、不同的季节而不断地变化。幼嫩的器官含氮量多，在春季生长开始阶段，叶片和新梢的氮含量最多，其氮的来源为头年贮存于植株内的氮素化合物。以后随着生长阶段的变化，叶片、新梢以及果实内的相对含氮量逐渐减少，但其总含量则急剧增长。秋季落叶前，部分氮以及其他营养物质（磷、钾）则由叶片内转移到枝条、树干和根内贮存起来，以备来年植株再次利用。氮是一种非常活跃的元素，它还可以由老组织内向新生的组织转移。

（二）钾

钾虽不是植物主要有机物的组成部分，但其在植物的生命过程中同样具有重要作用。钾在植物体内是以其离子状态存在的，它是植物新陈代谢中和蛋白质及碳水化合物的合成中酶的催化剂；由于它以离子状态存在，因此，它是细胞溶液和各种有机酸的积极缓冲者。钾还是一种渗透剂，这对于植物体内水的运动具有特别重要的作用；它可以自由地进入植物的木质部和韧皮部，在碳酸的同化过程中以及碳水化合物的运转中都具有重大作用。在钾供应不足时，植物的合成过程受阻，碳水化合物，特别是蔗糖和淀粉量明显减少，可塑性物质由叶片内向外输出而减少，而由于呼吸作用使碳水化合物遭到消耗。

植物对氮素的吸收和蛋白质的合成都需要在有钾的情况下才能顺利进行。

在钾供应不足的情况下，苹果果实变小、色泽发暗、风味变淡、抗寒能力降低，春季芽和花易遭受霜害。钾在根系生长和叶片气孔的开张与闭合上起关键作用。钾的含量不应低于1.4%。

钾在苹果植株体内以叶片和果实内含量最多，其在叶片内的含量约占其干重的1.6%，在果实内的含量约占1.2%。

（三）磷

磷在植物的生命活动中也起有非常重要的作用，是核蛋白和各种核酸的组成部分，而核蛋白在细胞核及其他有机合成物（磷脂、辅酶、肌醇）的建造中具有重要的作用。此外，磷还参与碳水化合物的代谢，它可以促进许多发酵过程的进行，在缺少磷的情况下，淀粉就不能转化为糖。有材料报道，磷还能提高植物的氧化还原反应。

磷在大多数土壤中的可溶性低。因此，磷在土壤溶液里的可利用性很小。

磷供应不足，植株的新梢和叶子的生长将行削弱，由于分生组织的活动减弱，可溶性氮和花青素的积累减少，致使浓绿色的叶片在秋季落叶前出现紫色和带有红色的斑点。磷若供应不足，花芽形成和结果将受到严重影响。

磷在植株内以叶片内含量最多，占其干重的0.45%；其次为结果枝，约占其干重的0.28%；果实内的磷含量占其干重的0.09%~0.20%。

磷在植物体内的含量像氮一样，随着植株及其不同器官年龄的增长和年周期中物候期的进程不断在发生变化。

（四）镁

镁是叶绿素的组成部分，镁对植物碳酸的同化作用有重要作用。随着叶片的衰老和叶绿素的破坏，镁流向种子，与磷一起以肌醇状态贮存起来。少量镁则参与果胶物质的合成。

镁在植物体内如果供应不足，则导致生长停止，叶片色泽变淡、出现褐斑，仅沿叶脉处色泽正常；如果镁严重不足，则将导致叶片及早脱落，仅在新梢顶端留有少量叶片。

（五）钙

钙对植物体中营养溶液的生理平衡起重大的作用，当某些阳离子物质，如一价的氢离子、钾离子、铵离子、钠离子，二价的镁离子和三价的铁离子、铝离子偏多时，钙则可起平衡作用。在营养液中钙离子的存在对于根系的生长，特别是在酸性土壤中尤为重要。

植物体在吸收和利用铵态氮方面，钙的充足供应是个十分重要的条件，如果钙供应不足，植物体就不能充分地吸收铵态氮。

钙与氮、磷、钾最大的差别是不能被植物体重复利用，这是因为钙主要积累在植物体衰老的或缺乏生命力的部分。钙在果实内的含量很少。近些年来，各国学者的研究指出，一些苹果品种果实的苦痘病主要就是钙的供应不足引起的。实践证明，在生长季节中定期喷施氯化钙可以减轻苦痘病的发病率。

（六）铁

铁在植物体内含量极少，虽然铁不是叶绿素的组成部分，但在植物正常形成叶绿素的过程中铁是必不可少的。

在缺乏铁的情况下，常会导致植物体生长停止，叶片出现黄化。由于铁在植物体内不能被重复利用，因此铁不能由原来的绿色叶片内转移到黄化的叶片内。

（七）硫

硫是构成蛋白质的氨基酸的组成部分。植物体仅能从硫酸盐中吸收处于氧化状态下的硫。植物体内很少发现有硫供应不足的情况，因为土壤内常含有很多的硫，而且由于硫酸铵、过磷酸钙的不断施用，因此土壤内的硫可以不断得到补充。

（八）硼

硼在植物体内处于不活动的状态，像钙和铁一样，不能被植物重复利用。植物对硼的需要量很少，植物的花含硼较多。一旦硼不足，则花不能正常受精。在缺硼的情况下，根系发育受阻，果实内常会出现木栓化斑点病。

（九）锰

锰与铁一样，对叶绿素的形成有一定的作用。另外，对植物的氧化还原过程也有一定的影响。

由于植物体对硼、锰的需要量很少,因此,人们通常称其为微量元素。除硼、锰外,在微量元素中对苹果重要的还有锌、铜、铝等。苹果植株在春季常出现一些先端轮生一簇小而质地较硬的叶片新梢的现象,这就是缺锌所致。

四、施肥时期与方法

(一)秋施基肥

基肥是果树施用的最主要肥料。果树生长结果所需要的营养物质主要来源于基肥,通常都是在秋末冬初土壤结冻前施用,因此,又称秋施肥。

用于基肥的都是有机肥料,如厩肥、堆肥等。秋末施用基肥时所切断的根系,在次年春季植株开始生长前可以得到很好的恢复,而施入土中的肥料经过一个冬季的分解,到了次年春天可以被果树的根系很好地吸收利用。

在施基肥时,可以同时掺入一些磷肥或钾肥。

矿物质磷肥和钾肥在土壤内活动性差,基本上停留在被施入的土层内。为了有利于果树根系的吸收利用,必须把它们施入到20~30cm深的土层内,因此,在秋施基肥时混入磷、钾肥料对日后发挥其作用是有益的。

基肥的施用方法一般都采用环状沟施法,即在沿树冠外围挖掘一条深40cm、宽30~40cm的轮状沟,将基肥施入其中,而后用土埋好。施肥的位置每年应随树冠的扩大而向外扩展。

除轮状沟施法外,也可以在树冠的两侧挖掘条沟施,沟深与宽度可与轮状沟相同,挖沟的位置应隔年轮换。

此外,还可采取放射沟施,即在树冠下,离主干1m以外的地方挖掘5~6条放射状沟,然后将基肥施入沟内,用土埋好。

生产实践证明,基肥的效用维持时间较长,因此,可每隔两年施一次基肥,以解决果园基肥来源不足的问题。

(二)生长季节内追肥

追肥的目的主要是补充基肥的不足,大多是在果树生长季节,根据果树植株生长发育的需要而及时追施。作为追肥的肥源一般都是氮肥,如硫酸铵、硝酸铵、尿素等。

氮素肥料由于施入土中容易流失,因此,一次追肥量不宜过多,一般5~6年生初结果树每株每次可追施100~150g,10年生以上开始大量结果的植株每株可追施250~500g。尿素的含氮量高,其追施量可少些。

除了硫酸铵、硝酸铵和尿素适宜作追肥外,腐熟的人粪尿也可以作追肥用。

通常一份人粪尿可对上5份水来施用。

禽粪（鸡粪、鸽粪等）在加水受溶后也可作苹果的追肥用。一般追施的浓度可为1kg禽粪对水10kg施用。

追肥一般采用放射沟施法，即在靠近树冠的外围用锄钩几条浅沟，将化肥均匀地撒入其内，或将人粪尿、禽粪汁施入其内，然后用锄推土加以覆盖。追肥最好在雨后土壤湿润时进行，若追用后能结合灌水则效果更好。

根据苹果植株一年内生长发育的客观规律，追肥应采用分期多次进行，其效果更佳。

在管理正常的果园内，通常每年需要进行以下若干次追肥：

（1）花前（开花前两周）追肥。适用于当年花芽量少、隔年结果习性严重的品种。

（2）花后（落花后）追肥。果树开花、坐果需要大量的营养，花后正是新梢旺盛生长时期，也需要耗费大量的营养，此时追施肥对提高坐果率、减少落果、保证新梢旺盛生长有良好的作用。

（3）生理落果后追肥。此次追肥大约在落花后一个月。在黄河故道地区为5月下旬，此时坐果已成定局，幼果正在迅速膨大，花芽临近分化前夕，为了保证幼果正常生长、促进花芽分化、并为次年丰产打下基础，此次追肥极为重要。

在上述三次追肥中，其中第2次和第3次，不论在什么情况下都必须追施。追肥时，除追用氮素化肥外，人粪尿、鸡粪和羊粪对水稀释后都可用来追肥。

（4）秋季追肥。除上述三次追肥外，于8月、9月中旬和9月下旬进行秋季追肥对植株的生长发育和连年优质高产效果更为明显。此时苹果的中晚熟和晚熟品种果实正处于采前果实膨大期，追肥有利于果实的增长，同时还可以促使叶片的同化作用、增强根系活动、提高植株秋季贮藏物质的积累，为果树的越冬和次年的开花结果、生长发育创造良好的物质基础。

生长季节追肥还应考虑到树龄和当年的结果情况。在幼龄苹果园内追肥的时期应重点放在早春和新梢的旺盛生长期。对于已经进入大量结果期的果园，在大年时追肥的时期应重点放在花后和生理落果时前，其目的在于促使新梢生长、增大叶营养面积，以满足当年产量的养分积累，又能形成足够的花芽，保证次年的产量；小年时追肥时期应重点放在早春和新梢旺盛生长时期，以延长新梢旺盛生长时期，使花芽形成量不致过多，避免大小年。

（5）根外追肥。在苹果生长季节内，除了可以采取土壤追肥外，还可以利用喷雾的方法进行根外追肥，即利用喷雾器将矿物质肥料的溶液喷洒到苹果树的叶面上，矿物质通过叶背的海绵组织细胞渗入叶内，被吸收和利用，同样也可以起到追肥的作用。

叶面追肥通常结合打药掺在药液内喷洒。凡是经过根外追肥的苹果树表现叶色深绿、果实着色好、枝条成熟度也好。

在作为叶面追肥的化肥中，以尿素最好。尿素易被叶片吸收。喷洒尿素的时期以花瓣脱落时效果最好，6月落果前不宜喷用，否则易引起大小年。

尿素通过叶片背面的气孔或直接通过叶片的角质层进入叶内；喷后最初几个小时内，尿素被吸收得最快。叶背由于具有气孔，其吸收尿素的效果较叶片表面为强。幼嫩叶片较老叶吸收快。叶片吸收尿素的时间可延续两个昼夜。

除尿素可作叶面追肥外，其他一些矿质肥料也可作为叶面追肥用，通常喷用的浓度如下：

尿素　0.5%

硫酸铵　0.3%

硝酸铵　0.3%

过磷酸钙　1.0%（此浓度对某些品种，特别是弱树，易产生药害）

氯化钾　0.5%

早春根外追氮（喷枝干）对苹果树以后的生长、坐果率提高和果实前期发育都有好处。早春根外喷氮的植株叶色浓绿，全氮量由1.4%~1.8%提高到2.6%，增加了30%~40%，叶绿素增加5%~10%，叶干重增加15%~25%，光合强度增加20%~35%，叶面积增大15%~30%。

五、生产上通常进行灌溉的时期

我国苹果产业的广大果农，在长期栽培苹果的生产实践中，根据本地区的土壤气候条件和苹果树一年四季生长、开花、结果、休眠的要求，积累了丰富的苹果灌溉经验，对苹果生产的发展起了重大的作用。根据灌溉时期的不同，一般分为冬春灌溉和生长期灌溉。

冬春灌溉在秋末冬初土壤封冻前和早春芽萌动前进行。我国东北、华北、西北以及黄河故道地区，冬季和春季降水量少、空气湿度低，且常刮寒风、蒸发量大，因此，进行冬春灌溉非常必要。

秋末冬初的灌溉，群众称之为"封冻水"或"越冬水"，对苹果树的越冬有良好影响，可以显著减轻苹果树一年生枝条或苹果幼树越冬期间因天气干旱、常吹寒风造成的枝条干枯和冻害现象，而且还可减轻早春干旱给苹果树造成的不良影响。早春时节正是这些地区易于干旱的季节，此时灌水有利于苹果植株的及早萌动和生长。各地的生产实践证明，冬春灌溉是保证苹果丰产稳产的一个重要因素。

根据前苏联摩尔达维亚蔬菜马铃薯灌溉试验站的试验，同一苹果品种，在不进行灌溉的地块上，每公顷的产量为 9 795kg 而在生长季节内进行 3 次灌溉的地块，每公顷的产量为 17 100kg，进行 3 次生长期灌溉和 1 次冬灌的产量则为 20 490kg，从而可以清楚地看出冬春灌溉对提高单位面积产量的重大作用。

我国苹果产区通常进行灌溉的具体时期和作用如下：

（1）早春发芽前。此时灌水可以促使芽及早萌动，减轻春旱对果树的不良影响。

（2）发芽后或开花前。此时灌水可以使花正常开放，有利于受精和坐果。

（3）开花后。此时幼果开始膨大，新梢处于旺盛生长时期，植株需要水分多，及时进行灌溉可以提高坐果率、减少落果、促进幼果和新梢旺盛生长。

（4）生理落果后（落花后约 1 个月）。此时灌水可以加速果实增长，并有利于花芽分化。

（5）采收前 2~3 周。此时灌溉有利于果实最后的增长，可提高单位面积产量与品质，并为植株的生长发育创造良好的水分条件。

（6）落叶后封冻前。此时应充分灌溉，以提高苹果植株的越冬能力、减轻寒地的冻害，并为次年的丰产打下良好的基础。

在苹果园内，灌溉一般都是结合施肥或追肥进行的。灌溉次数的多少，因具体情况而定。

灌水量的多少通常以水分充分渗透根系集中分布层为标准。晚秋冬初和早春的灌溉应使苹果植株整个根系分布层的土壤全部被水湿润为宜。

根据 1955 年苏联学者召开的苹果丰产稳产学术讨论会的材料报道，夏季干旱，而特别是在花芽分化阶段水分供应不足是引起大小年的主要原因之一。此一时期，大量的水分供应给幼果的生长，因而果台副梢不能分化花芽，这样也就导致次年小年。因此，在花芽分化阶段，对果树进行灌溉是一项保证丰产稳产的重要措施。

六、水肥一体化的灌溉方法

（一）喷灌法

喷灌是通过水泵加压或自然落差形成有压力的水，用管道将水送到田间，再经喷头喷射到空中形成细小的水滴，像自然落雨一样均匀地落到树体上，以达到灌溉目的的一种灌溉方法。喷灌是一种经济用水、效果良好的灌溉方法，整个果园灌溉比较均匀，而且可以在坡地或地形复杂的果园或沙质、砾质的土壤上应用，另外固定的喷灌系统还可以用来喷药、施肥和防霜，在特别炎热的天气还可用来喷水降温。喷灌与地面灌溉相比，一般可节水 30%～50%。

喷灌的缺点是喷灌机具的购买和安装成本较高，增加了能源费用；而且树体上方喷灌增加果园湿度，易引起病虫发生；再是，喷灌受风的影响较大。

（二）滴灌法

滴灌是利用塑料管道将水通过直径 10cm 毛管上的孔口或滴头将水送到树体的根部进行局部灌溉的一种灌溉方法，是干旱缺水地区最有效的节水灌溉方式。滴灌在水资源利用方面更优于喷灌，其利用率可达 95%。滴灌还可以结合进行施肥，使肥效提高 1 倍以上，能省工、减少污染、保持土壤结构、增加产量。

滴灌能使果树植株的全树盘土壤或部分土壤达到或接近田间最大持水量，增加果树新梢的生长量，延长花期和授粉期，有利于丰产稳产和增大果实、改善品质。

滴灌的缺点是造价高。另外，由于杂质和矿物质的沉淀，容易使毛管和滴头堵塞。

滴灌系统由控制系统、主管、侧管、阀门和滴头所组成。控制系统中有过滤器、流量计、控制阀、化学药物、肥料注射器和自动控制器的水泵。这些部件的质量必须有良好的保证，否则整个系统运转不佳，常收不到所期望的效果。

第二节 梨树水肥一体化技术应用

一、梨树的生长特性

（一）生长结果习性

1. 根系的分布与生长

梨根系分布较深较广，垂直分布为树高的 0.2～0.4 倍，水平分布约为冠幅的 2 倍，个别可达 5 倍。根系分布的状况受树种、品种、砧木、土壤、地下水位及栽培管理的影响。据俞德浚等在河北定兴调查，鸭梨的根系分布随地势、土质、土层和地下水位的高低而有变化。因此，深翻改土、降低地下水位等措施可诱导根系纵、横向生长，使根系生长发育良好。

梨根系的活动及生长与树龄、树势、土壤理化性质、水分及温度有密切的关系。一般在萌芽前土壤温度为 0.5℃ 时便开始活动，15~25℃ 生长较好，超过 30℃ 或低于 0℃ 则停止生长。在适宜的条件下可周年生长，而无明显的休眠期。

梨根系在周年活动中有 2~3 次生长高峰。在武汉地区，梨幼树根系有 3 个生长高峰：3 月下旬至 4 月下旬为第一个生长高峰，延续时间最长，根系生长量最大；5 月中旬至 7 月下旬为第二个生长高峰，延续时间较长，根系生长量较大；10 月上中旬至 11 月上旬为第三个生长高峰，这次延续的时间短，生长量亦小。结果树由于开花结果的影响，一般只有 2 个生长高峰：第一个生长高峰在新梢生长停止、叶面积形成后至高温来临之前（即 5 月下旬至 7 月上旬），由于地上部同化养分供应充足，地下土温又适宜根系生长，故这个时期根系生长最快，以后生长逐渐缓慢；第二个生长高峰在果实采收后，土温不低于 20℃，这时由于土温适宜，养分积累较多，根系又迅速生长（在 9—10 月），以后根系生长逐渐缓慢，至落叶后进入相对休眠期。如结果过多、树势很弱、管理粗放、病虫危害较重、受旱受涝的树，根在一年中无明显的生长高峰。

2. 枝梢生长

梨芽属于晚熟性芽，在形成的当年不易萌发，所以除个别品种或树势很强的树（尤其是幼树）以外，一般每年只有上一年的芽抽生 1 次新梢。梨芽的萌发率高，但成枝力低，除基部盲节外，几乎所有明显的芽都能萌发生长，但抽生成长枝的数量不多。因此，梨的绝大多数枝梢停止生长较早，枝与果争夺养分的矛盾较小，花芽比较容易形成，坐果率也较高。由于各枝的生长势差异较大，第一次枝生长特强，第二、三次枝依次减弱，所以顶端优势特强，易出现树冠上部强、下部弱和主枝强、侧枝弱的现象。

梨新梢自萌芽即开始生长，其生长强弱因树种、品种、树龄、树势、营养状况而不同。在条件相同的情况下，主要决定于芽内幼茎分化时间的长短及其发育程度。据莱阳农学院研究，叶芽的分化从萌芽露叶时开始，至休眠前形成 3~7 片叶原基，休眠后再分化 3~10 片叶原基，故短枝具有 3~7 片叶，中长枝不超过 14 片叶。只有少数生长强旺、有芽外分化的新梢，因生长后继续分化，其叶片数可超过 14 片。

一年生枝具有花芽的是结果枝，无花芽的为生长枝。生长枝根据发育特点和长度分为短枝（5cm 以下）、中枝（5~30cm）和长枝（30cm 以上）3 种。果枝根据长度也分为短果枝（5cm 以下）、中果枝（5~15cm）和长果枝（15cm 以上）3种。

在广东、广西、福建、云南等亚热带地区，由于气温高、雨水多、夏季长、冬季短，而夏季因受海风调节，一般较长江中游地区凉爽，故梨树枝梢生长次数较多。如广东惠阳的梨，一年可发新梢3次，除温暖的华南外，梨秋梢抽出后未及老熟即进入低温休眠期，因此，抽生秋梢对梨树生长和结果都不利。

3. 叶的生长

叶片随着新梢的生长而生长，基部第一片叶最小，自下而上逐渐增大。短枝上的叶片以最上的一两片叶最大；中、长枝上的叶片因生长期的营养状况、外界条件的不同而有大小不同的变化。在有芽外分化的长枝上，自基部第一片叶开始自下而上逐渐增大，当出现最大叶片后，有1~3片叶较小，以后渐次增大，再渐次减小。第一次自基部由小至大之间的叶片称为第一轮叶，为芽内分化的叶，在此以上的叶是芽外分化的叶，称为第二轮叶。叶片从萌芽到展叶需10d左右，一叶片自展叶到停止生长需16~28d。凡叶片小的品种及树势弱的单株，单叶的生长期短，反之则较长。

不同类型的新梢，叶面积的大小不同，以长梢最大，中梢次之，短梢最小；但就枝梢单位长度所占有的叶面积而言，则情况相反，以短梢的单位长度面积最大，中梢次之，长梢最小，故短、中梢的营养物质积累最多，有利于花芽分化和果实的肥大。但是，没有一定数量的长梢，便没有产生中、短梢的基础，也就不利于梨树的生长和结果。同时，长梢叶面积大，合成的营养物质较中、短梢多，除供给自身需要外，还有剩余外运他用。因此，各类枝梢合理的比例是丰产、稳产的生物学基础。

4. 花芽分化

梨是易形成花芽的树种，不仅能由顶芽发育形成顶花芽，也能由侧芽发育形成腋花芽，有的品种腋花芽数量大，坐果较可靠。梨花芽形态分化可分为花芽分化始期、花原基及花萼出现期、花冠出现期、雄蕊出现期及雌蕊出现期5个时期。梨的花芽开始分化一般在新梢生长停止后不久，或在迅速生长之后芽的生长点处于活跃状态的时期，若树体内外条件适宜便开始分化。若新梢停止生长过早，叶片小而少，营养不良，或停止生长过迟，气候不良，就不能形成花芽。据佐藤研究，绝大多数日本梨的早熟品种，花芽开始分化在6月中旬，中熟品种在6月下旬，晚熟品种在6月下旬至7月中旬。我国长江流域的气温比日本梨的主产地高，故武汉等地的日本梨花芽分化比在日本略早，长十郎在5月中旬，明月在6月中旬已开始分化。一般至10月花期已基本形成，此后气温降低，树体逐步进入休眠，花芽暂停分化。开春后，气温回升再继续分化，直到4月上中旬雌蕊内的胚珠发育完成，花的其他部分也生长发育直到开花。

5. 开花与授粉受精

（1）开花。花芽经过冬季休眠后，当日平均气温0℃以上时，花器内的器官即缓慢生长发育，随着气温的升高，生长发育逐渐加快，花芽的体积也相应膨大而萌动、开绽。长江流域梨花芽的开绽期一般在3月中旬。花芽开绽以后，经过现蕾、花序分离，最后花瓣伸展而开花。

梨的花芽是混合芽，芽内具有一段雏梢，萌芽后由雏梢发育为结果新梢，其顶端着生一伞房花序。一般每花序有5~8朵花，但不同品种间差异很大，少的有5朵以下，多的在10朵以上。结果新梢的叶腋内可发生果台副梢，发生果台副梢的多少、强弱与品种特性及营养状况有关。一般发生1~2个居多，个别也有不发或多达3个的。

梨的多数品种是先开花后展叶，少数品种是花叶同时开放或先展叶后开花，前者如康德，后者如京白梨。开花的顺序，就单个花序而言，基部的花先开，中心的花后开，先开的花发育较好，易于授粉受精，坐果可靠。同一朵花的花药花粉散放的顺序也是先外后内，呈向心生长。

梨开花的迟早及花期的长短因品种、气候变化、土壤管理不同而异。一般秋子梨和白梨的花期较早，西洋梨的花期偏晚，砂梨居中。同一品种因年份不同，开花期迟早差异亦大，但不同年份、不同品种的花期迟早仍相对一致，因此可归纳为早、中、晚3个类型，以作为选配授粉树的参考依据。

早花类型包括茌梨、鸭梨、苍溪梨、金川雪梨、二宫白、金水1号、黄蜜、康德等。

中花类型包括砀山梨、新高、湘南、黄花、长十郎、二十世纪、金水2号等。

晚花类型包括晚三吉、江岛、太白、王冠、伏梨、巴梨、日面红等。

根据浙江农业大学在杭州观察，不论哪个类型，初花期时间较短，大致为2d，盛花期较长，大致为4d，至于末花期各品种差异较大，有的品种3~4d，有的长达10d左右。总之，各类型的花期从始花至终花只有5~14d，且以5~6d占多数，从初开转入盛开只需1~3d。同一地方同一品种开花的迟早和花期长短则主要受气温的影响。

在一般情况下，梨树都是春季开花，但也有特殊情况，例如二次开花和晚期花。

秋季开花常称为二次开花。二次开花是当年分化的花芽在当年秋季就开放。这种现象常常是在花芽分化期由于天旱或病虫等自然灾害造成早期落叶，蒸腾作用大幅度降低，而根系吸收的水分和无机盐类仍不断上运，细胞液浓度降低，已通过质变期的花芽发育进程加快，并促使花芽提前萌发而开秋花。二次开花也常能授粉受精而结果，但因季节已晚，温度不够，不能正常成熟或果实过小，无商品价值。故生产上必须加强夏秋季的肥水供应和病虫害防治，防止早期落叶而引起二次开花。

春季正常开花后，一部分果台副梢的顶端开花。这种花的开花期比结果新梢上的花晚 10~20d，故称为晚期花。晚期花是在冬眠前已分化的花芽果台副梢顶部发生，休眠后至萌芽前继续分化，由于分化时间短，花器发育不完善，花的形态变异很大，如花瓣多、雄蕊少、花萼像叶片等，一般不能正常结实膨大。

（2）授粉受精。梨是异花授粉的果树，一般自花授粉不能结实或结实率很低。因此，生产上需配置授粉树。选择授粉树时，宜选亲和力强、花粉量大、花粉发芽率高，开花期与主栽品种相同的品种。一株树上的头批花或同一花序中的第 1~2 朵花养分较足，分化较好，结实率最高，而且结实后果实发育最快，果型最大，品质最好；盛花期开的花次之；最后开的花一般最差。

6. 坐果、落果及果实发育

（1）坐果及落花落果。梨是坐果率较高的树种，凡树势较强、能正常授粉而又管理得当的情况下，一般都能达到丰产的要求。但不同品种间坐果能力差异很大。按 1 个花序坐果 3 个以上者为强、2 个为中、1 个为弱的标准，坐果率高的品种有华梨 2 号、丰水、二十世纪、鸭梨等；坐果率中等的品种有湘南、砀山酥梨、长十郎等；坐果率低的品种如苍溪梨等。生产实践表明，坐果率高的品种往往由于坐果过多，负载过重，在大量结果的年份抑制了新梢的正常生长，影响了花芽分化而造成大小年结果或隔年结果现象。坐果率低的品种，又往往由于坐果量少而达不到丰产的要求。

在年周期中梨有 3 次生理落果高峰期。另外，在果实成熟前有一次轻微的落果。梨的果实较大，果柄较脆，遇 5~6 级以上的大风，如无防风设施会造成大量落果而减产。

（2）果实发育。梨果是经授粉受精的花发育而成，花托发育为果肉，子房发育为果心，胚发育为种子。在梨果发育过程中三者之间有着密切的相互制约的关系。根据果实发育的快慢可分为 3 个时期：

果实迅速膨大期：此期从子房受精后膨大开始，至幼嫩种子出现胚为止。胚乳细胞大量增殖，占据种皮内绝大空间。花托及果心部分细胞迅速分裂，细胞数目增加，果实迅速增大，其纵径比横径增大快，故幼果多呈椭圆形。

果实缓慢增大期：此期自胚出现起至胚发育基本充实止。这个时期胚迅速发育增大，并吸收胚乳而逐渐占据种皮内全部胚乳的空间。在胚迅速增大时期，花托及果心部分体积增大较慢。果肉中的石细胞团开始发生，并达到固有的数量和大小。

果实迅速增大期：此期自胚占据种皮内全部空间起至果实成熟止。由于果肉细胞体积和细胞间隙容积的增大，果实的体积和重量迅速增加，但这时种子的体积增大很少，甚至不再增大，只是种皮由白色变为褐色，进入种子成熟期。

果实大小决定于细胞的多少和体积的大小，凡细胞多而大的，果实亦大；细胞数目少而小的，果实亦小；细胞数目多而体积小，或体积大而数目少的，则果实大小居中。

所有活跃生长的植物器官在生长速率上都具有生长的昼夜周期性，影响植物昼夜生长的因子主要有温度、植物体内水分状况和光照。一般一天内有两个生长高峰，通常一个在午前，另一个在傍晚。果实生长昼夜变化主要遵循昼缩夜胀的变化规律。梨果实在晴天，由于白天高温干燥，果实收缩，晚上则膨大，其中光合产物在果实内的积累主要是前半夜，后半夜果实的增大主要是吸水。因此，夏季高温期晴天采果要赶在早晨露水干前，果既大，水分又足，脆凉适口。雨天由于空气湿度处于饱和状态，果实完全不收缩，甚至由于降雨，果实异常膨大而发生裂果，特别是当高温干旱，果实渗透压高的时候突降暴雨或大量灌水后会加重裂果。套袋有防止裂果的作用，因为套袋果对由于降雨而促进果实膨大的影响比未套袋果迟几小时出现，膨大的速度也比未套袋果慢一半，故很少发生裂果现象。

(二) 对环境条件的要求

1. 温度

梨因种类、品种、原产地不同，对温度的要求也不同，分布范围也有差异。砂梨原产我国长江流域，要求温度较高，多分布于北纬 22°~32° 之间，一般可耐-23℃以上的低温。

梨的不同器官耐寒力不同，以花器、幼果最不耐寒。江浙一带常因早春天气回暖之后骤然降温而发生冻花芽现象。梨是异花授粉果树，传粉需要昆虫媒介，8℃蜜蜂开始活动，其他昆虫则要在15℃以上。梨花粉发芽要求10℃以上温度，18~25℃最为适宜。在16℃的条件下，日本梨从授粉至受精约需44h，若温度升高，授粉受精过程可缩短，反之则要延长。故花期天气晴朗，气温较高，授粉受精一般良好，可望当年增产；若连续阴雨，或温度变化过大，会导致授粉受精不良，落花落果严重，必然造成减产。

温度对梨的品质也有影响，原产冷凉干燥地区的鸭梨引入高温多湿地区栽培，果型变小，风味变淡。长十郎果实成熟期（7—8月）的气温与品质呈正相关。

2. 水分

梨的生长发育需要充足的水分，若水分供应不足，枝条生长和果实发育都会受到抑制。但降水过多，湿度过大，亦非所宜，因为梨根系生长需要一定的氧气，土壤内空气含氧量低于5%时根系生长不良，降至2%以下时则抑制根系生长，土壤空隙全部充满水时根系进行无氧呼吸，会引起植株死亡。

梨对降水量的要求和耐湿程度因种类和品种不同而异。砂梨耐湿性强，多分布于降水量1 000mm以上的地区。降水量及湿度大小对果实皮色影响较大，在多雨高湿气候下形成的果实，果皮气孔的角质层往往破裂，一般果点较大，果面粗糙，而缺乏该品种固有的光洁色泽，尤以绿色品种如二十世纪、菊水等表现明显。同时，4—6月当新梢生长和幼果发育期间，若雨水过多，湿度过高，病害必然严重。

3. 日照和风

梨是喜光果树，若光照不足，往往生长过旺，表现徒长，影响花芽分化和果实发育；如光照严重不足，生长会逐渐衰弱，甚至死亡。

风对梨的影响很大，强大的风力不仅影响昆虫传粉，刮落果实，而且使梨的枝叶发生机械损伤，甚至倒伏。风还能显著增加梨的蒸腾作用，使叶内水分减少，从而影响光合作用的正常进行。一般无风时叶的水分含量最高，同化量最大，随着风力的增加，水分逐渐减少，同化量亦相应降低。但若果园长期处于无风状态，空气不能对流，二氧化碳含量必然过高，恶化环境，同化量便会下降。因此，适于梨树生育的以微风（风速$0.5\sim1m/s$）为好。建立防风林，可改善果园的小气候，减少风害，增强同化作用。

4. 土壤与地势

梨对各种土壤都能适应，无论沙土、壤土还是黏土都可栽培。但由于梨的生理耐旱性较弱，故以土层深厚、土质疏松肥沃、透水和保水性能较好的沙质壤土最为适宜。梨对土壤酸碱度的适应范围较广，pH值在5~8.5的土壤均可栽培，但以pH值在5.8~7为最适。梨树耐盐碱能力较强，土壤含盐量不超过0.2%时均能生长正常。

梨对地势选择不严格，不论山地、平原、河滩都可栽培。平原、河滩土壤水分条件较好，管理较为方便，但在高温多雨的地区，往往因雨水过多，地下水位过高，影响树势和果实品质，且真菌性病害容易蔓延。山地丘陵果园，排水、光照及通风条件较好，树势易于控制，病害也可减少，易获得高产优质的果实，但水土易于流失。同时，随着海拔的升高，气温逐渐降低，在高山地区栽培梨树，往往花期易受霜害。

二、梨树的水肥一体化管理

（一）施肥时期

梨树年周期生长中，根系和枝梢生长、开花、果实发育及花芽分化都有一定的顺序性及相互制约的关系。其中由依靠贮藏养分生长过渡到新叶同化养分供应的转折时期和果实发育及花芽分化的生殖时期是栽培管理上的关键时期，在确定施肥时期时应注意这些特点。

（1）基肥在采果后落叶前施用。这时土温较高，树体又在活动时期，有利于根系愈合和生长。秋施基肥的新根增长量比春施基肥的多3倍左右。同时，秋施基肥对恢复树势、加强同化作用、增强树体营养贮备有显著的影响，故有利于提高坐果率及产量和品质。如因劳力等原因不能秋施基肥，采后也应及时追施氮肥，维持强壮的树势。南方梨产区秋季气温尚高，蒸发量较大，秋施基肥后应及时灌透水。

（2）追肥梨树在不同生长时期对营养元素的吸收量是不同的。据日本佐藤对20世纪梨的养分吸收量的研究，氮的吸收以新梢生长期及幼果膨大期最多，生理落果期极少，果实第二次膨大期又增多，果实采收后则吸收减少；钾的吸收动态基本上与氮相同，不过第二次果实膨大期对钾的吸收远比氮为高；而磷的吸收较氮、钾为少，且各生长期比较均匀。仅施基肥不能及时满足各个时期对不同养分的需要，必须根据梨树需肥特点合理追肥。全年追肥3次，天旱时要结合灌水进行。

①花前肥：于萌芽后开花前进行，施速效性氮肥，若用人粪尿、腐熟的饼肥，应提前半个月左右施下。花前肥对提高坐果率、促进枝叶生长和提高叶果比有一定的作用。弱树、幼树以促进枝叶生长为主要目的可以单施氮肥，施用量约占全年施用量的20%。初结果的树和成年的旺树一般不宜单施氮肥。

②壮果肥：于新梢生长盛期后果实第二次膨大前进行。以施速效氮肥为主，

配合磷、钾肥料,氮肥用量约占全年施用量的20%。如施肥过早,会促进枝梢旺长,果实糖分下降,影响品质。

③采前肥:于采果前进行。施速效氮肥,用量占全年的20%。对树势较弱和结果多的树,采果后若不能及时施基肥,还可适当补施速效氮肥,对恢复树势、防止早期落叶起作用。

(二) 施肥方法

梨的根系强大,分布较深。基肥应采用环状、条沟、扩槽放窝,分层深施。沟宽0.8m左右,深0.6m左右。轮换开沟,每1~2年一次,逐步将果园深翻施肥一遍,引导根系深入扩展。过4~5年后,可分期分批深耕,以更新根系,活化土壤。

追肥方法应根据肥料种类、性质,采用放射沟、环状沟或穴施,深10~15cm,施后及时覆土。如土壤水分不足,要结合灌水,氨水要先对水稀释,否则肥效不易发挥,甚至起破坏作用。

根外追肥目前已普遍采用,随着机械化的发展,今后将广泛应用管道灌溉。特别是4—5月梨树由贮藏养分到当年同化养分的转变时期,采用根外追肥效果更明显。常用浓度为尿素0.3%~0.5%,人尿5%~10%,过磷酸钙2%~3%,硼0.2%~0.5%,硫酸重铁0.5%,锌0.3%~0.5%等。

(三) 施肥量

许多试验证明,施肥要适量,必需元素既不能缺少,又不应过量,否则不仅会对树体产生肥害,还会污染土壤环境。所以,国内外都很重视对施肥量的研究。

果树的需肥量受品种、树龄及砧木的影响较大。例如,20世纪与长十郎相比,前者较后者需肥量要大,特别是对磷的需要量大。美国将巴梨嫁接在不同的砧木上,研究不同砧木吸收营养物质的情况,结果表明:氮的吸收,西洋梨实生砧比榅桲和山楂砧为多;磷的吸收,以花楸砧最多,雪梨砧最少。又据国外研究材料指出,营养器官(根、茎、叶)与果实的氮、磷、钾含量的比率是不同的,营养器官的氮、磷、钾比率为10:(3.5~5):(6.4~6.9),果实中的比率为10:(4.5~4.8):(21~25)。两者间氮、磷含量基本一致,而钾的含量果实比营养器官高3倍以上。因此,对成年树和幼树施肥应有所区别。

梨施肥量确定的依据是正确的营养诊断,目前常用的诊断方法有形态(树相)诊断、叶片分析、土壤分析等,还有近十多年来兴起的肥料综合诊断法(DRIS法)。一般是根据树体大小、花芽多少、产量高低,以产定肥,并根据丰产、稳产的规律性指标酌量增减。

在施肥的种类上，应有机肥料与化肥相结合。一般有机肥应占施肥所需营养要素的 1/3~1/2，以改善土壤的理化性状。在化肥方面，应逐步由使用单一化肥过渡到使用复合肥料，并依据营养诊断结果配方施肥。

（四）灌溉

梨树较耐湿，但土壤过湿，通气不良，根的生理机能减退，树体生长恶化，尤其是温度高，梨树旺盛生长时这种湿害的表现更为严重。生产上常见到地势低洼、地下水位高的平原梨园和不透水的山地梨园，在降水多或降水集中的季节，叶色黄绿，生长不良，一旦进入旱季，便发生黄叶早落的现象。因此，南方建园时提倡山地抽槽改土，平地深沟高畦，心土黏重必须抽槽抬高，要深翻改土，打通不透水的硬土层，四周围沟要深，沟河相通，既能排明水又要排暗水，做好排水防涝工作。长江流域及其以南的大部分梨产区的气候特点：春夏季阴雨连绵，降水较多；夏、秋季高温干旱，降水较少。概言之，春、夏之际要注意排水，夏、秋乃至冬季要注意灌水，个别春旱年份还要进行春灌。不同的树种品种、生育时期和立地条件对需水要求和保水性能是不同的，灌水时期、次数应区别对待。灌水的方法以沟灌、滴灌、喷灌、微喷灌和穴灌等为好，但高度密植的梨园采用高喷头喷灌易造成小气候湿度过大而引起真菌性病害。灌水要一次灌透，应使根系主要分布层内土壤含水量达到田间最大持水量的 70%~80%。

第三节　桃树水肥一体化技术应用

一、桃树对环境条件的要求

（一）温度

桃的适应范围广，在年平均气温为 8~17℃ 的地区均可栽培，北方品种群适宜的年平均气温为 8~14℃。生长期平均气温为 19~22℃，开花期需 10℃ 以上。果实生长发育的适宜温度为 20~25℃。据美国资料介绍，生长期月平均温度在 18.3℃ 以下时，果实品质差，达 24.9℃ 时，则产量高、品质好。我国桃产区 6—8月温度一般均在 24℃ 以上，所以有利于果实的生长发育。

桃树在冬季休眠期需一定量的低温才能正常萌芽生长、开花结果。如果冬季温度过高，则不能顺利完成休眠，造成翌春萌芽晚，开花不整齐，授粉受精不良，产量降低。休眠期需冷量以日平均温度≤7.2℃的温度累积时数为500~1 200h。所以，我国冬季气温较高限制了桃树的发展。

桃的不同品种对低温抵抗力不同，一般品种可耐-25~-22℃的低温。有的品种在-18~-15℃时，花芽和幼树发生冻害。-25℃以下时树体受害，-30~-28℃时严重受害，甚至整株死亡。花芽萌动后的花蕾变色期受冻温度为-6.6~-1.1℃，开花期为-2~-1℃，幼果期为-1.1℃。

(二) 水分

桃在年周期发育中，需要适量的水分。试验证明，桃树要求适宜的土壤田间持水量为60%~80%。当田间持水量在20%~40%时还能正常生长，降到10%~15%，叶片出现凋萎，严重影响桃树的生长发育。

桃树耐旱怕涝。排水不良和地下水位过高，会引起根系早衰，叶片变薄，叶色变淡，生长降低，进而落叶、落果、流胶以至死亡。因此，雨季要注意排水防涝。

(三) 土壤

桃根系呼吸旺盛，需氧量多。据测定，土壤含氧量在10%~15%时，根系生长正常；当含氧量降至7%~10%时生长不良；5%以下时，根变褐，不能发生新根；2%时，细根开始死亡，新梢停止生长甚至落叶，所以桃树宜栽在土质疏松、排水良好的沙壤土或壤土地上。黏重土壤通气不良，易患流胶病、颈腐病等。桃在土壤pH值为4.5~7.5时生长正常，在碱性土壤中易患缺铁黄叶病。不同砧木耐碱力不同，其中，山桃较耐碱，所以在pH值稍高的北方多用山桃砧。桃较耐盐，土壤含盐量为0.13%时，对生长无不良影响，当盐的浓度达0.28%时，生长不良或部分致死。

(四) 光照

桃原产我国西北部光照很强的大陆性气候，形成了喜光的特性。当光照不足时，树体同化产物显著减少，根系发育差，枝叶徒长，花芽分化少，质量差，落花落果严重。小枝易枯死，结果部位上移，树冠下部光秃。因此桃园应选在通风透光良好的地方，栽植密度要适宜，树形要合理，留枝量要适度。

二、桃树的水肥一体化管理

(一) 施肥

1. 桃树的需肥规律

桃果实膨大生长有三个主要时期。第一生长期从落花至果实硬核前，细胞开始迅速分裂，品种间有一定差异。第二生长期从核开始硬化至硬化完成，细胞数迅速增加，早熟与晚熟品种所需的时间差异较大。第三生长期从核层硬化结束至果实成熟，该期细胞增长迅速，果实在采收前15d增长十分明显。新梢生长从果实生长的第一期至第二期的前半期都处于急速生长状态，到果实逐渐成熟时，新梢生长逐步缓慢，渐至停止。

合理施肥至关重要，桃树对氮、钾的需求量大，对磷的需求量较少。每生产1kg鲜果，需氮1.52g、磷0.79g、钾3.64g。

在一年中，桃树对营养元素的吸收随生长期而变化。从硬核期开始，对氮、磷、钾的吸收量迅速增加，大约到采收前20d达最高峰。在这段时期，磷、钾的吸收量增长较快，尤其是钾。

微肥作基肥施用时，每亩硼砂用量为0.25~0.5kg、硫酸锌2~4kg、硫酸锰1~2kg、硫酸亚铁5~10kg（应配合优质有机肥一起施用，有机肥与铁肥用量比为5:1）。微肥可叶面喷施，喷施的浓度根据叶的老化程度控制在0.1%~0.5%，叶嫩时宜稀，叶衰老时可浓一些。

2. 基肥

基肥施用的时期可分为春施和秋施，施基肥最好在秋季，以9月上旬至10月中旬为宜。此时新梢已停止生长，但根系仍在活动，尤其9月份正处于根系的秋季生长高峰，施用基肥所造成的伤根经过20d左右即可愈合并恢复生长。秋季施入的肥料被根系吸收后，为地上部叶片的光合作用提供大量营养原料，提高树体营养贮藏水平，为翌年生长结果打下物质基础。如果秋季没有施基肥，翌春必须在解冻后及时施入，过晚，因施肥造成的伤根对开花、坐果和春梢生长都会产生不利影响。

基肥的施用方法有撒施法、环状沟施法、条沟施法、放射沟施法等。幼龄桃树，结合全园土壤改良，多采用环状法施肥。沟的深度40~60cm，宽50~60cm。成龄桃树因结合环沟施肥已把土壤全部深翻改良完成，施基肥时可采用条沟、放射沟、大穴和地面铺施法交替进行。如果地面铺肥或杂草等有机物达10~20cm，并坚持每年续铺10cm以上，可每隔3年进行一次大穴换土。这样，可以节省大量劳动力，效果也好。大穴换土的方法：在树冠外缘垂直地面挖2~4个40~50cm见方的坑穴，表土和生土分放，把含有大量有机物的表土填入穴内，生土填在坑表。每3年进行一次，每次更换坑穴的位置。未覆盖桃园深沟施肥时，要把挖出来的生土和表土分放，将腐熟的有机肥均匀地掺入表土后填入下层，生土放在表层，每株施优质有机肥30~40kg，次质肥要多施些，施肥后及时灌水。

3. 追肥

追肥的次数和数量，与品种、地力、树势和产量等因素有关。如早熟品种，或土壤肥沃、树势强健、结果又不过量的树，施肥量和次数要少，反之要多。土壤追肥的方法：采用直沟或放射沟法。沟深 20cm，肥料施入后，将沟两侧的土扒入沟内。桃园主要有如下几次追肥：花前肥、坐果肥、果实膨大肥。

4. 叶面喷肥

为了及时补充桃树对养分的急需，也常采用根外追肥，尤其在盐碱地上栽培的桃树，进行根外追肥效果更好。喷施的肥液通过叶片气孔进入叶内。吸收的强度和速度与叶龄、肥料成分、肥液浓度有关。幼嫩的叶片生理机能旺盛，气孔所占面积比成龄叶片大，因此比老叶吸收快；肥液浓度越高，叶片吸收量越大；叶片背面气孔多、细胞间隙大，吸收能力强于正面。

在桃树上常喷施的肥料较多，如尿素、磷酸二氢钾、磷酸二铵、硝酸钾等。生长季喷施 0.3%~0.5%，落叶前 10d 喷施浓度为 1%。

在桃树的年生长周期中，从开花前至落叶前，浓度要逐渐提高，一天中最好在 10:00 以前和 16:00 以后喷施。大风天和炎热的中午不要喷肥，因为此时喷肥不仅肥效差，而且还因水分蒸发过快容易发生肥害。

（二）灌水与排水

桃树在生长季节的各个发育阶段都需要一定的水分供应，生长季依据土壤干旱情况灌水，一般追肥后应灌水，促进肥料分解、树体及时吸收。灌水后应及时松土保墒。只有在适宜的水分供应条件下，树体才能正常进行生理活动。在年生长周期中，以春季需水最多。北方地区春天多风少雨，如果供水不足，树体因缺水造成根系生长缓慢、新梢生长不良、坐果率低、果个小；即便是桃树在休眠期间，树体的各类枝条仍在蒸腾水分，此时缺水也会造成翌春树体衰弱，甚至发生嫩枝抽条。桃树对水分的需要有一定限量，其耐水力远不如苹果、梨等其他果树，过多供水或较长时间浸泡，也会影响桃树生长发育，土壤积水，还会造成整株死亡。因此，适宜灌水、及时排水，是保证桃树丰产、稳产、优质的重要条件。

1. 灌水

桃树灌水的方法有畦灌、沟灌、穴灌、滴灌、喷灌等。因各地水源和生态条件不同而有各种各样的灌水方法，目前大多数桃园采用的是串树盘的畦灌法。这种灌水方法，虽然速度快、又省工，但沟渠占地多，灌水量不均匀，还容易造成病害蔓延。有条件的桃园最好采用喷灌，其优点是灌水均匀，可控制灌水量，有利于喷灌设备进行根外喷肥和打药。总的原则是要节约用水，并保证水分及时渗透到根系分布最多的土层，使土壤保持一定的水分。

桃树灌水时期分为萌芽前、硬核期、果实第二次迅速生长期和落叶后封冻前。①萌芽前灌水：为保证萌芽、开花坐果顺利进行，需灌透、灌足水，渗水深达80cm。②硬核期灌水：此期果实虽然生长缓慢，但种胚处在迅速生长期，桃树对水分敏感，是桃树需水临界期。水分过多，枝叶生长过旺，影响坐果；缺水，也造成落果，影响产量。此期灌水应浅灌，尤其对初果期的树更应慎重。③果实第二次迅速生长期灌水：大约在采前2周，此时果实迅速膨大生长，充足的水分可以明显增产。④封冻前灌水：在桃树落叶后、土壤结冻前灌水，保证土壤有充足的水分，以利桃树安全越冬，但不能过晚，以免因根颈部积水或水分过多、昼夜冻融交替而导致颈腐病发生。

2. 排水

桃树怕涝，桃园一旦积水，土壤中的空气被挤走，就会使桃树进行无氧呼吸，产生乙醇、甲烷等有害物质，造成根系毒害，导致桃树死亡。一般情况下，雨季淹水一昼夜便死亡。因此，雨季做好排水防涝是栽培桃树的一项重要工作。

第四节　葡萄水肥一体化技术应用

一、葡萄对环境条件的要求

（一）温度

温度是影响葡萄生长和结果最重要的气象因素。在春季当气温达到7~10℃时（地温10℃左右），葡萄根系开始活动；在25~30℃时生长最快，35℃以上时生长受到抑制。10~12℃时开始萌芽。葡萄新梢生长、开花、结果和花芽分化的适宜温度为25~30℃。开花期间如出现低温天气（<15℃），葡萄就不能正常开花和授粉受精。鲜食葡萄和制干葡萄浆果成熟期的适宜温度为28~32℃，而酿酒葡萄则为17~24℃。不同成熟期的葡萄，要求达到一定的有效积温后，果实才能充分成熟。

低温对葡萄的伤害是世界葡萄栽培中常遇到的问题。不同种和品种之间抗寒力差异很大、不同组织和器官之间也有相当差别。通常情况下，美洲葡萄的抗寒力大于欧亚种。根系是抗寒性最弱的器官，大部分葡萄的根系在-5℃左右即受冻致死。但山葡萄能耐-15.5℃。为了减轻根系冻害、采用山葡萄和贝达（Beta）作抗寒砧木，使葡萄通过埋土防寒能在较寒冷地区栽培，有重要的经济价值。一般认为，多年平均最低温度在-15~-14℃的地方，葡萄可不埋土越冬，而在低于-15℃的地方只有进行程度不等的覆土，葡萄才能安全越冬。葡萄的冬芽抗寒能力比较弱，其次是成熟的一年生枝条，多年生枝条、主干最抗寒。欧亚种葡萄的芽眼，在冬季能耐-20~-18℃的低温。但如果枝条成熟度较差，在-15~-10℃时，芽就会受冻，在-18℃的低温持续3~5d，不仅芽眼受冻，枝条也会受冻。北方地区冬季低温造成的伤害，往往是与干旱缺水相关联。

缺水导致葡萄的耐寒性下降。春季的嫩梢和幼叶在-1℃时即开始受冻，0℃时花序受冻。

（二）光照

太阳光是葡萄进行光合作用唯一的能源，是葡萄进行能量和物质循环的动力，葡萄产量和品质的90%~95%来源于光合作用。葡萄是喜光果树，长日照植物。在葡萄生长季节、充足的光照使花芽分化良好，叶片生长色绿、肥厚，新梢粗壮，果实着色良好，尤其是对光照特别敏感的欧洲种葡萄，只有在阳光直射条件下才能着色正常。葡萄对光照的需求，也并不是光照越强越好。夏季中午高温伴随着强烈的光照，果面温度可达50℃以上，常会发生日烧病。叶片在中午光照条件最好的时候，则又会发生"午睡现象"，抑制生长。

在我国一般葡萄园太阳能的利用率仅为0.5%左右，现代科学一直在追求利用太阳能，提高转化率，挖掘增产潜力以达到高产优质。

（三）水分

自然降水的多寡和降水量的季节分配，强烈地影响着葡萄的生长和发育、产量和品质。一般认为，生育期内至少需要250~350mm的降雨量。春季芽眼萌发新梢生长，如果雨量充沛，有利于花序原始体继续分化和新梢生长。葡萄开花期需要晴朗温暖和相对较为干旱的天气。天气潮湿或连续阴雨低温会阻碍正常的开花和授粉受精，引起幼果脱落。成熟期雨水过多或阴雨连绵都会引起葡萄糖分降低，病害滋生，果实烂裂，严重影响葡萄的品质。葡萄生长后期多雨，新梢成熟不良，越冬时容易受冻。

（四）土壤

影响葡萄生长发育的土壤因素包括土壤通气、土壤水分、土壤养分及土壤的酸碱度等因素。葡萄对土壤的适应力较强，从沙土到黏土都能生长，但以轻松的沙土或沙质土和带有大量粗沙和石砾的山根土为好。黏重土壤的土层厚，保肥保水能力强，但容易引起葡萄旺盛生长，进而影响结果和品质。此外，微酸性土（pH值不低于5）和微碱性土壤（pH值不超过8.5）都可种葡萄。盐碱地经过改良和排盐，土壤盐分降到0.2%以下时，栽培葡萄也能成活良好。不同的葡萄品种只有在适合自身条件的土壤中才能生产出优质的果实。

二、葡萄的水肥一体化管理

（一）施肥

1. 肥料种类和作用

葡萄植株必需的矿质营养元素主要从肥料中获得。肥料可分为有机肥和无机肥两大类。生产上经常使用的有机肥如圈肥、厩肥、禽粪、饼肥、人粪尿、作物秸秆等，一般腐熟后施入土壤，多作基肥使用。有机肥中除含有大量元素外，还含有各种微量元素，故称完全肥料。多数有机肥要通过微生物作用，才能被葡萄根系吸收利用，因此也称迟效性肥料。有机肥不仅供给葡萄植株生长发育所需的营养元素，而且还能调节土壤通透性，提高土壤保肥、保水能力，增加土壤肥力，为微生物活动创造物质基础条件，对改良土壤结构起到重要作用。

我国土壤有机质含量普遍较低，要进行优质葡萄生产，提高土壤有机质含量是非常关键的。

无机肥料也称化学肥料，可作为有机肥的补充。无机肥所含营养元素单一，但纯度高、易溶于水，多数无机肥可直接被植物根系吸收，也称速效性肥料。无机肥施用后见效快，多用作葡萄生长期追肥。

2. 施肥量

依据葡萄本身的需肥特点、土壤状况、立地条件以及肥料利用率等确定合理的施肥量，以便充分满足果树对各种营养元素的需要。叶分析是一种确定葡萄施肥量比较科学的方法，当分析发现某种营养成分处于亏缺状态时，就要根据亏缺程度进行补充。生产中多凭经验和试验结果确定施肥量。

3. 秋施基肥

基肥通常以迟效性的有机肥料为主，可以混合一定数量的化肥如过磷酸钙等，按照全年施肥量的60%~70%施用。时间一般在葡萄采收后至土壤封冻前的秋季进行，此期正值葡萄根系第二次生长高峰，吸收能力较强，伤根容易愈合；加上叶功能尚未衰退，光合能力增强，有利于树体贮藏营养的积累，提高葡萄的抗寒能力。施用时，要距离根系分布层稍深、稍远些，以诱导根系向深广生长，扩大根系吸收范围。

4. 追肥

追肥主要追施速效性化肥，按全年总施肥量的 30%~40% 施用。生产中要根据负载量及土壤状况，合理确定葡萄追肥的时期和比例。对结果多的园区，需增加钾、氮、磷、镁、铁、硼等肥料，尤其是增加氮肥和钾肥的数量。施肥时期，氮、硼肥在果实生长前期，磷、镁肥在果实生长中期，钾、铁肥在果实糖分积累期。葡萄一年中需进行 3 次追肥。①花前追肥以氮肥为主，辅以氮、磷结合的速效性肥料，主要目的是促进枝叶生长及花序分化。②谢花后幼果开始生长，是需肥较多的时期。花后追肥应及时补充速效性氮肥，配合适量磷、钾肥，以促进新梢生长，保证幼果膨大，减少落果。③催熟肥于浆果开始着色时进行，此时追肥以磷、钾肥为主，主要促使枝条充实、果实成熟与着色。

（二）灌水与排水

葡萄耐涝性及抗旱性均较强，应在干旱季节和需水时期适时灌水、果园积水时及时排水，可获得更高的产量和更优质果品。一般成龄葡萄园在葡萄生长的萌芽期、花期前后、浆果膨大期和采收后浇水 5~7 次，而在花期和浆果成熟期则要注意控水，以防落花落果和降低果实品质。

1. 灌水

花前期灌水在树液流动、萌芽至开花前进行。此时正值春旱，土壤急需补充水分。萌芽前应灌一次透水，以促进萌芽及新梢生长。

果实膨大期灌水在生理落果至果实着色前进行，此期新梢、幼果均旺盛生长，且气温不断升高，叶片水分蒸发量大，对水分和养分最为敏感，是葡萄需水、需肥的临界期。要结合施催果肥浇水，之后再根据天气、土壤情况决定浇水多少。一般干旱少雨时，可每隔半个月浇一次，以促进果粒膨大。

封冻前，在秋施基肥后先灌一次水，有利于伤根愈合和沉实土壤，土壤结冻前再灌一次越冬水，防越冬时枝条抽干，以利安全越冬。

2. 控水

花期控水在初花至末花期，在 10~15d。花期遇雨，影响葡萄授粉受精，出现大小粒现象。花期浇水引起枝叶徒长，营养消耗过多，严重时将影响花粉发芽，落花落蕾，造成减产。花期适当控水，可促进授粉受精，提高坐果率。

浆果着色成熟时，水分过多或降雨多，影响浆果着色，降低品质；易发生炭疽病、白腐病等；有些品种还可能出现裂果。因此，着色期控水可提高浆果含糖量，加速着色和成熟，防止裂果，提高果实品质。

3. 排水

葡萄生长季节土壤水分过多时，易引起枝蔓徒长，降低果实含糖量，严重时根系缺氧，抑制呼吸，造成植株生理干旱甚至死亡。因此，要注意涝、雨季节及时排水。

第六章 蔬菜水肥一体化技术的应用

我国是一个农业大国，每天需要大量的蔬菜供应市场，为了提高我国蔬菜的产量，农业领域使用了蔬菜水肥一体化技术。蔬菜水肥一体化技术是借助压力系统（或地形自然落差），按土壤养分含量和蔬菜需肥规律和特点，将可溶性固体或液体肥料配对成的肥液与灌溉水一起相融后，通过管道和滴喷头形成滴喷灌，均匀、定时、定量地浸润蔬菜根系，满足蔬菜生长需要。该技术具有"水肥均衡、节水省肥、省工省时、减轻病害、控温调湿、增加产量、改善品质、效益显著"等特点。

第一节 番茄水肥一体化技术应用

番茄原产南美热带高原原始森林中，为多年生草本植物，在有霜地区栽培为一年生。别名西红柿、洋柿子，古名六月柿、喜报三元。在秘鲁和墨西哥，最初称之为"狼桃"。果实营养丰富，具特殊风味。具有减肥瘦身、消除疲劳、增进食欲、提高对蛋白质的消化、减少胃胀食积等功效。番茄为成熟多汁浆果，每100g鲜果含水分94g左右、碳水化合物2.5~3.8g、蛋白质0.6~1.2g、维生素C 20~30mg，以及胡萝卜素、矿物盐、有机酸等。图6-1所示为膜下滴灌在番茄上的应用。

一、番茄水肥管理

1. 水分管理

我国用于食用或加工的番茄基本都是一年生的。从定植到采收末期保持根层土壤处于湿润状态是水分管理的目标。一般保持0~40cm土层处于湿润状态。可以用简易的指测法来判断。用小铲挖开滴头下的土壤，当土壤能抓捏成团或搓成泥条时表明水分充足，捏不成团散开表明土壤干燥。通常每次滴灌1~2h，根据滴头流量大小来定。微喷带每次5~10min。切忌过量灌溉，淋失养分。番茄生育

第六章 蔬菜水肥一体化技术的应用

图 6-1 膜下滴灌在番茄种植上的使用

期长，耗水量较大。在番茄的生长用水中，要对灌溉量进行合理设计，多次少量为最佳用水，直至根部土壤湿润。灌溉种植时要多加观察田间灌溉水管是否良好，出现冒水、漏水现象时要及时处理，更换新的水管。定植水，灌水定额 15~20m³/亩，滴灌或沟灌；缓苗水（定植后 7d），灌水定额 10~12m³/亩；然后进行中耕蹲苗，至第一穗果膨大，视情况滴水一次或不滴水，灌水定额 10m³/亩；第一穗果膨大至 5cm 后，5~7d 滴水一次，滴水 2~3 次，每次灌水定额 10~12m³/亩；进入盛果期，4~5d 滴水一次，每次灌水定额 12~15m³/亩。定植至拉秧生育期 160d 左右，滴水 20~22 次，总灌水量 260~300m³/亩。番茄定植后，由于土壤性质比较干燥，需及时用水滴灌浇透，灌水量多少要根据墒情和秧苗生长情况来确定。秧苗生长期间，每天要进行 3~6 次的滴灌，每次灌水时间约为 1min，选择灌水量为 16mL/min 的滴头，每次灌水量约为 500L/hm²；在快速生长阶段，用水量会逐渐增加，每天需要进行 6~8 次的灌溉，单次灌溉时间为 2min，每次灌水量约为 520L/hm²。番茄在生长过程中的用水量还要根据当天的天气情况来决定。

2. 养分管理

番茄的根系发达，分布深且广，再生能力强，因而吸收养分的能力较强。番茄需肥量大，也比较耐肥，对钙、镁的需求量也较大。番茄为持续生长和结果的蔬菜，生长与结果期长，除施足底肥外，还要求充足的追肥。苗床施肥，一般施腐熟有机肥 3~5t/亩、过磷酸钙 30kg/亩、硫酸钾 10kg/亩、尿素 5kg/亩，混匀后翻施到 25cm 耕层内，充分混匀播种。出苗后，如养分不足，幼苗长势差，可随水浇施稀薄人粪尿或喷施 0.1%~0.2% 的尿素溶液和 0.2% 的磷酸二氢钾溶液。

大田基肥一般按每公顷（5~10）×10^4kg 厩肥或堆肥，（4~6）×10^2kg 过磷酸钙，150kg 硫酸钾，移栽前均匀地翻入耕层土壤。追肥以滴肥为主，肥料应先在容器溶解后再放入施肥罐。滴肥与滴水交替进行，即滴一次肥后，再滴一次水。第一次追肥在定植后 1 周内按每公顷 10×10^3kg 左右腐熟人粪尿或 80~120kg 尿素进行浇施，称为"催苗肥"；第二次追肥在第 1 簇果实开始膨大时进行，按每公顷 15×10^3kg 人粪尿或 100~120kg 尿素进行浇施；第 3 次追肥在第 1~2 簇果实收获后，第 3~4 簇果实正在迅速膨大时进行，施肥种类和数量与第 2 次追肥相同，同时可按每公顷 40~60kg 增施硫酸钾。在果实采收期还可用 0.2%~0.4% 的磷酸二氢钾和 0.2%~0.5% 的尿素溶液进行叶面喷施。

二、滴灌带的铺设和播种

播种前对播种机具进行全面维修调试，并安装好铺设滴灌管装置，达到安全使用状态。

1. 种子处理

为了防治早期病害，如猝倒病，对购买回的种子，播种前要用 70% 代森锰锌进行拌种，药剂用量为种子量的 0.2%，处理后的种子要进行发芽试验。

2. 喷洒除草剂

药剂可用禾耐斯或都尔，用量：禾耐斯 1 500g/hm^2，都尔 2 250g/hm^2，喷洒药剂一定要均匀一致，到头到边，喷洒后用之字耙对角耙一遍。

3. 做好机具准备

播种前对播种机具进行全面维修调试，并安装好铺设滴灌毛管装置，达到安全使用状态。

4. 播种方式

膜下条播可以播种、铺地膜、铺滴灌带一次完成。一是干播湿出，播量 1 200~1 500g/hm^2，播种深度 1.5~9cm，株距 25cm。两膜八行或三膜十二行，膜上行距 30cm，结合部 60cm，膜下条播，滴灌毛管铺设在地膜下，一根毛管管两行番茄。铺管播种覆膜一次完成。移栽番茄在定植前一天晚上就开始滴水，定植时保证土壤湿润，在滴孔附近栽苗，边滴水边栽苗，栽苗后封土。二是育苗移栽。在温室内育苗，播期一般比同地区直播提前 40~60d，提前 15~20d 采收。育苗栽培可以抵御早春低温和霜冻的危害，增加丰产的稳定性。由于移栽时切断主根，致使侧根大量萌发。使耕作层内根系数量大大增加，增强了根的吸收能力，从而可提高产量。加工番茄落花现象比较普遍，水分不足、根系发育不好、低温、植株徒长等均可引起落花。防止落花的主要措施：一是促进苗壮；二是使用浓度为（30~50）×10^{-6}mg/L 的番茄灵喷花，当 50%以上的植株第一穗花有 2~3 朵花开放时，开始喷花，隔 1 周喷 1 次，连续喷 3 次。

5. 播种

条播一般在 4 月中旬，当 5cm 地温稳定通过 12℃时，开始播种，播后铺设支管，接通毛管，准备滴出苗水。播种一定要达到质量要求，深度 1.5cm，播种均匀，无缺行断垄现象，膜边封土量要求要严，并按 6m 距离在膜上用碎土横压一道线，防止风害。干播湿出的地块，苗出齐后要及时放苗封洞。幼苗在 2 片真叶时应间苗，4~6 片真叶时定苗。定苗后再封土护根，缺苗时应及时补苗。如有缺苗现象，可在旁边留些双株，第一滴水后立即补苗。

6. 滴水

①出苗水。播种后在膜下地温达到 12℃时开始滴出苗水，播后 5d 内必须滴完水，滴水量 230m³/hm²；育苗移栽，滴水后，等霜期过后移栽，膜上行距 30cm，结合部 60cm，株距 30cm，栽苗后再滴一次缓苗水。②生育期滴水。从出苗后 15d 开始滴第一次水，以后根据天气、土壤含水量和作物长势等综合因素判定，1~5d 滴一次水，每次滴水量 180m³/hm²，全生育期滴水 15 次，采摘前 7d 停水。总滴水量 3 000m³/hm² 左右。

7. 放苗、封洞、定苗

干播湿出的地，苗出齐后要及时放苗封洞，随后进行定苗，株距 30cm，如有缺苗断垄现象，可在旁边留些双株，第一水后立即补苗。

8. 施化肥

全生育期化肥全部随水滴施。出苗水随水滴施专用肥 45kg/hm²，以后每次滴水随水滴施专用肥 60kg/hm²，番茄开始采收前 5~7d 停止滴肥，其中在第一穗花序浆果膨大期，随水滴施一次硝酸钙，施用量 45kg/hm²，可有效防止番茄脐腐病的发生，从而提高番茄的品质。

9. 调控

实施水调化控相结合，尽可能减少药剂用量，保证产品符合国家卫生标准。化学药剂，随用 15%多效唑，4~6 片叶随水滴施，用量 45~75g/hm²。

第二节 辣椒水肥一体化技术应用

辣椒，别名为海椒、辣子、辣角、番椒等。茄科辣椒属多年生或一年生植物。辣椒的营养价值较高,维生素 C 含量在蔬菜中名列前茅,且味辣,是我国人民喜食的鲜菜酱菜及调味品,特别是西北的甘肃、陕西,西南的四川、贵州、云南,华中的湖南、江西等省,几乎每餐必备。辣椒果内含有辣椒素($C_{18}H_{97}NO_3$)和辣椒红素,有促进食欲、帮助消化等功效。我国产的辣椒干、辣椒粉远销新加坡、菲律宾、日本、美国等地。

一、辣椒的需水特性

辣椒植株全身需水量不大,但由于根系浅、根量少,对土壤水分状况反应十分敏感,土壤水分状况与开花、结果的关系十分密切。辣椒既不耐旱也不耐涝,只有土壤保持湿润才能高产,但积水会使植株萎蔫。一般大果类型的甜椒品种对水分要求比小果类型辣椒品种更严格。辣椒苗期植株需水较少,以控温通风降湿为主,移栽后为满足植株生长发育应适当浇水,初花期要增加水分,坐果期和盛果期需供应充足的水分。如土壤水分不足,极易引起落花落果,影响果实膨大,果实表面多皱缩、少光泽、果形弯曲。灌溉时做到畦土不积水,如土壤水分过多、淹水数小时,植株就会萎蔫,严重时成片死亡。此外,对空气湿度要求也较严格,开花结果期空气相对湿度以 60%~80% 为宜,过湿易造成病害,过干则对授粉受精和坐果不利。

辣椒是一种需水量不太多,但不耐旱、不耐涝,对水分要求较严格的蔬菜。苗期耗水量最少,定植到辣椒长至 3cm 左右大小时,滴水量要少,以促根为主,适当蹲苗。进入初果期后,加大滴水量及灌水次数,土壤湿度控制在田间持水量的 70%~80%;进入盛果期,需水需肥达到高峰,土壤湿度控制在田间持水量的 75%~85%。

定植水,灌水定额 15m³/亩;定植至实果期(7月至8月上旬)4~6d 滴水一次,灌水定额 6~8m³/亩;初果期(8月中下旬)5d 滴水一次,灌水定额 8~10m³/亩;盛果期(9月)5d 滴水一次,灌水定额 10~15m³/亩;植株保鲜(10—11月)10月上旬滴水一次,灌水定额 8~15m³/亩。定植至商品上市生育期 130d 左右,滴水 20~30 次,总灌水量 190~230m³/亩。

二、辣椒的需肥特性

辣椒为吸肥量较多的蔬菜类型,每生产 1 000kg 鲜辣椒需氮 3.5~5.5kg、五氧化二磷 0.7~1.4kg、氧化钾 5.5~7.2kg、氧化钙 2~5kg、氧化镁 0.7~3.2kg。

辣椒在各个不同生育期,所吸收的氮、磷、钾等营养物质的数量也有所不同。从出苗到现蕾,由于植株根少叶小,干物质积累较慢,因而需要的养分也少,约占吸收总量的5%;从现蕾到初花植株生长加快,营养体迅速扩大,干物质积累量也逐渐增加,对养分的吸收量增多,约占吸收总量的11%;从初花至盛花结果是辣椒营养生长和生殖生长旺盛时期,也是吸收养分和氮素最多的时期,约占吸收总量的34%;盛花至成熟期,植株的营养生长较弱,这时对磷、钾的需要量最多,约占吸收总量的50%;在成熟果收摘后,为了及时促进枝叶生长发育,这时又需较大数量的氮肥。

辣椒对氮的吸收随生育进展稳步增加,对硝态氮的吸收量与果实的产量相平衡,直到采收结束。对磷的吸收虽然随生育进展而增加,但吸收量变化的幅度较小。钙的吸收也随生育期的进展而增加,若在果实发育期间钙的供给跟不上,易出现脐腐病。对钾、镁的吸收量,同样在生育初期较少,从结果起不断增加,盛果期吸收得最多。钾素缺乏,易造成植株落叶;缺镁会造成叶片叶脉间黄化。

一般每生产1 000kg辣椒果实,需要吸收纯氮3~5.2kg、磷0.6~1.1kg、钾5~6.5kg、钙1.5~2kg、镁0.5~0.7kg。

三、辣椒水肥一体化栽培技术

(一) 品种选择

日光温室冬季栽培的环境特点是低温、弱光、通风不良、温室内湿度大,因此日光温室越冬茬辣椒栽培宜选用耐低温弱光、抗病性强的高产品种,如中椒108、国禧105、红塔系列、海丰系列等。

(二) 茬口安排

华北地区日光温室辣椒一般选择越冬茬栽培,7月上旬育苗,8月下旬定植,10月下旬开始采收,翌年5月下旬拉秧结束。

(三) 定植

(1) 整地施肥。辣椒越冬茬栽培生长期长,必须施足基肥,一般每$667m^2$施优质腐熟厩肥7 500~10 000kg、过磷酸钙75~100kg、硫酸钾20~30kg、碳酸氢铵50~75kg、饼肥50~100kg。基肥宜采取地面普施和开沟集中施相结合的方法,2/3基肥耕地普施,1/3基肥沟施,人工深翻2遍,把粪和土充分混匀,而后在沟内浇水。

(2) 定植密度。辣椒栽植密度较大,可采用双行单株定植,大行距70~80cm,小行距35~45cm,株距30cm。起垄栽培,垄不宜太高,一般相对高度为20~25cm。

(3) 定植方法。选择晴好天气，于 9:00—15:00 定植最好。按株距成"丁"字形挖穴，每穴浇灌 65%甲霜灵可湿性粉剂 1 000 倍液和 72%硫酸链霉素可溶性粉剂 4 000 倍液共 100mg，防止土壤中疫病、根腐病和疮痂病病菌对植株的侵染。定苗时苗坨与垄面平齐或略高，子叶的方向与行垂直，定植后用滴灌浇透扎根水。

（四）主要病虫害防治

(1) 疫病。在定植前每 667m^2 用 4%疫病灵颗粒剂 5kg 进行土壤消毒。盛果期根施 2 次 4%疫病灵颗粒剂进行早期预防，用量为每穴 2g。发病后可用 25%甲霜灵可湿性粉剂 800~1 000 倍液，或 75%百菌清可湿性粉剂 600 倍液，或 20%噻菌铜悬浮剂 400~600 倍液喷雾防治。

(2) 白粉病。发病初期，可选用 50%硫黄悬浮剂 200~300 倍液防治，隔 7~15d 喷 1 次，连续 2~3 次，防效明显。

(3) 灰霉病。发病初期用 50%腐霉利可湿性粉剂 1 500~2 000 倍液，或 40%嘧霉胺可湿性粉剂 800~1 200 倍液喷雾防治。在番茄灵溶液中加 0.1%的 50%腐霉利可湿性粉剂或 50%异菌脲可湿性粉剂蘸花或涂果柄，既防止落花又兼治灰霉病。

(4) 病毒病。可用 1.5%烷醇·硫酸铜乳剂 500 倍液，或 10%混合脂肪酸水剂 200 倍液，或 20%吗胍·乙酸铜可湿性粉剂 500 倍液喷洒防治。

(5) 蚜虫、白粉虱。可选用 25%噻虫嗪水分散粒剂 1 500~2 500 倍液，或 2.5%氯氟氰菊酯乳油 4 000~5 000 倍液喷雾防治。

第三节　黄瓜水肥一体化技术应用

一、黄瓜栽培的生物学基础

黄瓜又称胡瓜、王瓜，葫芦科 1 年生攀缘性草本植物，起源于喜马拉雅山南麓的热带雨林地区。汉代张骞出使西域时带回我国中原，经过 2 000 多年的栽培和培育，形成了果实细长、皮薄、刺瘤多的华北系黄瓜。另外，黄瓜经越南传入我国南方，形成了果实短粗、刺瘤稀少的华南系黄瓜。

黄瓜是全球性的大众化蔬菜，其栽培面积仅次于番茄、甘蓝和洋葱，名列第四。其中亚洲的栽培面积最大，约占世界总面积的 50%，其次是欧洲及北美洲、中美洲。我国黄瓜栽培面积约为241 000hm²，占世界栽培面积的28%，居各国之首。黄瓜适应性较强，可进行多种形式和茬次的栽培，是我国北方地区保护地栽培的主要蔬菜之一。

（一）黄瓜的植物学特征

（1）根。黄瓜是浅根性作物，根系主要分布在25cm的土层中，以5cm土层最多，侧根多集中于半径30cm范围内，根量少，吸收能力弱，而且根系木栓化早，再生能力差，断根后不易发新根。生产中应采用营养钵、营养土方等护根措施育苗。另外，黄瓜根系好气，要求土壤疏松透气。

（2）茎。黄瓜为蔓生攀缘茎，其茎中空、五棱、生有刚毛，5~6节后节间开始伸长。茎的高度、节间长短、分枝多少受品种、环境条件和栽培技术的影响。

（3）叶。黄瓜的真叶为单叶互生、掌状五角形，叶片大而薄，蒸腾量大。叶表面着生刺毛和气孔，刺毛在叶正面密、背面稀，气孔则是正面少而小、背面多而大，因此喷药时应侧重叶背面。

（4）花。黄瓜一般为雌雄同株异花，花的着生和开花顺序由下而上进行。主蔓上第一雌花节位高低直接影响采瓜的早晚，生产上也常以此作为品种鉴别的重要指标。早熟栽培，应选用第一雌花节位低的品种。

（5）果实和种子。黄瓜果实为假果，表皮部分由花托的外皮发育而成，有棱瘤刺或无；皮层由花托皮层和子房壁发育而成。有单性结实特性，若环境条件不适、栽培管理不当，会出现畸形果实。黄瓜种子披针形、扁平、黄白色，每条瓜有种子150~300粒，种子着生于侧膜胎座上，千粒重22~42g。

（二）黄瓜的生育周期

黄瓜生育周期大致分为发芽期、幼苗期、甩条发棵期和结果期。

（1）发芽期。从种子萌动到第一片真叶出现，需5~6d。发芽期主要靠种子贮藏的养分，生产中应选用饱满的新种子，并给予较高的湿度、温度和充足的光照条件。

（2）幼苗期。从第一片真叶出现到5~6片叶展开，茎蔓开始伸长，需20~30d。此期幼苗直立生长，分化大量叶芽和花芽。黄瓜幼苗1~2片叶时，甚至真叶充分展开后即进行花芽分化，花芽在分化之初，既有雌蕊又有雄蕊。但在花芽发育成花的过程中，有的雌蕊退化，发育成雄花；有的雄蕊退化，发育成雌花，这一过程受到内外因素的影响。雌、雄花的数量和比例是可塑的，目前认为黄瓜性型分化的内在因素主要有植株体内的碳氮比（C/N）、代谢水平和植物激素。

植株体内含氮化合物多可促进雄花分化；含碳化合物多可促进雌花分化。植株较低的代谢水平有利于雌花分化；茎叶生长旺盛，代谢水平高，有利于雄花分化。赤霉素可促进雄花分化；生长素和乙烯可促进雌花分化。外界条件中低温，特别是低夜温，短日照，充足的磷、钾肥营养，适宜的水分、乙烯利、脱落酸以及二氧化碳、一氧化碳等均可促进雌花分化；而长日照、氮肥过多、营养不良、使用赤霉素等均有利于雄花分化。生产中应创造适宜的环境条件，促进幼苗多分化雌花。

（3）甩条发棵期。从5~6片叶到第一雌花开放。此期植株转为蔓性生长，植株生长旺盛，并由营养生长为主过渡到生殖生长为主。

（4）开花结果期。从第一雌花开放到拉秧。结果期长短与栽培季节、栽培环境有密切关系，春黄瓜一般50~60d；日光温室越冬黄瓜可达6~8个月。

二、黄瓜品种选择

（一）黄瓜品种类型

黄瓜栽培历史悠久，分布广泛，品种类型十分丰富。根据其分布区域和生物学性状分为以下类型。

（1）南亚型。分布于南亚各地。植株茎叶粗大，易分枝。果实大，单瓜重达1~5kg，果实短圆筒形或长圆筒形，皮色浅，瘤稀，刺黑色或白色，皮厚，味淡。喜湿热，严格要求短日照。地方品种群很多，如锡金黄瓜、中国版纳黄瓜及云南昭通大黄瓜等。

（2）中国华南型。分布在我国长江以南及日本各地。茎叶较繁茂，耐湿、耐热，为短日照植物。果实较小，瘤稀，多黑刺，幼嫩果实绿色、绿白色、黄白色，味淡；成熟果实黄褐色，有网纹。代表品种有昆明早黄瓜、广州二青、上海杨行、武汉青鱼胆、重庆的大白及日本的青长、相模半白等。

（3）中国华北型。分布于我国黄河流域以北及朝鲜、日本等地。植株长势中等，喜土壤湿润、天气晴朗的自然条件，对日照长短反应不敏感。幼嫩果实棒状、绿色，瘤密、多刺；成熟果实黄白色，无网纹。代表品种有山东新泰密刺、北京大刺瓜、津研系统黄瓜、津杂系统黄瓜和津优系统黄瓜等。

（4）欧美露地型。分布于欧洲及北美洲各地。茎叶繁茂，果实圆筒形、中等大小，瘤稀，白刺，味清淡，成熟果实黄褐色。有东欧、北欧、北美等品种群。

（5）北欧温室型。分布于英国、荷兰。茎叶繁茂，耐低温弱光，果面光滑、浅绿色，果长达50cm以上。有英国温室黄瓜、荷兰温室黄瓜等。

(6) 小型黄瓜。分布于亚洲及欧美各地。植株矮小，分枝性强，多花多果，以盐渍加工为主。代表品种有扬州乳黄瓜等。

我国栽培的黄瓜主要有华北型和华南型2种生态类型。华北型黄瓜按栽培季节分为春黄瓜类型、春夏黄瓜类型和秋黄瓜类型。春黄瓜类型比较耐寒和早熟；春夏黄瓜类型生长势和适应性强，耐热抗病，多为中熟品种；秋黄瓜类型多为中晚熟品种，叶片厚叶色深绿，适应性强。近年来随着黄瓜产业的发展，出现了一些适应冬季生产需要的品种，如新泰密刺、津春3号、津优2号、津优3号等。

(二) 黄瓜主要优良品种

(1) 津优2号。天津市黄瓜研究所育成的早熟杂交一代，1998年通过天津市农作物品种审定委员会审定。植株生长势较强，主蔓结瓜为主，瓜码密，几乎节节有瓜，回头瓜多。瓜长棒形，瓜长34cm左右，瓜把短，瓜色深绿，有光泽，刺瘤中等，白刺。瓜肉深绿色，口感脆，品质优。耐低温弱光，高抗霜霉病、白粉病和枯萎病。播种至始收60~70d，采收期为80~100d，单瓜重约200g。适合我国北方各地日光温室冬春茬和大棚春提早栽培。

(2) 津优30号。天津市黄瓜研究所育成的杂交一代。植株生长势强，主蔓结瓜为主。瓜条顺直，长35cm左右，单瓜重220g左右。瓜绿色、有光泽，瘤显著，密生白刺。瓜肉淡绿色，口感脆，味甜。耐低温弱光，抗枯萎病、霜霉病和白粉病。适合华北、东北、西北和华东地区日光温室越冬茬和冬春茬栽培。

(3) 津优3号。天津市黄瓜研究所育成的早熟杂交一代。植株生长势强，较耐低温和弱光，株型紧凑，叶色深绿，叶片中等大小。以主蔓结瓜为主，分枝性弱，瓜码密，回头瓜多，在良好栽培条件下，可反复多茬结瓜。瓜条顺直，瓜把短，瓜长30~33cm，瓜色深绿，瘤显著，密生白刺。单瓜重200g左右。该品种是目前抗病性强、丰产性好的越冬茬日光温室栽培的高产品种之一。

(4) 新泰密刺。山东省新泰市地方品种，1987年通过山东省农作物品种审定委员会审定。植株生长势较强，茎粗，节间短，主蔓结瓜，第一雌花着生在4~5节，一节多瓜，回头瓜多。瓜长棒形，瓜长25~35cm，横径3~4cm，瓜把较长、深绿色，刺瘤密，白刺，棱不明显。口感脆嫩，微甜，品质好。耐低温、弱光，抗枯萎病，不抗霜霉病、白粉病，早熟，单瓜重250克左右。适合日光温室及大棚栽培。

(5) 中农202。中国农业科学院蔬菜花卉研究所最新育成的极早熟杂交一代。植株生长势强，生长速度快，以主蔓结瓜为主。瓜长棒形，瓜把短，瓜条直，瓜长 35cm 左右、横径 4cm 左右。瓜色深绿，有光泽，瓜表面无棱，刺瘤小，稀密中等，白刺。瓜肉厚，心腔小，质脆，味微甜，商品性好。第一雌花节位 2~3 节，单瓜重 250g 左右。该品种只在保护地区种植。

(6) 中农 9 号。中国农业科学院蔬菜花卉研究所新近育成的中早熟少刺型杂种一代。植株生长势强，第一雌花始于主蔓 3~5 节，雌花间隔 2~4 节，前期主蔓结瓜，中后期侧蔓结瓜为主，雌花节多为双瓜。瓜短筒形、长 15~20cm，瓜色深绿一致，有光泽，无花纹，瓜把短，刺瘤稀，白刺，无棱，品质中上等。抗霜霉病、枯萎病、黑星病等病害。适合春大棚、春日光温室及秋延后栽培。

(7) 中农 12 号。中国农业科学院蔬菜花卉研究所新近育成的中早熟杂种一代。植株生长速度快，以主蔓结瓜为主，瓜码密。瓜长棒形，瓜长 30cm 左右，瓜色深绿一致，有光泽，瘤小，白刺，口感脆甜，品质佳。抗霜霉病、白粉病、花叶病毒病，中抗黑星病、细菌性角斑病、枯萎病。早熟，单瓜重 150~200g。适合北方早春保护地、早春露地和秋延后栽培。

(8) 中农 19 号。水果型雌型杂种一代品种。瓜短筒形，瓜长 15~20cm，瓜色亮绿一致，无花纹，果面光滑，口感脆甜。具有很强的耐低温、弱光能力，抗枯萎病、黑星病、霜霉病和白粉病等。连续坐果能力强，单瓜重约 100g。适合日光温室越冬茬和大棚春早熟栽培。

三、种子处理

播种前进行种子处理是蔬菜育苗的重要技术环节之一，主要目的和作用是提高种子使用价值、消毒、促进萌发、促进生育等。种子处理的方法有多种，根据处理的目的可分为以下几种。

(一) 种子清选

除掉杂质和不成熟的种子，提高种子纯净度，也就提高了种子的使用价值。比较简易可行的方法有风选、水选、筛选和人工手选等。

(二) 种子消毒处理

许多蔬菜的病害是通过种子传播的，其中多数病原菌寄生在种子的表面，种子消毒处理可以杀死病原菌，避免病害的发生。种子消毒的方法较多，常用的方法有以下 5 种。

(1) 温汤浸种。温汤浸种所用水温为 55℃ 左右，用水量是种子体积的 5~6 倍。方法是先用常温水浸 15min，后转入 55~60℃ 热水中浸种，期间不断搅拌，并保持该水温 10~15min，然后将水温降至 30℃，继续浸种 4~6h。

（2）热水烫种。此法一般用于难于吸水的种子，水温为70~75℃，水量不宜超过种子量的5倍，种子要经过充分干燥。烫种时要用2个容器，将热水来回倾倒，最初几次动作要快而猛，使热气散发并提供氧气。一直倾倒到水温降至55℃时，再改为不断搅动。倾倒烫种7~8min，用55℃温水搅动浸种10~15min，当水温降至30℃时停止搅拌，再浸种3~4h。

（3）药液浸种。种子消毒常用的药剂有1%高锰酸钾溶液、10%磷酸三钠溶液、1%硫酸铜溶液、40%甲醛100倍液等，一般经过5~10min药液浸种，再用清水反复冲洗种子至无药味为止。用10%磷酸三钠溶液浸种20~30min，或40%甲醛100~200倍液浸种15~20min，捞出后清水洗净，还可预防黄瓜病毒病。

（4）药剂拌种。用干种子播种的，可将药剂与种子混合均匀，使药剂黏附在种子的表面，然后再播种，药剂的用量一般为种子重量的0.2%~0.3%，常用的药剂有70%敌磺钠可溶性粉剂、50%多菌灵可湿性粉剂、40%福美·拌种灵可湿性粉剂、25%甲霜灵可湿性粉剂。50%多菌灵可湿性粉剂用种子量的0.4%拌种，可预防黄瓜真菌性病害。

（5）干热处理。干热处理是将充分干燥（含水量低于4%）种子放在75℃以上的高温条件下处理，可钝化病毒，还可提高种子活力。适用于较耐热的蔬菜种子，如瓜类和茄果类蔬菜种子。在70℃高温条件下处理2d，可使黄瓜绿斑花叶病毒完全丧失活力而死亡。

（三）催芽处理

催芽是在消毒浸种之后，为了促进种子萌发所采取的技术措施。催芽过程中应满足种子萌发所需要的温度、湿度、通气和光照条件，促使种子的营养物质迅速分解转运，供给种子幼胚的生长需要。温度管理，应初期低后期逐渐升高，当种子露白时再降低，使胚根苗壮。湿度管理，以种皮不发滑又不发白为宜。催芽适宜温度为25~28℃，一般经24h即可达到播种要求。催芽过程中注意每隔一段时间翻动1次使种子受热均匀，保证种子水分需求和氧气供应。催芽期间还可进行胚芽锻炼，方法是将刚破嘴的种子连同包布置于-2~-1℃低温条件下12h，然后再置于18~22℃条件下催芽，如此反复2~3d，可显著提高种子耐寒能力，增强幼苗对低温的适应性。生产中，浸种后若遇天气状况不适宜播种时，也可以采用胚芽锻炼方法控制种子出芽时间。

（四）播种

每667m^2苗床播种量为120~200kg，播种前先浇足底水，待水分渗下后，按8~10cm见方在苗床预先计划方格，然后于每方格的中央播1粒种子。采用容器育苗时，每个营养钵点播 1 粒种子，播后需用过筛细土进行覆盖。覆盖厚度一般为 1~2cm，过厚不利于种芽拱土，过薄则易导致"戴帽"出土。为利于苗床或营养钵增温、通气和种芽拱土，覆土时最好形成以种子为中心、直径 5cm 左右的小土堆。然后再于整个畦面均匀铺撒厚 5~7mm 的过筛细土，利于畦面保墒和防止出现裂缝。最后，再用塑料薄膜封严密闭，并于夜间盖草苫。

四、育苗技术

（一）常规育苗

播种前 7~10d，对日光温室进行消毒处理，方法是每 100m³ 空间用硫黄粉 250g、锯末 500g 混合熏烟 12h 左右。在日光温室中柱前 50cm 以南，做宽 1~1.5m 的畦，有条件的可铺设地热线。用 60%田土和 40%腐熟有机肥混合配制营养土，每立方米土加 50%多菌灵可湿性粉剂 180g、58%甲霜·锰锌可湿性粉剂 50g，过筛后用农膜盖好闷 24h。采用直径 10cm、高 10cm 的营养钵，内装营养土 8cm 高，浇透水，水渗后在每个营养钵内播发芽种子 1 粒，覆土厚约 1cm，平盖地膜，以利于保墒。出苗后除雨天和夜间外，尽量使幼苗通风透光。幼苗期气温高，蒸发量大，苗床必须保证水分供应，缺水要及时补充。育苗期间，为促进雌花形成可用乙烯利处理，一般在 2 片真叶期用 40%乙烯利水剂 4 000 倍液在傍晚进行喷雾。幼苗 2 叶 1 心至 3 叶 1 心时定植。

（二）嫁接育苗

黄瓜是日光温室种植的主要蔬菜，轮作倒茬困难，连作引起的枯萎病日趋严重，采用嫁接育苗是防治枯萎病的有效措施；而且砧木根系发达，耐旱、耐寒和吸收水肥的能力较强，利于黄瓜高产优质。黄瓜砧木以南瓜为主，主要有黑籽南瓜、南砧 1 号、新土佐、壮士、共荣等品种。

（1）嫁接方法。黄瓜嫁接方法主要是劈接、靠接、插接和断根插接、芯长接、二段接等。

一是靠接。通常接穗黄瓜比砧木南瓜早播 2~5d，黄瓜播种后 10~12d、第一片真叶始露至半展，砧木南瓜子叶全展，第一片真叶显露时即可嫁接。嫁接过早，幼苗太小操作不方便；嫁接过晚，成活率低。嫁接前先将砧木苗和接穗苗的基质喷湿，从育苗盘中挖出后用湿布覆盖。嫁接时，取接穗在子叶下部 1~1.5cm 处呈 15°~20°角向上斜切一刀，深度达胚轴直径的 3/5~2/3；去除砧木生长点和真叶，在其子叶节下 0.5~1cm 处呈 20°~30°角向下斜切一刀，深度达胚轴直径的 1/2，砧木和接穗切口长度均为 0.6~0.8cm。最后，将砧木和接穗的切口相互套插在一起，用嫁接夹固定或用塑料条带绑缚。将砧穗复合体栽入营养钵中，两者根茎距离保持1~2cm，以利于成活后断茎去根，如图6-2所示。

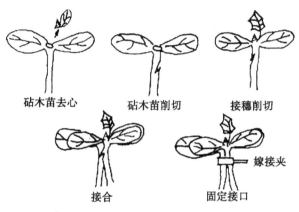

图6-2 黄瓜靠接技术

黄瓜靠接苗易管理，成活率高，生长整齐，操作简易。但此法嫁接速度慢，接口需要固定物，成活后需要断茎去根工序，而且接口位置低易受土壤污染和发生不定根，幼苗搬运和田间管理时接口部位易脱离。

二是插接。接穗子叶全展，砧木子叶展平、第一片真叶显露至初展为嫁接适宜时期。根据育苗季节与环境，南瓜砧木比黄瓜早播2~5d，黄瓜播种后7~8d嫁接。嫁接时先喷湿接穗和砧木苗钵（盘）内的基质。剔除砧木苗生长点，用竹签从顶心呈40°角向下斜插0.5cm，以手指能感觉到其尖端压力为度。然后将接穗苗在子叶下0.5cm处削成楔形。拔出砧木上的竹签，将削好的接穗插入砧木小孔中，使两者密接，砧、穗子叶伸展方向呈"十"字形，利于见光，如图6-3所示。

插接法嫁接，砧木苗不需起出，还减少了嫁接苗栽植和嫁接夹使用等工序，也不用断茎去根，嫁接速度快，操作方便，省工省力；嫁接部位紧靠子叶节，细胞分裂旺盛，维管束集中，愈合速度快，接口牢固，砧、穗不易脱裂折断，成活率高；接口位置高，不易再度污染和感染，防病效果好。但插接对嫁接操作熟练程度、嫁接苗龄、成活期管理水平要求严格，技术不熟练时嫁接成活率低，后期生长不良。

三是断根插接。从砧木胚轴适当长度切断后进行嫁接，促其生根长成完整植株，嫁接时采用插接法即为断根插接。用新土佐作砧木嫁接时常用此法，黑籽南瓜胚轴太短，子叶太大，应用较少。根据嫁接时温度条件，砧木比接穗提前2~

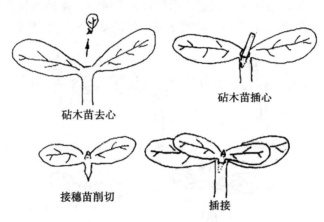

图 6-3　黄瓜插接示意图

3d 播种或砧、穗同时播种，砧木第一片真叶 0.5~1cm、接穗第一片真叶 0.2~0.5cm 时嫁接。嫁接前 1~2d 适当降温控水，促使胚轴硬化，嫁接当日苗床充分浇水，使植株吸足水分，最好喷洒 1 次低浓度杀菌剂。嫁接时在砧木子叶节下留 5~6cm 将胚轴切断（越靠近根部胚轴生根能力越强），接穗于子叶节下 2~3cm 处切断，将两者分别放入湿润的容器中，用湿布覆盖防止萎蔫。注意一次性剪取砧木、接穗数量不要过多，最好随剪断随嫁接。插接完毕后将砧穗复合体插栽入装有基质的育苗钵（盘）中，插栽深度 2~3cm，然后扣棚密闭遮光。断根插接法操作简单、省力，砧木和接穗完全不附着泥土，嫁接效率高。幼苗重新生根，侧根数量多，植株长势旺盛，利于提高产量。但嫁接后管理要求精细，比普通插接费工。

　　四是劈接。适宜嫁接时期为接穗 2 片子叶充分展开，砧木第一片真叶出现，砧木比接穗提早 3~8d 播种。嫁接时将砧木心叶摘除，然后用刀片在胚轴正中央或一侧垂直向下纵切，切口长 1~1.5cm，再把接穗胚轴削成楔形，削面长短与砧木切口长度相对应，最后将接穗插入砧木切口并用嫁接夹固定或用塑料袋缠绑。黄瓜劈接后管理比较困难，成活率也较低，生产中应用较少。

　　（2）嫁接后的管理。嫁接后 3d 内苗床不通风、不见光，苗床温度白天保持在 25~28℃、夜间 18~20℃，空气相对湿度保持 90%~95%。3d 后视苗情，以幼苗不萎蔫为度进行短时间少量通风，以后逐渐加大通风量。1 周后接口愈合，即可逐渐揭去草苫，并开始大通风，苗床温度白天保持 22~26℃、夜间 13~16℃，若苗床温度低于 13℃ 应加盖草苫。育苗期视苗情浇 1~2 次水，采用靠接法的，

第六章 蔬菜水肥一体化技术的应用

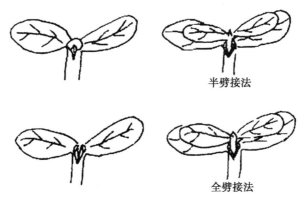

图6-4 黄瓜劈接法

在接口愈合后，及时剪断接穗的根。

为提高黄瓜的抗逆性，培育适龄壮苗，应进行大温差管理。嫁接苗成活后，白天温度保持25~30℃，不超过35℃不通风，前半夜温度保持15~18℃、后半夜11~13℃、早晨揭苫前10℃左右，有时可短时间降至5~8℃，地温保持在13℃以上。水分不需过分控制，以适宜的水分、充足的光照、昼夜温差大来防止幼苗徒长。冬春茬黄瓜苗龄不宜太大，以3~4叶1心、株高10~13cm时定植为宜，日历苗龄35d左右，不宜超过40d。

（三）壮苗标准

幼苗子叶完好，茎粗壮，叶色浓绿，无病虫危害；4~5片真叶，株高15cm左右，根系发达，苗龄25~30d。

五、水肥一体化管理

定植后至坐瓜前不追肥，可结合喷药，用0.2%磷酸二氢钾+0.2%尿素或0.3%三元复合肥叶面喷肥。第一次灌水在定植时进行，用水量为15~20m³/667m²。当植株有8~10片叶、第一瓜长约10cm时，进行第二次灌水并结合施肥，每667m²施尿素10~15kg、硫酸钾10~12kg，用水量为15m³/667m²。入冬前，每15~20d追肥1次，除结合追肥浇水外，从定植到深冬季节，应以控为主，如果植株表现缺水现象，可浇小水。2月下旬后，黄瓜需水肥量增加，要适当增加浇水次数和浇水量，每隔7~10d浇1次水。进入盛果期后，可适当减少灌水次数，盛果期结合浇水每10~15d追施1次化肥，每次每667m²用尿素8~10kg、硫酸钾12~15kg。黄瓜全生育期可以灌水12~15次，追肥8~10次。生育后期可用0.2%~0.3%尿素溶液或磷酸二氢钾溶液进行叶面追肥,以壮秧防早衰。

六、主要病虫害及防治

(1)白粉病。发病初期在叶片上生圆形粉斑,迅速扩展,重病时整个叶片布满白粉。病后期叶片变黄干枯。病菌在病残体上越冬,借气流传播。植株生长不良,光照不足或干燥时病情发展迅速。防治方法:用25%三唑酮可湿性粉剂1 500倍液喷雾防治。

(2)枯萎病。主要危害根和茎,受害根系呈褐色腐朽,茎部皮层有时呈纵裂状。潮湿时产生粉红色霉,病部维管不变褐色,全株萎蔫枯死。土壤病菌从根部伤口侵入,连作地、氮肥过多或排水不良的地块发病严重。防治方法:实行轮作,选用无病种子,用无病土育苗,合理浇水施肥,拔除病株并在病穴内及周围撒石灰,与黑籽南瓜砧木嫁接。发病初期,可用50%多菌灵可湿性粉剂800倍液,或70%甲基硫菌灵可湿性粉剂1 000倍液灌根防治,每株灌药液0.25kg。

(3)炭疽病。叶片受害发生黄褐色圆形病斑,边缘明显,外缘淡黄色,易破裂,高温高湿时病斑上产生粉红色糊状物,茎、果受害不严重。病菌随病残株在土中越冬,种子带菌,引起子叶发病。病菌借雨水(或大棚内滴水)传播,阴雨天气、暴风雨及高温高湿时易发病。防治方法;用无病瓜留种或进行种子消毒,用无病土育苗。注意通风排湿和防涝排水,采用高畦地膜覆盖栽培。可用2.5%嘧菌酯可湿性粉剂1 000~1 500倍液喷雾防治。

(4)害虫。黄瓜虫害主要有蚜虫和白粉虱,主要危害叶片,受害叶片卷缩,严重时整片卷曲直至整株萎蔫死亡。老叶受害不卷曲,但提前干枯死亡。防治方法:蚜虫,用50%抗蚜威可湿性粉剂2 000倍液雾防治。白粉虱,用25%噻嗪酮可湿性粉剂2 500倍液喷雾防治。

第四节 茄子水肥一体化技术应用

一、茄子的需水特性

茄子在高温高湿环境条件下生长良好,对水分的需要量大。但是,茄子不同生长发育阶段对水分的要求不同,在幼苗初期,在光照度和温度等条件适宜的情况下，苗床水分充足，能促进幼苗健壮生长和花芽顺利分化，并能提高花的质量。所以育苗时，应选择保水能力强的土壤作床土，同时浇足底水，以减少播种后的浇水次数，稳定苗床温度。开花坐果期，由于茄子处于从营养生长向生殖生长过渡的阶段，为了维持营养生长与生殖生长的平衡，避免营养生长过盛，在水分管理上，应以控为主，不干旱不浇水。结果期，在门茄"瞪眼"以前需要水分较少，门茄迅速生长后需水量逐渐增多，对茄收获前后需水量最大。茄子坐果率和产量与当时的降雨量及空气湿度成负相关。空气相对湿度以70%~80%为宜，长期超过80%，容易引起病害发生。土壤相对含水量以保持在60%~80%为好，一般不能低于55%，否则会出现僵苗、僵果。生产中要尽量满足茄子对水分的需求，否则就会影响其生长发育，水分不足，结果少、果实小、果面粗糙、品质差。茄子不耐通气不良和过于潮湿的土壤，因此要防止土壤过湿；否则，易出现沤根现象。

二、茄子的需肥特性

茄果类蔬菜是一类根系发达的植物，而且茄子采摘期长，产量高，养分吸收量大。茄子对土壤要求不太严格，但以富含有机质、土层深厚、保水保肥能力强、通气排水良好的沙质土壤生长为好，土壤pH值以6.8~7.3为好。

（一）茄子对不同营养元素的需求

（1）对氮肥的需求。茄子以采收嫩果为主，氮素营养对产量的影响特别明显。氮肥充足时，植株生长旺盛，花芽发育良好，结实率高。茄子从定植到拔秧均需要氮肥供应，定植后茄子耐高浓度氮肥的能力比番茄强，不太容易因氮肥过多引起植株徒长。

（2）对磷的需求。磷对茄子花芽分化发育有着直接的影响，苗期磷肥充足不仅有利于根系发展，而且可以分化出优良的花芽；如果磷肥不足，会造成花芽发育迟缓或不发育，或形成不能结实的花。开花结果后，对氮、钾的需求总量增多，但对磷的吸收开始减少；磷过多会使果皮变硬，影响果实品质。

（3）对钾的需求。钾对花芽发育虽无明显影响，但缺钾少钾或钾过量，均会使花芽分化推迟。在茄子生育中期以前，茄子对氮、钾的需求基本是一致的，盛果期对钾的需求明显增多。茄子整个生育期缺钾均会影响产量，因此，在茄子全生育期均应注意钾肥的施用。

（4）对镁、钙的需求。茄子对镁的需求在结果以后开始明显增加，缺镁会使花芽发育迟缓或不发育或形成不能结实的花，叶片主叶脉附近褪绿变黄，叶片早落而影响产量。土壤太湿或含氮、钾、钙过多，均会引起缺镁症，表现为果实或叶片网状叶脉褐变而产生铁锈状。茄子对缺钙的反应不如番茄敏感。

（二）茄子不同生育期对养分的吸收

茄子幼苗期对养分的吸收量不大，但对养分的丰缺非常敏感，养分供应状况影响幼苗的生长和花芽分化。从幼苗期到开花结果期对养分的吸收量逐渐增加，开始采收果实后茄子进入需要养分量最大的时期，对氮、钾的吸收量急剧增加，对磷、钙、镁的吸收量也有所增加，但不如钾和氮明显。茄子对各种养分的吸收特性也不同，氮对茄子各生育期都非常重要，任何时期缺氮，都会对开花结果产生不良影响。茄子对钾的吸收量到生育中期都与氮相当，到果实采收盛期钾的吸收量显著增高。在盛果期，氮和钾的吸收增多。氮、磷、钾肥配合施用，可以相互促进。

一般每生产1 000 kg 茄子果实需氮 3.65kg、磷 0.85kg、钾 5.75kg、钙 1.8kg、镁 0.4kg。生产中茄子栽培基肥应以有机肥为主，同时适量施入氮、磷、钾化肥。结果前期注意使用氮、磷肥，结果后期氮、磷、钾肥配合施用，并酌情追施镁、钙等微肥。

三、茄子水肥一体化栽培技术

（一）品种选择

日光温室茄子冬春栽培的环境特点是低温、弱光、通风不良、温室内湿度大，因此应选择株型紧凑、耐寒性和抗病性较强、连续结果能力强、肉质致密细嫩的品种。主要栽培品种有青选长茄、沈茄一号、齐茄一号、齐杂茄二号、绿圆茄、圆杂二号、天津快圆茄等。

（二）茬口安排

华北地区一般选择越冬茬栽培，8月下旬至9月上旬育苗，10月上中旬定植，日历苗龄50~55d，12月上中旬始收，翌年6月下旬拉秧。

（三）定植

（1）整地施肥。茄子忌连作，在前茬作物收获后要进行深翻晾晒。基肥一般每667m² 施腐熟农家肥5 000kg、尿素46kg、过磷酸钙100kg、硫酸钾30kg。基肥2/3撒施，1/3施于定植沟内。

（2）定植方法与密度。定植要选阴天过后晴天开头时进行。定植时先按50cm窄行、70cm宽行交替开沟，沟深5~6cm，然后按45cm株距摆苗，埋土浇水。当土壤见干见湿时中耕松土，培土使垄高超过垄面4cm，使苗行形成10cm

左右的垄台,再在两垄上覆一幅80~90cm宽的地膜,地膜开纵口把苗引出膜外,实施膜下暗灌。每667m²栽苗2 500余株。

(四) 主要病虫害防治

日光温室茄子病害主要有黄萎病、绵疫病、褐纹病等,害虫主要有棉红蜘蛛和茶黄螨。

(1) 黄萎病。播种前可用0.2%的50%多菌灵可湿性粉剂浸种1h。苗期和定植前喷洒50%多菌灵可湿性粉剂600~700倍液。定植后发病初期浇灌15%混合氨基酸铜·锌·锰·镁水剂500倍液,也可用2%嘧啶核苷类抗生素水剂200倍液灌根,每株灌150~250g,每10d 1次,连续2~3次。

(2) 绵疫病。又叫掉蛋、烂茄、水烂,是茄子的重要病害之一。发病初期可用75%百菌清可湿性粉剂500~600倍液,或64%噁霜·锰锌可湿性粉剂400~500倍液,或90%乙铝·锰锌可湿性粉剂600倍液,每7~10d 1次。还可用百菌清烟剂或粉尘剂防治。

(3) 褐纹病。发病初期喷洒75%百菌清可湿性粉剂600倍液,或40%甲霜灵可湿性粉剂600~700倍液,或64%噁霜·锰锌可湿性粉剂500倍液,交替使用,视天气和病情隔10d左右1次,连续防治3~4次。

(4) 虫害。红蜘蛛用25%灭螨猛可湿性粉剂1 000倍液喷洒防治,连喷2~3次。茶黄螨用73%炔螨特乳油2 000~3 000倍液喷洒防治,每7~10d喷1次,连喷3次。

第五节 西葫芦水肥一体化技术应用

一、西葫芦的需水特性

西葫芦又称荬瓜,原产于热带干旱地区。西葫芦根系发达,侧根数多,吸收能力强,表现为耐旱力强。西葫芦根系主要分布在耕层土壤中,由于一般耕作层较浅,蓄水蓄肥能力有限,往往容易干燥、脱肥;而且西葫芦茎叶繁茂,叶片大而多,蒸腾作用强,耗水量大。因此,生产中需要加强灌溉,保持适宜的土壤含水量,方能获得高产。

西葫芦要求较干燥的空气条件,温室栽培时必须通过减少地面水分蒸发和通风来调节空气湿度,适宜的空气相对湿度为45%~55%。高温、干旱条件下易发生病毒病；高温、高湿条件下易发生白粉病。在日光温室内种植西葫芦要特别注意控制温度和湿度，防止病毒病、白粉病等病害的发生和蔓延。

二、西葫芦的需肥特性

西葫芦根系发达，直播栽培主根入土深达 2m 以上，育苗移栽主根入土深达 1m 以上，根系横向扩展范围 2m 左右。侧根的分枝能力也很强，大部分侧根分布在 30cm 深的耕作层内。西葫芦抗旱、抗瘠薄能力强，对土壤要求不严格，黏土、壤土、沙壤土均可栽培。但因其根群发达，宜选用土层深厚、肥沃疏松的沙壤土，以利根系在低温条件下保持较强的生长势和吸收能力，提早收获和延长结果期。适宜的土壤 pH 值为 5.5~6.8。

西葫芦不同生育期对肥料种类、养分比例需求有所不同。前期植株生长缓慢，对养分吸收量较少，出苗后到开花结瓜前需供给充足氮肥，促进植株生长，为果实生长奠定基础。西葫芦对厩肥和堆肥的反应良好，生产上应多施有机肥作基肥。增施优质有机肥和三元复合肥，适当控制氮肥的用量，有利于平衡营养生长和生殖生长，提高产量。若氮肥用量过大，容易引起茎叶徒长，导致落花落瓜和病害的发生。

西葫芦吸肥能力强，对矿质养分的吸收，以钾最多，氮次之，其次为钙，磷最少。每生产 1 000kg 西葫芦果实，需要吸收纯氮 3.92~5.47kg、磷 2.13~2.22kg、钾 4.09~7.29kg、钙 3.2kg、镁 0.6kg，除钾的需要量低于黄瓜外，需氮、磷、钙的数量均高于黄瓜。

三、西葫芦水肥一体化栽培技术

（一）品种选择

冬季日光温室的环境特点是低温、弱光、通风不良、湿度大，因此日光温室越冬茬西葫芦宜选用抗病、耐低温弱光、抗逆性强、优质丰产商品性好的品种，如早青一代、中葫三号、冬玉西葫芦、寒玉西葫芦等。

（二）茬口安排

华北地区一般选择日光温室越冬茬栽培，10月上中旬播种育苗，10月下旬至11月上旬定植，11月下旬开始采摘，翌年5月下旬拉秧结束。

（三）育苗技术

1. 种子处理

（1）晒种。播种前进行种子精选，选择有光泽、籽粒饱满、无病斑、无虫伤、无霉变的新种子，将选好的种子摊在簸箕内晒1~2d，以提高发芽率、发芽势，加速发芽和出苗，同时还兼有杀菌消毒作用。若用陈种子播种，播前晒种尤为重要。

（2）种子消毒。用10%磷酸三钠溶液浸种20min，或用50%多菌灵可湿性粉剂500倍液浸种30min，然后用清水冲洗干净，再用温水浸种。

（3）浸种催芽。每667m^2需种子400~500g。在容器中放50~55℃的温水，将种子投入水中后不断搅拌，待水温降至30℃时停止搅拌，浸泡3~4h。浸种后将种子从水中取出，摊开晾10min，再用洁净湿布包好，置于28~30℃条件下催芽，经1~2d、70%左右的种子发芽时即可播种。

2. 播种

在日光温室内建造宽1.2m、深10cm的平畦苗床。营养土用肥沃大田土6份、腐熟农家肥4份，混合过筛。每立方米营养土加捣细的腐熟鸡粪15kg、过磷酸钙2kg、草木灰10kg，或三元复合肥3kg、50%多菌灵可湿性粉剂80g，充分混合均匀。将配制好的营养土装入营养钵或纸袋中，营养钵密排在苗床上。也可购买商品基质。播前浇足底水，将种子点播于营养钵内，播后覆1.5~2cm厚的干细土。播后苗畦覆盖地膜并插拱覆膜。

3. 育苗期管理

（1）温度管理。育苗期白天温度保持25~28℃、夜间15~20℃，阴天温度适当低些。

（2）光照管理。光照较强时，应覆盖遮阳网进行遮阳处理和叶片喷水降温。幼苗出土后尽可能提供充足的光照条件，防止光照不足引起徒长。

（3）水分管理。出苗期间保持床土湿润，以后视墒情适当浇水。

（4）炼苗。定植前1周不浇水，加强通风，进行低温炼苗，以利于缩短缓苗期。幼苗3叶1心、株高12~15cm时即可定植。

（四）主要病虫害防治

日光温室西葫芦主要病害有猝倒病、白粉病、灰霉病、疫病等，主要害虫有蚜虫、白粉虱、红蜘蛛、斑潜蝇等。

（1）猝倒病。发病初期用64%噁霜·锰锌可湿性粉剂500~600倍液，或72.2%霜霉威水剂5 000倍液，或15%噁霉灵水剂450倍液喷施防治。

（2）白粉病。发病初期用25%三唑酮可湿性粉剂2 500~3 000倍液，或40%氟硅唑乳油8 000~10 000倍液，或15%嘧菌酯乳油2 000~3 000倍液喷施防治。也可用4%嘧啶核苷类抗生素水剂600~800倍液喷施防治。

（3）灰霉病。发病初期用40%甲基嘧菌胺悬浮剂1 200倍液，或65%甲硫·乙霉威可湿性粉剂1 000~1 500倍液，或50%腐霉利可湿性粉剂1 000~2 000倍液喷施防治。

（4）疫病。发病初期用58%甲霜·锰锌可湿性粉剂750~1 500倍液，或90%三乙膦酸铝可湿性粉剂500倍液，或20%噻菌铜悬浮剂400~600倍液，或100万单位新植霉素可湿性粉剂2 000~3 000倍液喷施防治。

（5）蚜虫。在蚜虫始盛期用10%吡虫啉可湿性粉剂2 000~3 000倍液，或3%啶虫脒乳油1 500~2 000倍液，或0.3%印楝素乳油1 000~1 500倍液喷施防治。

（6）白粉虱、红蜘蛛、斑潜蝇。在危害初期用10%联苯菊酯乳油4 000~8 000倍液，或1.8%阿维菌素乳油3 000~4 000倍液，或20%氰戊菊酯乳油1 500~2 500倍液，或0.3%印楝素乳油1 000~1 500倍液喷施防治。

第六节　西瓜水肥一体化技术应用

西瓜属葫芦科，原产南非热带沙漠地区，属耐热性作物，喜光、耐高温干燥。果实属于瓠果，味甜多汁甘美，富含维生素、矿物质与糖分，为夏季重要的消暑果品，在我国栽培普遍。我国西瓜种植面积占世界总面积的55%以上，总产量占世界总产量的70%以上。西瓜在世界园艺业中占有重要地位，其生产规模居第5位，仅次于葡萄、香蕉、柑橘和苹果。

西瓜生长速度快，要及时供应养分，但在西瓜种植中，瓜农在浇水施肥上多采用大水畦灌冲施肥的传统方法，盲目、超量施肥和灌溉，不但造成了肥水的大量浪费，且易使土壤污染和酸化板结，降低肥料利用率，严重影响西瓜的产量和品质。因此，西瓜栽培宜采用滴灌、渗灌、喷灌等水肥一体化灌溉技术，实现西瓜优质高产。此外，西瓜多种植于轻质或沙质土壤，而这类土壤往往保肥保水能力差。应用少量多次的水肥一体化技术也正好可解决这一问题。我国关于西瓜滴灌以及微喷的研究始于20世纪70年代，近年来，许多西瓜种植专业户采用滴喷灌水肥一体化技术，对西瓜进行水肥管理，收效显著，其应用也越来越受到人们的重视。

第六章 蔬菜水肥一体化技术的应用

一、滴灌网管的铺设

对于西瓜滴灌来说,铺设网管时,工作行中间铺设送水管,输水管道一般是三级式,即干管、支管和滴灌毛管,其中毛管滴头流量选用每小时2.8L,滴头间距为30cm。进水口处与抽水机水泵出水口相接,送水管在瓜行对应处安装1个带开关的四通接头,直通续接送水管,侧边分别各接1条滴管,使用90cm宽的地膜,每条膜内铺设一条滴灌毛管,相邻2条毛管间距2.6m,用量为390m/hm^2。滴管安装好后,每隔60cm用小竹片拱成半圆形卡过滴管,插稳在地上半圆顶距滴管充满水时距离0.5cm为宜,这样有利于覆盖薄膜后薄膜与滴管不紧贴、泥沙不堵塞滴管出水孔。最后覆盖地膜,春季为防寒要加小拱棚。

二、水分管理

西瓜适宜的灌溉模式以膜下喷水带最常用,也有膜下滴灌等。通常一行西瓜安装一条喷水带,孔口朝上,覆膜。沙土质地疏松,对水流量要求不高,但黏土上水流量要小,以防地表径流。喷水带的管径与喷水带的铺设长度有关,以整条管带的出水均匀度达到90%为宜。如采用间距40~50cm,流量1.5~3.0L/h,沙土选大流量滴头,黏土选小流量滴头。西瓜灌水按照"中间丰两头控"的原则实施,灌水量可采用灌水时间控制,并结合天气、植株长势等因素决定灌水时间的长短。西瓜全生育期共滴水9~10次,滴水量350~400m^3/hm^2。西瓜播种后滴水20m^3/hm^2。出苗水要充足,浸透播种带以确保与底墒相接,滴水量为45m^3/亩。出苗后根据土壤墒情蹲苗。在主蔓长至30~40cm时滴水1次,滴量为40m^3/hm^2。开花至果实膨大期共滴水6次,每隔5~7d滴水1次,每次滴水量为40m^3/亩,其中开花坐果期需水量较大,为45~50m^3/hm^2,瓜果膨大期保持在50m^3/hm^2。果实成熟期滴水1次,为保证西瓜的品质、风味要减少灌水量,根据瓜蔓长势保持在35~45m^3/hm^2。灌水时入沟流量以不漫垄为宜,果实采收前7~10d停止滴水。

三、施肥管理

西瓜生育期有发芽期、幼苗期、伸蔓期和结果期,不同生育期对氮、磷、钾的需求量和吸收比例不同。发芽期吸收量最少,仅占总吸肥量的0.01%,幼苗期约占总吸肥量的0.54%,伸蔓期吸肥量增多,约占总吸肥量的14.67%,结果期吸肥量最多,约占总吸肥量的85%。西瓜生育期对氮、磷、钾的吸收量以钾肥最多，氮肥第二，磷肥最少，其比例是 3.3：1：4.3（氮：磷：钾）。钾肥要充足，以提高西瓜含糖量。据研究，每生产 1 000kg 果实，需 N 4.6kg，P_2O_5 3.4kg，K_2O 3.4kg。西瓜整个生育期主要施用基肥、种肥和追肥，追肥主要是提苗肥、伸蔓肥和结果肥，全生育期共施肥 7 次，施肥量为 35kg/亩。苗期和开花期，随水滴施西瓜营养生长滴灌肥 75kg/hm²。坐瓜后，随水滴施西瓜生殖生长滴灌肥 3~4 次，每次 60kg/hm²，成熟期不滴施肥料。施肥要做到"足、精、巧"，即底肥要足，种肥要精，追肥要巧。

（1）提苗肥。在团棵期，即 5 叶期施肥，主要是促进生长，迅速伸蔓，扩大同化面积，为花芽分化奠定物质基础。西瓜苗期根系范围尚小，宜选择少量速效肥进行施用，以促进根系发育，加速地上部生长。幼苗期追肥：尿素 50g、复合肥 50g、水 50kg，或尿素 100g、水 50kg，或复合肥 125g、水 50kg，人粪尿可对 2 倍水淋苗。中苗期追肥：尿素 100g、复合肥 50g、水 50kg，或复合肥 100g、尿素 75g、水 50kg，麸肥要提前用水浸泡半个月，施肥浓度不要超过 0.8%，人粪尿中可加 1/3 水淋苗。

（2）伸蔓肥。应根据各地气候条件、植株长势，以促使茎叶快速生长但不徒长为原则，巧施伸蔓肥。伸蔓肥在植株"甩龙头"前后施用，于两棵瓜苗中间开一条深 10cm、宽 10cm、长 40cm 左右的追肥沟，每株施腐熟饼肥 100g 或腐熟农家肥等优质肥料 500g 左右。

（3）结果肥。在田间大部分植株已结果，幼瓜鸡蛋大小（落花后 7d）时即可追施结果肥，目的是促进果实膨大，保持植株的生长势。在植株一侧距根部 30~40cm 处开沟，结合浇水每亩施入复合肥 7.5~10kg、硫酸钾 5.0~7.5kg，或单用复合肥 10~15kg，也可用 0.5% 尿素加 0.1% 磷酸二氢钾进行叶面喷施。在采收前 5~7d，停止滴灌施肥，以保证内含色素和糖分的正常转化，以便贮运过程中不易烂瓜。

第七节　白菜水肥一体化技术应用

　　白菜原产于我国北方，俗称大白菜。引种南方，南北各地均有栽培。黄河流域一年可栽培春茬、夏茬和秋茬，东北地区可栽培春茬和秋茬。青藏高原和大兴安岭北部地区一年只栽培一茬，华南地区可以周年栽培。目前我国各地多以秋季栽培为主。也有利用设施进行越夏白菜栽培。

一、白菜灌溉类型

白菜多为露地栽培，水肥一体化技术应用较少。最适宜的灌溉方式为微喷灌。微喷灌可分为移动式喷灌、半固定式喷灌和固定式喷灌。在水源充足的地区（畦沟蓄水），采用船式喷灌机。一些农场采用滴灌管，滴头间距20~30cm，流量1.0~2.5L/h，用薄壁灌带。

微喷器的喷水直径一般为6m，为保持其灌溉的均匀性，应采用喷水区域圆周重叠法，可将微喷器安装间距设定为2.5m，使相邻的两个喷水器的喷水区域部分相重叠。

二、白菜水分管理

（一）白菜需水规律

大白菜叶片多，叶面角质层薄，水分蒸腾量很大。大白菜不同生长期对水分需求是不同的，幼苗期土壤持水量要求65%~80%（土壤湿润），莲座期是叶片生长最快的时期，但需水量较少，一般土壤含水量15%~18%即可。结球期是大白菜需水最多的时期，必须保持含水量19%~21%，不足时需要灌水。

（二）白菜水分管理

白菜发芽期和幼苗期需水量较少，但种子发芽出土需有充足水分；幼苗期根系弱而浅，天气干旱应及时浇水，保持地面湿润，以利幼苗吸收水分，防治地表温度过高灼伤根系。莲座期需水较多，掌握地面见干见湿，对莲座叶生长即促又控。结球期需水量最多，应适时浇水。结球后期则需控制浇水，以利储存。

（1）经验法。在生产实践中可凭经验判断土壤含水量。如壤土和沙壤土，用手紧握形成土团，再挤压时土团不易碎裂，说明土壤湿度大约在最大持水量的50%以上，一般不进行灌溉；如手捏松开后不能形成土团，轻轻挤压容易发生裂缝，证明水分含量少，及时灌溉。夏秋干旱时期还可根据天气情况决定灌水时期，一般连续高温干旱15d以上即需开始灌溉，秋冬干旱可延续20d以上再开始灌溉。

（2）张力计法。白菜为浅根性作物，绝大部分根系分布在3cm土层中。当用张力计检测水分时，一般可在菜园土层中埋1支张力计埋深20cm。土壤湿度保持在田间持水量的60%~80%，即土壤张力在10~20cm时有利于白菜生长。超过20cm表明土壤变干，要开始灌溉，张力计读数回零时为止。当用滴灌时，张力计埋在滴头的正下方。

(3) 适时浇水。白菜定植后及时灌足定根水，随后结合中耕培土 1~2 次之后根据天气情况适当灌水以保持土壤湿润。每次灌水时间为 3~4h，土壤湿润层 15cm，喷灌时间一般选在上午或下午。这时进行灌溉后地温能快速上升。喷水时间及间隔可根据蔬菜不同生长期和需水量来确定，大白菜从团棵到莲座期，可适当喷灌数次，莲座末期可适当控水数天。大白菜进入结球期后，需水分最多。因此，刚结束蹲苗就要喷水一次，喷灌时间为 3~4h，然后隔 2~3d 再接着喷灌第二次水以后，一般 5~6d 喷水一次，使土壤保持湿润，前期灌水的水量要比后期小才能保证高产。

三、白菜养分管理

（一）白菜需肥规律

大白菜生长迅速，产量很高，对养分需求较多，见表 6-1。每生产 1 000kg 大白菜需吸收氮 1.3~2.5kg，五氧化二磷 0.6~1.2kg，氧化钾 2.2~3.7kg。三要素大致比例为 2.5 : 1 : 3。由此可见，吸收的钾最多，其次是氮，磷量少。

表 6-1　不同产量水平下大白菜氮、磷、钾的吸收量

产量水平（kg/亩）	养分吸收量（kg/亩）		
	N	P	K
5 000	12.1	1.6	8.6
6 000	14.5	2.3	9.3
8 000	19.3	2.9	10.4
10 000	22.5	3.4	13.9

大白菜的养分需要量各生育期有明显差别。一般苗期（自播种起约 31d），养分吸收量较少，氮吸收占吸收总量的 5.1%~7.8%，磷吸收占总量的 3.2%~5.3%，钾吸收占总量的 3.6%~7.0%。进入莲座期（自播种为 31~50d），大白菜生长加快，养分吸收增长较快．氮吸收占总量的 27.5%~40.1%，磷吸收占总量的 29.1%~45.0%，钾吸收占总量的 34.6%~54.0%。结球初、中期（自播种为 50~69d）是生长最快养分吸收最多的时期，吸氮占总量的 30%~52%，吸磷占总量的 32%~51%，吸钾占总量的 44%~51%。结球后期至收获期（自播种 69~88d），养分吸收明显减少，吸氮占总量的 16%~24%，吸磷占总量的 15%~20%。而吸钾占总量已不足 10%。可见，大白菜需肥最多的时期是莲座期及结球初期，也是大白菜产量形成和优质管理的关键时期，要特别注意施肥。

（二）白菜养分管理

大白菜忌与十字花科蔬菜连作，前作最好是西瓜、黄瓜、豆类、葱蒜类和水稻。秋冬大白菜生长期长，产量高，需肥量大。在其营养长过程中，经历幼苗期、莲座期，后进入结球期形成产品。不同的生长期施肥方式不同。发芽期生长量小，生长速度快，所需的养分主要靠种子供给，不需施肥。幼苗期为保证迅速发苗，需追施提苗肥，用尿素或磷酸一铵等 30~60kg/hm²。莲座期生长速度较快，生长量逐渐增加，对养分和水分的口及收量增多，充分施肥浇水是保证莲座期叶健壮生长和丰产的关键。采用滴灌方式的话，遵循"见干浇水"的原则。一般莲座期需施入氮肥 100~150kg/hm²。结球期的生长量最大，此期氮肥可施入 175~300kg/hm²，硫酸钾及过磷酸钙各 100~150kg/hm²。要始终保持土壤的湿润，可每天进行滴灌 1h 或喷灌 30min。大白菜总的吸肥特点是：苗期吸收养分较少，氮、磷、钾的吸收量不足总吸收量的 1%；莲座期明显增多，约占总量的 30%；包心期吸收养分最多，约占总量的 70%。全生育期对钾吸收量最大，氮、钙次之，磷、镁吸收量较小。每生产 1 000kg 大白菜约需 N 1.82kg、P_2O_5 0.36kg、K_2O 2.82kg、CaO 1.6lkg、MgO 0.21kg，其比例为 5 : 1 : 7.8 : 4.5 : 0.6。

第八节 莴苣水肥一体化技术应用

一、莴苣的需水特性

莴苣为菊科莴苣属，1~2 年生草本植物，分为叶用和茎用两类。叶用莴苣又称春菜、生菜，其中的结球莴苣 20 世纪 80 年代后期在北京及一些沿海城市开始发展。叶用莴苣叶片多，叶面积较大，蒸腾量也大，消耗水分较多，需水分较多。莴苣生育期 65d 左右，每亩需水量为 215m³ 左右，平均每天需水量为 3.3m³。叶用莴苣在不同生育期对水分有不同的需求，种子发芽出土时，需要保持苗床土壤湿润，以利于种子发芽出土。幼苗期适当控制浇水，土壤保持见干见湿。土壤水分过多幼苗易徒长，土壤水分缺乏幼苗易老化。发棵期要适当蹲苗，促使根系生长。结球期要供应充足的水分，缺水易造成结球松散或不结球，同时造成植株体内莴苣素增多，产品苦味加重。结球后期浇水不能太多，防止发生裂球，并导致软腐病和菌核病的发生。

二、莴苣的需肥特性

莴苣根系吸收能力较弱，对氧气的需求量高，沙土或黏土栽培根系生长发育均不良。因此，生产中应该选择有机质含量较高、通透性较好的壤土或沙壤土栽培。莴苣喜欢微酸性土壤，适宜的土壤pH值为6左右，pH值大于7或小于5均不利于莴苣生长发育。

叶用莴苣生长期短，食用部分是叶片，对氮的需求量较大，整个生育期要求有充足的氮素供应，同时要配合施用磷、钾肥，结球期应充分供应钾素。莴苣生长初期，生长量和需肥量均较少，随着生长量的增加，对氮、磷、钾的需求量逐渐增加，特别是结球期需肥量迅速增加。莴苣全生育期对钾需求量最高，氮次之，磷最少。每生产1 000kg叶用莴苣，需纯氮2.5kg、磷1.2kg、钾4.5kg、钙0.66kg、镁0.3kg。

叶用莴苣在幼苗期缺氮，会抑制叶片的分化，使叶片数量减少；在莲座期和结球期缺氮，对产量影响最大。幼苗期缺磷，不但叶片数量少，而且植株矮小，产量降低。缺钾影响叶用莴苣结球，叶球松散，叶片轻，品质下降，产量减少。叶用莴苣还需要钙、镁、硼等中微量元素，缺钙常造成叶用莴苣心叶边缘干枯，俗称干烧心，导致叶球腐烂。缺镁导致叶片失绿。生产中微量元素可以通过叶面施肥补充，一般在叶用莴苣莲座期补施，后期喷施效果较差。

三、莴苣水肥一体化栽培技术

（一）品种选择

日光温室秋冬茬结球莴苣宜选用优质、抗病、适应性强的"五湖"结球生菜、大湖659、圣利纳、美国PS等品种。

（二）茬口安排

华北地区日光温室结球莴苣一般选择秋冬茬栽培，多采用育苗移栽，一般苗龄30~35d，定植后60~65d即可收获。9月中下旬播种育苗，10月下旬定植，翌年元旦前后供应市场。

（三）播种育苗

(1) 苗床准备。苗床土壤宜选择保水、保肥性能好的肥沃沙壤土，深翻后充分暴晒。每10m² 苗床施优质腐熟农家肥10~15kg、硫酸铵0.3kg、过磷酸钙0.5kg、氯化钾0.2kg，各种肥料与土壤充分混匀，整平床面，准备播种。

(2) 浸种催芽。将种子用纱布包裹，置于20℃清水中浸种3h，取出后在

15~20℃条件下催芽2~3d，种子露出白色芽点后即可播种。

（3）播种多采用撒播法，每亩温室需育苗畦50m²，用种量为30g。为使播种均匀，播种时种子中可拌入适量细土粒，播后覆土厚0.5~1cm。低温季节可在苗床上覆盖地膜，以提温保湿。播种后苗床温度保持20~25℃，出苗后白天温度保持18~20℃、夜间8~10℃，幼苗长至4叶1心时即可定植。

（四）定植

定植前结合整地每亩施腐熟有机肥4 000kg、三元复合肥30~40kg，深翻25cm。作畦，畦宽1m左右，多采用平畦栽培。栽培密度依品种而定，一般株行距为30cm×40cm。早熟品种，植株开展度小，可适当密植，株行距以30cm×30cm为宜；中晚熟品种植株开展度较大，行株距以35cm×40cm为宜。带土坨定植，定植深度以土坨与地面平齐为宜，栽后及时浇水，促使迅速缓苗。

（五）定植后的环境管理

缓苗期，加强保温，密闭棚室不通风，白天温度保持20~25℃、夜间10℃以上，10cm地温15℃以上；莲座初期，适当控水，加大通风，白天温度保持15~18℃、夜间10~15℃；结球前期，加大水肥管理，及时中耕除草，发现干旱及时浇水，一定要保持土壤含水均衡，否则易裂球影响品质；结球后期，加大通风，保持土壤湿润，发现干旱及时浇水，防止裂球。

（六）主要病虫害防治

结球生菜病害主要有软腐病、霜霉病、菌核病和灰霉病等，害虫主要有蚜虫、地老虎等。

（1）软腐病。发病初期可用72%硫酸链霉素可溶性粉剂200~250mg/L溶液喷雾防治，每隔7~10d喷1次，连喷2~3次。

（2）霜霉病。可用25%甲霜·锰锌可湿性粉剂或25%甲霜灵可湿性粉剂600倍液，或75%百菌清可湿性粉剂或65%代森锌可湿性粉剂600倍液喷雾防治，7~10d喷1次，连喷2~3次。

（3）菌核病。可用50%多菌灵可湿性粉剂600倍液，或50%腐霉利可湿性粉剂或40%菌核净可湿性粉剂1 000~1 500倍液喷雾防治，隔7~10d喷1次，连喷2~3次。

（4）灰霉病。可用50%腐霉利可湿性粉剂1 000~1 500倍液，或50%硫菌灵可湿性粉剂或50%多菌灵可湿性粉剂500~600倍液喷雾防治，每7~10d喷1次，连喷2~3次。

（5）蚜虫。可用10%吡虫啉可湿性粉剂1 500倍液喷雾防治。

参考文献

曹新芳，姜召涛．2010．现代苹果高效栽培实用新技术［M］．北京：中国农业出版社．

陈清，陈宏坤．2016．水溶肥料生产与施用［M］．北京：中国农业出版社．

程季珍，巫东堂，蓝创业．2013．设施无公害蔬菜施肥灌溉技术［M］．北京：中国农业出版社．

程季珍，巫东堂，姚小平．2010．蔬菜灌溉施肥技术问答［M］．北京：金盾出版社．

郭彦彪，邓兰生，张承林．2007．设施灌溉技术［M］．北京：化学工业出版社．

何龙，何勇．2006．微灌工程技术与装备［M］．北京：中国农业科学技术出版社．

李保明．2016．水肥一体化实用技术［M］．北京：中国农业出版社．

李建明．2015．温室番茄甜瓜水肥技术研究［M］．北京：中国科学技术出版社．

李久生，张建君．薛克宗．2005．滴灌施肥灌溉原理与应用［M］．北京：中国农业科学技术出版社．

李俊良，金圣爱，陈清，等．2008．蔬菜灌溉施肥新技术［M］．北京：化学工业出版社．

刘学卿．2018．图解设施樱桃高产栽培与病虫害防治［M］．北京：化学工业出版社．

罗金耀．2003．节水灌溉理论与技术［M］．武昌：武汉大学出版社．

彭世琪，崔勇，李涛．2008．微灌施肥农户操作手册［M］．北京：中国农业出版社．

宋志伟，等．2016．设施蔬菜测土配方与营养套餐施肥技术［M］．北京：中国农业出版社．

宋志伟，等．2016．蔬菜测土配方与营养套餐施肥技术［M］．北京：中国农业出版社.

宋志伟，邓忠.2018.果树水肥一体化实用技术［M］.北京：化学工业出版社.

宋志伟，翟国亮.2018.蔬菜水肥一体化实用技术［M］.北京：化学工业出版社.

隋好林，王淑芬.2015.设施蔬菜栽培水肥一体化技术［M］.北京：金盾出版社.

王克武，周继华.2011.农业节水与灌溉施肥［M］.北京：中国农业出版社.

王少敏，刘涛.2010.苹果、梨、桃、葡萄套袋栽培技术［M］.北京：中国农业出版社.

王玉宝.2018.甜樱桃栽培彩色精解［M］.北京：中国农业出版社.

吴普特，牛文全.2001.节水灌溉与自动控制技术［M］.北京：化学工业出版社.

徐坚，高春娟.2014.水肥一体化实用技术［M］.北京：中国农业出版社.

徐卫红.2014.水肥一体化实用新技术［M］.北京：化学工业出版社.

严以绥.2003.膜下滴灌系统规划设计与应用［M］.北京：中国农业出版社.

杨国，王廷忠，王立平.2016.果树栽培实用技术［M］.北京：中国农业科学技术出版社.

张承林，邓兰生.2012.水肥一体化技术［M］.北京：中国农业出版社.

张承林，郭彦彪.2006.灌溉施肥技术［M］.北京：化学工业出版社.

张洪昌，李星林，王顺利.2014.蔬菜灌溉施肥技术手册［M］.北京：中国农业出版社.

赵习平.2017.杏实用栽培技术［M］.北京：中国科学技术出版社.

Chapter I Introduction

Integrated water-fertilizer technology is a new agricultural technology which integrates irrigation and fertilization. It makes use of micro-irrigation system to timely, appropriately and accurately provide the crops with fertilizer and water, according to the rule of crop water demand, fertilizer requirement, soil moisture and nutrient conditions. The investment of the integrated water-fertilizer system (including pipeline, fertilizer pool, power equipment, etc.) is about 1 000 yuan per mu, which can be used for about 5 years, saving 50%~70% of fertilizer, compared with conventional fertilization, and increasing the yield by more than 30%. At the same time, the problem of water pollution caused by excessive fertilization can be greatly reduced. The following is a detailed description of the development and application of the integrated water-fertilizer technology.

Section I Overview of the Integrated Water-Fertilizer Technology

I . Concepts

The integrated water-fertilizer is a new agricultural technology which integrates irrigation and fertilization to realize simultaneous control of water and fertilizer, which is also called "coupling of water and fertilizer", "fertilization with water", "irrigation and fertilization" and so on. In a narrow sense, it is to dissolve fertilizer in irrigation water, and to be brought to every crop in the field by irrigation pipes. In a broad sense, water and fertilizer supply crop needs at the same time. The integration of water and fertilizer is to mix the soluble solid or liquid fertilizers with irrigation water, and to supply water and fertilizer through controllable plumbing system, with the aid of pressure system (or natural topography), according to different soil environment and nutrient content, the characteristics of different crops' need of fertilizer, and the law of water and fertilizer requirement in different growing period. After the water and fertilizer are mixed, drip irrigation was formed through pipes and drips, which infiltrated the crop root region for the growth and development in a uniform, regular and quantitative way, to make sure the soil always kept loose and having suitable amount of water and fertilizer. In the 1990s in China, the irrigation is mainly canal irrigation, which is the simplest and less requirement for fertilizer, but this type of irrigation does not avail for the requirement of water saving. But, drip irrigation is the most accurate water supply according to crop demand, fertilizer demand and root distribution, and it is not limited by wind and other external conditions; sprinkler irrigation is relatively not widely adapted compared with drip irrigation. So the narrow sense of the integrated water-fertilizer technology is also known as drip irrigation fertilization.

In recent years, with the global warming and drought aggravation, the shortage of water resources has become the bottleneck of the sustainable development of our national economy. At present, the annual total water resources in China is about 2.81 trillion m^3, accounting for only 6% of the world's total water resources, and the per capita water resources is only $2\,800m^3$, which is 1/4 of the world's. At the same time, there is a problem of uneven distribution of water resources in China. The northern area north of Huaihe River has 60% of the cultivated land, but only 15% of the water resources. Agriculture is an industry consuming the most water in China, the total amount of water used being 400 billion m^3, accounting for 70% of the total water consumption. Among them, the water consumption of farmland irrigation is 360 billion to 380 billion m^3, accounting for 90% to 95% of agricultural water consumption. The average annual water shortage of agricultural irrigation is more than 30 billion m^3, and the average annual dry farmland area is 2.1467 million hm^2, accounting for 1/5 of the country's cultivated land, and the drought farmland area is 8.6 million hm^2, which lead to the reduction of 10 billion to 15 billion kg per year due to the drought. Moreover, under the condition of water shortage of, the problem of waste of water resources is very serious. The traditional irrigation method in our country is mainly canal irrigation, and the traditional soil canal results in large water leakage, which generally accounts for 50% to 60% of the total water flow. Some poor soil canal can

Chapter Ⅰ　Introduction

reach as high as 70% of water loss. Compared with the traditional irrigation methods, 20% to 30% of water can be saved by pipelines and seepage preventing canals; 50% can be saved by sprinkler irrigation; 60% to 70% by micro-irrigation, over 80% by drip irrigation under mulch, and over 95% by subsurface drip irrigation. At the same time, the utilization rate of irrigation water in our country is a little low, only about 43%, with the grain productivity per cubic meter of water being about 1kg, while the utilization rate of irrigation water in developed countries is 70% and 80%, with the grain productivity being more than 2kg. Due to the unreasonable irrigation ways and low efficiency of irrigation water use, the problem of the shortage of water resources has been aggravated more seriously. According to the forecasts of the Ministry of Water Resources, the Chinese Academy of Engineering and other departments, the use of China's agricultural water must maintain zero or negative growth in order to ensure the safety of water use and ecological security. And one of the important ways to alleviate the contradiction between supply and demand of water resources is to develop water-saving irrigation. We can not only save irrigation water, but also improve land use efficiency, reduce labor intensity, and improve crop yield and quality. China is a big country in fertilizer production, and a big one of fertilizer consumption. According to the data of the International Association of Fertilizer Industry and the statistical data in China, the use of fertilizer in China accounted for about 35% of the global consumption in 2007, and use is still growing at 3.5% annually. The utilization rate of fertilizer in the current season is low, as a result of unscientific fertilization technology and unqualified fertilizer products, the utilization rate of nitrogenous fertilizer, phosphatic fertilizer and potassic fertilizer in China is 15% to 35%, 10% to 20% and 35% to 50% respectively, which is lower than that of Japan, the United States, Britain, Israel and other developed countries. The large amount use and unreasonable application of fertilizer resulted in the change of soil structure, the decline of fertility, the aggravation of heavy metal pollution, and the aggravation of salinization and alkalization, especially the aggravation of water quality pollution of surface runoff, which results in a series of hazards such as the eutrophication of water body, groundwater pollution, and produce quality, etc. Therefore, reducing the use of fertilizer, rational fertilization, and raising the utilization rate of fertilizer have become an important issue for the sustainable development of agriculture and the safety of agricultural produces in our country.

II. Characteristics

Traditional irrigation methods include flood irrigation, furrow irrigation, border irrigation and so on. Fertilizer can be mixed in irrigation water and applied with irrigation water. Flood irrigation means not to make any ridges in the field, but to allow to flood the ground, through gravity infiltration into the soil, which is a more primitive irrigation method. This irrigation method has poor uniformity of irrigation, leading to great waste of water and serious pollution of groundwater by fertilizer washing. Border irrigation is to divide the field into a series of rectangular small ridges. When irrigating the field, the water enters the border and moves along the border, gradually wetting the soil by gravity. Border irrigation is simple and does not need additional irrigation facilities, but irrigation consumption is large, water waste is serious, and it is easy to cause soil consolidation and lead to the spread of crop diseases. Furrow irrigation is to excavate the irrigation ditch between crop rows, and the water enters into the ditch to moisten the soil through capillary action. Compared with border irrigation, furrow irrigation does not destroy soil structure near the root of crop, and has less damage to soil aggregate structure. It does not lead to soil consolidation, can keep loose soil and can reduce soil evaporation loss.

1. Increasing the utilization rate of irrigation water

Drip irrigation drips water oneby one into the soil, so the surface of the field does not have runoff. It changes from irrigating the land to irrigating the crops, thus reduces the evaporation of water between the crops. At the same time, by controlling irrigation quantity, the deep leakage of soil water is very little, reducing the loss of ineffective field water. In addition, the drip irrigation water system enters into a whole closed water transportation system starting from the water sources. The water is transported to the crop roots through multi-stage pipeline, and the water efficiency is high, reducing the irrigation quantity.

2. Improving the utilization rate of fertilizer

Water and fertilizer are directly transported to the most developed parts ofthe crop roots, which can fully ensure the rapid absorption of nutrients by crop roots. As for drip irrigation, fertilization efficiency is high because the wetting range is limited to areas where roots are concentrated, thus saving fertilizer.

Chapter Ⅰ Introduction

3. Saving labour

The traditional method of irrigation and fertilization is to dig holes or shallow grooves for each fertilization, and then irrigate it after fertilization. A large amount of labor can be saved by using water and fertilizer integration technology to realize synchronous management of water and fertilizer.

4. Controlling the quantity of fertilization conveniently, flexibly and accurately.

According to the law of crop fertilizer, fertilization can be targeted, achieving precise fertilization with the need and quantity.

5. Benefiting the protection of the environment

Unreasonable fertilization results inthe waste of fertilizer. A large amount of fertilizer can not be absorbed and utilized by crops, which lead to great waste as well as environmental pollution. The integrated water-fertilizer technology can avoid leaching fertilizer into deep soil by controlling irrigation depth, so as to avoid soil and groundwater pollution. Moreover, the integration of water and fertilizer can effectively utilize and develop the marginal land such as hilly land, mountainous area, sandy and stony area, and mild saline-alkali land, etc.

6. Benefiting the application of trace Elements

Metal trace elements are usually used inchelated form, which is expensive, and drip irrigation system can improve fertilizer utilization by supplying accurately.

7. Limitations of the integrated water-fertilizer technology

It is difficult to popularize rapidly and widely because this technique is a facility fertilization with much investment in the early stage and a higher requirement for the solubility of the fertilizer.

Section Ⅱ Research Progress of the Integrated Water-Fertilizer Technology

Ⅰ. The History of the Development of Integrated Water-Fertilizer Technology

Integratedwater-fertilizer technology is the crystallization of human wisdom and

◐ Modern Practical Integrated Water-Fertilizer Technology on Agriculture (Chinese-English)

continuous development of productivity. Its development has a long history, and the inspiration of integrated water – fertilizer technology originated from soilless cultivation technology. As early as the 18th century, John Woodward, a British scientist, compounded the first nutrient solution for hydroponic culture by using soil extracts. Later, integrated water-fertilizer technology went through three stages of development.

The first stage is from the end of the 18th century to the beginning of the 20th century, which is a stage of the cultivation of nutrient solution and the stage of soilless culture, the earliest stage of the integrated water-fertilizer technology. The cultivation of nutrient solution initially refers to the water culture without any fixed root system. In the middle of the 19th century, Wignen and Pollsloff made an important experiment, successfully culturing plants with distilled water and inorganic salt, which directly verified that the plant needed salt in the course of growth. The most representative figure in this period was Van Liebig because he found that C needed in plant growth was the CO_2 from the air and H and O come from NH_3、NO_3^-, and some of the minerals that are essential to plants are supplied by the soil environment. Van Liebig's discovery directly overturned the popular theory of humus nutrition theory at that time. His theory frame of mineral nutritious was formed, and his academic theory is still applied to "modern nutrition farming", which lays a solid foundation for the development of modern agriculture.

The second stage is from mid-19th century to mid-20th century, which is a stage of commercialized production of soilless cultivation and the preliminary formation of water-fertilizer integration technology. World War II accelerated the development of soilless cultivation, because the United States established large-scale soilless farming farms at various military bases in order to provide a large number of fresh vegetables to soldiers. Soilless cultivation technology is becoming mature and gradually commercialized. The commercial production of soilless cultivation began in the Netherlands, Italy, Britain, Germany, France, Spain, Israel and other countries. Later, countries and areas began to promote soilless cultivation, such as Mexico, Kuwait and Central America, South America, the Sahara Desert, and other regions with poor land and scarce water resources.

In 1950, the method of irrigation after mixing fertilizer and water was invented, in which fertilizer was dissolved in irrigated water for surface irrigation, flood irrigation and ditch irrigation, which was the embryonic form of water-fertilizer integration tech-

Chapter Ⅰ Introduction

nology. Nitrogen fertilizer was the most commonly used in the early days, and the ammonia water and ammonium nitrate were the two most common types of nitrogen fertilizer. Due to the volatile nature of nitrogen fertilizer in the soil and the low utilization rate of water caused by flood irrigation, the absorption rate of nitrogen fertilizer in crop growth was very low. The development of plastic pipes and plastic containers promoted the development of integrated water-fertilizer technology, by which the nutrient solution with reasonable formula was transported to plants. So, pipe irrigation system can greatly improve the utilization rate of fertilizer, but irrigation through pipe will increase the quantity of using fertilizer, which further promoted the design and development of mixed irrigation of water pump and fertilizer for fine supply of nutrients.

The third stage is from the middle of the 20th century to date, which is a stage of rapid development of integrated water-fertilizer technology. In the 1950s, Farmers in-Hatzerim Kibuz in the Negev desert in Israel stumbled upon the fact that crops near the leaking water pipes grew better than those elsewhere. Later, experiments concluded that drip irrigation could effectively reduce evaporation of water. It improved the utilization rate of irrigation water and was the most effective method to control water fertilizer and pesticide. With the maturity of the technology, the Israeli government thus increased the promotion of drip irrigation technology, and the famous Netafim drip irrigation company was established in 1964. From backward agriculture country to modern industrial countries, Israel is benefited mainly from the drip irrigation technology. Compared with sprinkler irrigation and furrow irrigation, the yield of tomato and cucumber by drip irrigation was increased by one time and two times respectively. Since the application of drip irrigation technology in Israel, the country's agricultural output has increased fivefold, but the agricultural water consumption has not increased.

The first-generation drip irrigation system by the Netafim company used a flow meter to control the one-way water flow in plastic pipes; the second generation used high-pressure equipment to control water flow; and the third and fourth generation products began to work with computers. Since the 1960s, water-fertilizer integration technology has been popularized in Israel. About two hundred thousand hm^2 of the four hundred and thirty thousand hm^2 cultivated land in the country used pressurized irrigation system. Due to the success of pipeline and drip irrigation technology, the irrigation area increased from 1.65 billion m^2 to 2.2 billion-2.5 billion m^2, and the cultivated land increased from 1.65 billion m² to 4.4 billion m². it's alleged that the drip irrigation technology in Israel has developed to the sixth generation. Fruit trees, flowers and greenhouse crops planted in greenhouses all use integrated water-fertilizer fertilization and drip irrigation technology, while only a small proportion of open-air vegetables and crops used integrated water-fertilizer technology, and the coverage rate is not very high. Whether the integrated technique of water and fertilizer can be used or not depends on the properties of soil and the application of basic fertilizer. With the development of technology, water-fertilizer integrated technology has been used in micro-irrigation and micro-sprinkler irrigation system, and it has also played a significant role in the growth of crops. With the change of sprinkler irrigation system from mobile to fixed type, the integrated water-fertilizer technology was successfully used in the sprinkler irrigation system. In the early 1980s, the automatic propulsion mechanical irrigation system also began to promote the water-fertilizer integrated technology, which effectively improved the growth of crops and promoted the development of agricultural technology.

  The integrated water-fertilizer technology can not only distribute water reasonably, but also ensure the sufficient nutrient of crops. The wetted soil volume is only a small part of the farming layer, especially when it comes to sandy soil, which is only a small part of the tillage layer. In the initial stage of the integration of water and fertilizer, there are two ways to fertilize crops with irrigation system. The first type is to inject fertilizer into irrigation system by spray pump; The second is to pour water directly into containers with water and solid fertilizers and then introduce them into irrigation systems for fertilization. These two methods are simple and convenient, but their accuracy is not enough, and irrigation of water and fertilizer is not well-proportioned. In the 1970s, the innovation and development of chemical technology produced a new form of fertilizer, that is liquid fertilizer. The emergence of liquid fertilizer accelerated the research and development of hydraulic drive pump. Membrane pump is the first hydraulic drive pump developed in the early stage. First, fertilizer is poured into a large open vessel. After mixing evenly, the pump is used to inject the mixed water and fertilizer into the irrigation system. The pressure of this pump is high, about twice the pressure in the irrigation system. The second hydraulic pump is the piston pump, which uses the physical motion of the piston to absorb and inject fertilizer into the irrigation system. The ap-

Chapter Ⅰ Introduction

plication of these fertilizer pumps secures the simultaneous supply of water and fertilizer. At the same time, the Venturi fertilizing appliance also began to be used. The fertilizer can be evenly dissolved in irrigated water and the distribution of nutrition is relatively balanced. It is mainly used in nursery and potted greenhouse, whose application effectively solves the problem of poor accuracy of early fertilizer pump at low flow rate. With the development of computer technology, the automatic control of fertilizer consumption and flow rate has been realized gradually, and the control has become more and more accurate.

Ⅱ. The Development of Integrated Water – Fertilizer Technology in China

China's agricultural irrigation had a long history, but most of the traditional irrigation methods, such as flood irrigation, had a low utilization rate of water resources, which not only wasted a large amount of water resources, but also did not increase the yield of the crops. In the 1970s in China, the integrated water-fertilizer technology was introduced into the field of planting. Over the past 30 years, with the popularization and application of micro – irrigation technology, the integrated water – fertilizer technology has been increasing, and its development is generally divided into the following three stages.

The first stage is from 1974 to 1980, among which China began to introduce irrigation facilities and gradually realized the research and manufacture of domestic equipment. At the same time, a large number of micro-irrigation experiments were carried out, and relevant experimental data were collected. With the development of China's science and technology, the first generation of drip irrigation equipment in China was successfully developed and produced in 1980.

The second stage is from 1981 to 1996, among which the foreign advanced production technology was successfully introduced to help the domestic equipment to realize the large-scaleproduction. The large-scale production of irrigation equipment promoted the micro – technology in the field of planting, from the initial small module to the large-scale promotion. The micro-irrigation experiment has obtained the satisfactory answer, and the research on the content of the irrigation and fertilization is being carried out in the some micro-irrigation experiment.

◆ Modern Practical Integrated Water-Fertilizer Technology on Agriculture (Chinese-English)

    The third stage is from 1996 to 2019, among which the study of irrigation technology has gained more attention and recognition in the field of agriculture, and a large number of research, development and training were carried out in China. A large number of agricultural technical talents have been cultivated, which lays a foundation for the popularization of integrated water-fertilizer technology in China, and realizes the implementation of this technology on large-scale agricultural lands as soon as possible.

    Since the mid-1990s, micro-irrigation technology and integrated water-fertilizer technology have been popularized rapidly in China. With the development of agricultural technology, the integrated water-fertilizer technology in China has gradually matured, and has evolved from the small areas in the past to the large areas. And the integrated water-fertilizer technology has also begun to be applied in the arid areas of northwest China, and it has also been successfully applied in the cold areas of northeast China and tropical areas of southeast Asia. This technology covers a wide range of techniques, including facilities cultivation and soilless cultivation models. More and more crop cultivation has been implemented with integrated water-fertilizer technologies, such as vegetables, seedlings, and cash crops, etc. In the economically developed areas of our country, where science and technology are advanced and the economic level is relatively high, a good condition for the popularization of integrated water-fertilizer technology is available, promoting the improvement and excellence of this technology. Gradually this improved the large-scale water conservancy facilities, and designed a mature automation control system, a large-scale model projects emerging. In some areas, gravity drip irrigation was carried out in the mountainous areas; in the semi-arid and arid areas of the northwest China, they were allocated with solar greenhouse rain irrigation systems, drip irrigation in kilns, drip irrigation in hanging bottles of melon cultivation, and fertilization by drip irrigation in hanging bottles of melon cultivation; in South China, technique of injecting Organic Fertilizer by Irrigation was applied. All these techniques make the irrigation and fertilization technology become more and more abundant and perfect. There is little precipitation in Northwest China and the soil environment is relatively dry. So the introduction of integrated water-fertilizer technology has greatly improved the local planting benefit and changed the traditional planting mode. In the field crop planting, the drip irrigation of cotton under the mulch in Xinjiang is the most successful model of fertilization and irrigation in northwest Chi-

Chapter Ⅰ Introduction

na. Xinjiang is located in the plateau with large wind, much sand and scarce water resources, which is not suitable for field crop cultivation. The emergence of integrated water-fertilizer technology has improved the planting conditions in Xinjiang and improved the economic benefits of the local residents. The new planting mode has brought more planting options. In 1996, in order to improve the local planting conditions, Xinjiang has successfully introduced the drip irrigation technology. Through the research and experiment of three years, the drip irrigation technology is becoming more and more mature, and a set of low-cost drip irrigation belt suitable for large-area farmland planting has been successfully formed. Only irrigation technology is not enough for the crop cultivation, so in 1998, the techniques of fertilization and cultivation management were also studied, for the combined technology can improve the benefits. The use of high-power tractors in crop cultivation can complete the work at one time, such as ditching, sowing, fertilizing, coating, etc. In the later period of cotton growth, it is necessary to make full use of the good irrigation facilities to reasonably fertilize and irrigate.

With the deep theoretical study and application of the irrigation and fertilization, the integrated water-fertilizer technology in China has been developed from the backward in the 1980 to the middle stage. With the improvement of the nationaleconomy, science and technology, our research on micro-irrigation technology has reached the international advanced level, and some of the facilities and systems have been in the international leading ranks. These advanced technical equipment resulted from the hard struggle of the great technicians. In 1982, China joined the International Committee on Irrigation and Drainage (ICID) and became one of the members of the world's micro-irrigation organization. Our country strengthened the exchange of international technology, and paid attention to the practice of the managing, designing & planning of micro-irrigation technology. We cultivated a large number of talents who can manage and design the engineering with integrated water-fertilizer technology.

Water-fertilizer integration is a modern agricultural technology of saving water and saving fertilizer. There is no water shortage in many parts of Europe, but the integrated water-fertilizer technology is still applied, for they considered the other advantages of this technology, especially the protection effect on the environment. The serious shortage of water resources in China and the development of water-saving agriculture are challenging. At the same time, our country is a big country consuming fertilizer, and the amount of fertilizer in the unit area is in the forefront of the world. The fertilizer production consumes a large amount of energy, and all the measures to save the fertilizer are the ones to save energy. According to some experts, the large-area popularization and application of the integrated water-fertilizer technology is not only saving water and fertilizer itself, but leads to a far-reaching revolution of our country's agriculture from the traditional to the modernization.

However, from the angle of technology application, the integrated water-fertilizer technology in China is developing slowly. Firstly, attention is only paid to the water-saving irrigation equipment, but there are not much research results of the application of the theory; secondly, the management level of the irrigation and fertilization system in China is low, the training and popularizing are not in place, and the application of the grass-roots agricultural technicians and farmers to the integrated water-fertilizer technology is not well-mastered. Furthermore, the proportion of the field applied with integrated water-fertilizer technology is small and the depth is not enough. And finally, the quality of some micro-irrigation equipment, especially the head supporting equipment, still has a large gap compared with the foreign products of the same kind.

Section III The Applicable Prospect of the Integrated Water-Fertilizer Technology

The integrated water-fertilizer technology is one which combines fertilization with drip irrigation to control water and fertilizer synchronously. It uses irrigation facilities to supply nutrients and water for crops with the most accurate amount, so as to save water resources better. Therefore, the integrated water-fertilizer technology will be greatly supported and popularized by the country, and the development prospect is very promising.

I. The integrated water-fertilizer technology is developing towards a scientific direction

The integrated water-fertilizer technology is developing towards precision agriculture and formula fertilization. The land of our country is very large, which leads

Chapter Ⅰ Introduction

to the big difference of the climate environment in differentregion. For example, the annual rainfall in northwest China is very little, and the soil is mainly sandy, while the southwest China is in a subtropical climate with perennial rainfall, and the soil is moist and fertile. Therefore, the future development of integrated water-fertilizer technology in China should be determined comprehensively according to the different regions, different crop species and soils. When mixing the fertilizer with this integrated technology, it is necessary to consider comprehensively the difference between regions, choose different fertilizers for different soil. According to the characteristics of soil fertility and the rule of fertilizer requirement which were obtained after tests, a specific design formula was made. Finally, the suitable fertilizer was selected for irrigation and fertilization.

In addition, the future of fertilizer selection will also develop scientifically. The fertilizer will select drip irrigation special fertilizer and liquid fertilizer according to the characteristics of integrated water-fertilizer irrigation system.

Ⅱ. The integrated water-fertilizer technology is developing to great scale and industrialization

For a long time in the future, the market potential of integrated water-fertilizer technology in China lies mainly in the following aspects: to establish modern agricultural model zone, introducing advanced integrated water-fertilizer technology and equipment as production models by the government, and letting farmers follow; to set up a batch of urban agriculture, such as leisure agriculture, sightseeing orchard and so on, to further promote the application and development of integration technology; trade groups invest in agriculture, carry out large-scale production, establish special agricultural products base, develop export trade, and process agricultural products or serve the catering industry of the city, etc; to improve the urban environment, parks, sports grounds, and the development of green lawns in residential areas are also related with the integrated water-fertilizer technology; to increase the farmers' income and technical training on farmers to enable them willingly use irrigation and fertilization integrated techniques and equipment to conserve water, soil and labour resources and to obtain maximum agricultural economic benefits.

Chapter II The Main Equipment and System of the Integrated Water-Fertilizer Technology

A set of integrated water-fertilizer equipment includes the head hub, water distribution network and emitters. Integrated water-fertilizer technology is realized by means of irrigation system. In order to control the quantity and concentration of fertilization reasonably, proper irrigation equipment and devices must be selected. Commonly used irrigation facilities are sprinkler irrigation, micro-sprinkler irrigation and drip irrigation, the latter two being called micro-irrigation. The following is a brief introduction to each equipment and system.

Section I The Head Hub of the Integrated Water-Fertilizer Technology

I. Pressure Equipment

1. Pump room

The pressure equipment is usually installed in the pump house. The type and specification are determined according to the irrigation design requirements. In addition to the deep well water supply, it is needed to build a corresponding area of the pump house, which can provide a certain space. Water pump house's structure is generally brick-concrete, and it can also be mobile house, whose main function is to prevent rain and theft, being convenient to install irrigation and fertilization equipment.

A water pool is required, with an inlet that can be connected to the irrigated water source, which requires the installation of a trash rack (preferably galvanized or stainless steel) to prevent floating matter from entering the pool. The bottom of the pool

Chapter Ⅱ The Main Equipment and System of the Integrated Water-Fertilizer Technology

is slightly deep, about 0.5 meter deeper than the outside of the pool. The bottom of the pool is paved with steel mesh and hardened with concrete. The four sides are concreted with bricks, top covered with reserved pump suction and maintenance hole (adding a small cover). We should regularly clean the bottom of the pool, ensure the inlet can not be blocked, otherwise it will affect the operation of the whole system.

2. Pumps

(1) Selection of water pump.

The selection of pump plays an important role in the normal operation of the whole irrigation system. The principle of pump selection is as follows: under the design lift of delivery, the flow rate should meet the requirement of designed flow; in the long run, the working efficiency of the pump is high, and it often runs best on the right side of the highest efficiency point; it is easy to operate and manage.

(2) Centrifugal self-priming pump.

The self-priming pump has certain self-priming ability, which can make the pump easily start up and maintain the normal state in case that the pump can not absorb water. The present products of China's self-priming pump is basically self-priming centrifugal pump. According to the different ways of self-priming, the pumps can be divided into external self-priming pump, internal self-priming pump and vacuum-assisted self-priming pump with power provided by the pump itself. According to the liquid and material transferred, it can be divided into sewage self-priming pump, clean water self-priming pump, corrosion-resistant self-priming pump, stainless steel self-priming pump andand so. Its main parts are pump body, pump cover, impeller, axle, bearing and so on. Self-priming pump has a unique scientific structure; the pump is equipped with the suction chamber, the storage chamber, the return valve, the gas-liquid separation chamber and the pipeline do not need to be installed with low valves. Before working, the only necessity is to have some quantitative liquid in the pump body, so the pipeline system is simplified and the working conditions are improved. There is a vortex channel in the pump body, a large air-water separation chamber around the outer part of the channel, and a seat angle in the lower part of the pump body to fix the pump.

ZW horizontal centrifugal self-priming pump is a low-lift, large-flow sewage pump. Its motor and pump body are connected with a shaft, which can be separated from the pump, easy to maintain. The motor can be removed in the season without water or in the season with typhoonflood, and can be reconnected when in use. Its concentric degree is good, with little noise, mainly used for farms with single crops with unified fertilization time and with the square of more than 50 mu to 100 mu. When the flow of a rotation irrigation area is $50 \sim 70 m^3$, the efficiency of the pump can better. The pump has the function of strong self-priming, full lift, no overload, no need to add water after adding water once, as shown in Fig. 2-1.

Fig. 2-1　Self-priming pump

The working principle of the water pump is as follows: the pump's impeller is drove by the motor, drawing the irrigation water up from the sink, sending the water through the filter of the outlet to every part of the field. And there installs a three-way outlet at the intake pipe 20cm or so far with the pump. It connects with the valve and small filter, then connected to the wire hose for absorbing the fertilizer.

When the pump is absorbing water, the inside of the inlet pipe is in a negative pressure state. At this time, the suction tube is placed in the fertilizer liquid in the bucket (which can be saturated fertilizer liquid), and the valve of the fertilizer pipe is opened. The fertilizer is sucked into the pump and pipe and mixed with clear water, transported to the root of the crops for fertilization. Compared with the electric injection pump, there is no need for power supply, and the pressure is not too high or too low lest it should supply excessive fertilizer or no fertilizer.

(3) Submersible pumps.

Submersible pumps are different from ordinary pumps in that they work underwater, and ordinary pumps mostly work on the ground. The advantage of using

Chapter Ⅱ The Main Equipment and System of the Integrated Water-Fertilizer Technology

submersible pump as the head power system is that it is simple and practical, does not need to fill water, and does not cause such faults as air leakage and excess NPSH (net positive suction head). There are many types of pumps, and there is plenty of room for choice. The disadvantage is that the pump motor and wire are immersed in water, so when using the prevention of power leakage is the essential.

Submersible pumps for irrigation are usually in the DN32-DN150 (pipe diameter of 32 to 150mm, the same below) according to the outlet diameter. If one is not enough, two pumps can be installed in parallel, which can save electricity. Taking the QS series submersible pumps as an example, this part introduces the composition of the head part, as shown in Fig. 2-2. Pump parameters: QS submersible pump, specified flow rate $65m^3$, specified head 18m, power 7.5kW, voltage 380V, the inner diameter of the outlet DN80. The intake of the QS pump is in the middle and upper part of pump, when working, the bottom silt will not be sucked in. The cooling effect is good; the output power is large; the outlet is on the top of the pump body, which is easy to install.

Fig. 2-2 QS series submersible pumps

3. Negative pressure variable frequency water supply equipment

In general, greenhouse and field irrigation uses pumps to extract water directly from the water source for pressurized use. Regardless of the amount of water consumed, the pump operates at full load, so when the water consumption is small, the electricity consumption is the same as when the consumption is large. It is easy to cause a great waste.

The negative pressure variable frequency water supply equipment can automatically change the number of the pumps and the speed of the motor according to the instantaneous changing pressure and flow parameters in the water supply network, and realize the purpose of constant pressure and variable water supply, as shown in the Fig. 2-3. The power of the pump varies with the change of the water consumption, that is when the water consumption is large, the pump power automatically increases; when the consumption is small, the pump power automatically decreases, so the power can be saved by 50% to achieve high efficiency and energy saving.

Fig. 2-3 Variable frequency water supply system

The application of this equipment optimizes the way of crop water supply. LFBP-L series of variable frequency water supply equipment was optimized and upgraded on the original base, as shown in the Fig. 2-4. The third generation equipment already has the

Chapter Ⅱ The Main Equipment and System of the Integrated Water-Fertilizer Technology

function of automatically filling water, starting automatically, automatically closing the machine, automatically searching the fault. When the valve is open, the pressure sensing of the pipeline starts the pump through the PLC. After all the outlet valves are closed, the pump stops working to achieve the effect of energy saving.

Fig. 2-4 **LFBP-L series of variable frequency water supply system**

Negative pressure water supply equipment is composed of frequency conversion control cabinet, centrifugal pump (DL or ZW series), vacuum water diversion tank, remote pressure gauge, water diversion tube, bottom valve and so on. The power of the motor is usually 5.5 kW, 7.5 kW, 11 kW. The number of water pumps controlled by frequency conversion is from 2 to 4, which can be determined according to the actual water consumption in the field. We must pay attention to the supply voltage for the 11 kW series.

The control cabinet of the variable frequency constant (variable) pressure water supply equipment is a complete set of mechatronic equipment for closed-loop control of pumps in the water supply system, as shown in the Fig. 2-5. The equipment adopts industrial microcomputer variable program controller and digital frequency conversion adjustment technology, which automatically adjusts the speed of the pump and the number of working pumps, according to the instantaneous change of flow rate and corresponding pressure in the water supply system. Thus it changed the pressure and flow rate of the outlet, so the pressure in the water supply network system is kept constant with the set pressure, to improve the water supply quality and save energy efficiently.

Fig. 2-5 The control cabinet of the variable frequency constant pressure water supply equipment

The control cabinet is suitable for automatic control of all kinds of closed water supply without high-rise water tower. It has the characteristics of many functions such as constant pressure, simple structure, simple operation, long service life, high efficiency and energy saving, reliable operation, complete function and perfect protection, etc.

The control cabinet has three kinds of working ways: manual, frequency conversion and industrial frequency automation, and it can add the following additional functions according to the requirements of the users, such as small flow switch or stop pump, pump stop without water in the pool, regular start and stop of the pump, dual power supply, double frequency conversion and two - way water supply system switching, automatic inspection, change of the water supply pressure, digital display of water supply pressure and other functions required by users.

II. Filtration Equipment

The function of the filter equipment is to filter the solid particles (sand, fertilizer sediment, and organic matter) from the irrigated water to prevent the sewage from entering the system to block thethe system and the emitters.

Chapter Ⅱ The Main Equipment and System of the Integrated Water-Fertilizer Technology

1. Screen filter

Screen filter is one of the most widely used filtration equipment in micro-irrigation system, whose filter media include plastic, nylon or stainless steel screen.

(1) Applicable conditions. Screen filters are mainly used as terminal filters, connected to the main filter (gravel or hydraulic circular filter) and used as control filters when the quality of the irrigated water is poor. It is mainly used to filter dust, sand and dirt in the irrigation water. When the content of organic substance is high, the filtering effect of this type would be very poor, especially when the pressure is high, the organic substance will squeeze through the screen and enter the pipe, causing the system and the emitters to be blocked. Screen filters are generally used for secondary or tertiary filters (i. e. used in conjunction with gravel separators or sand filters).

(2) Classification. There are many kinds of screen filters, vertical and horizontal according to the installation, manual cleaning and automatic cleaning by the cleaning way, plastic and metal by manufacturing materials, closed and open by closing or not.

(3) Structure. The screen filter is mainly composed of screen, shell, and top cover, as shown in Fig. 2-6. The size of the screen (i. e. , the number of meshes) determines the filter capacity of the filter. The emitters are blocked by particles of dirt passing through the filter screen and squeezing into each other in the holes or channels of the emitters, therefore, it is generally required that the pore size of the selected filter should be 1/7 to 1/10 of the diameter of the irrigator used. See Table 2-1 for the relationship between the specification of the screen and the size of the orifice.

Table 2-1 Relationship between the specification of the screen and the size of the orifice

Filter specification	size of the orifice		Particle type	diameter (mm)
	mm	μm		
20	0.710	710	Coarse sand	0.50~0.75
40	0.420	420	Medium sand	0.25~0.41
50	0.181	181	Fine sand	0.15~0.20
100	0.151	151	Fine sand	0.15~0.20
120	0.126	126	Fine sand	0.10~0.15
150	0.105	105	Very fine sand	0.10~0.15

(续表)

Filter specification	size of the orifice		Particle type	diameter (mm)
	mm	μm		
200	0.075	75	Very fine sand	< 0.10
250	0.052	52	Very fine sand	< 0.10
300	0.045	45	Powder sand	< 0.10

Fig. 2-6　The appearance and the filter elements of the screen filter

The selection of filter size depends on the type of emitters used and the cross section of the flow. At the same time, because the filter reduces the flow and has certain head loss, the pressure loss range of the filter must be considered when calculating the pressure of the system design. Otherwise, when a certain degree of blockage occurs in the filter, it will affect the irrigation quality of the system. In the modern irrigation system of our country, the sprinkler irrigation system usually requires 45~85 meshes, the micro-irrigation system requires 85~105 meshes, and the drip irrigation system requires 105~155 meshes. But the greater the number of filter is, thegreater the pressure loss is, causing the greater energy consumption.

2. Laminated filter

You can get a simple idea of how this filter works from its name. The laminated filter is filtered by overlapping circular filters, and each filter has filter slots on both

Chapter Ⅱ The Main Equipment and System of the Integrated Water-Fertilizer Technology

sides, and as the water flows through the laminates, impurities and filth in the water will be blocked, so that the purification of water quality can be achieved. In principle, the laminated filter has a better purification effect than the screen filter, which can be used for primary and final filtration between 40~400 meshes, but when there are more contaminants in irrigation water, a laminated filter is not suitable as a primary filter device, for the device needs to be cleaned many times, which affects the efficiency of the irrigation system, see Fig. 2-7, Fig. 2-8.

Fig. 2-7 The appearance and lamination of the laminated filter

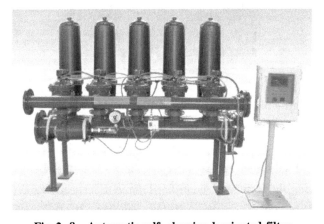

Fig. 2-8 Automatic self-cleaning laminated filter

3. Centrifugal filter

Centrifugal filter is also known as whirl water – sand separation filter or vortex water-sand separator. A centrifugal force is generated by high – speed rotating water flow, separating sand andother heavier impurities from the water, and there is no screen inside it. There are also no detachable parts, and the maintenance is very convenient. This kind of filter is mainly used to filter the water source with much sands. When the sand content in the water is large, the centrifugal filter should be chosen as the main filter. It consists of inlet, outlet, vortex chamber, separation chamber, sewage storage chamber and drain outlet, as shown in Fig. 2-9.

Fig. 2-9　Centrifugal filter

The working principle of the centrifugal filter is that when the pressure water enters the vortex chamber from the inlet, the water rotates and moves downward under the gravity at the same time, and it is spirally moving in the swirl chamber. The sand particles and other solid substance in the water are thrown onto the shell wall of the separation chamber under the action of centrifugal force. Because of gravity, the particles gradually move downward along the wall and converge into the sewage storage chamber. In the storage chamber, the cross-section increases, the flow velocity decreases, centrifugal force reduces, and the gravity increases. Finally, the particles subside and then drain out the filter through the sewage pipe. And the pure water velocity at the center of the vortex is lower, the potential energy is higher, so the spiral flow goes up through the outlet at the top of the separator into the irrigation pipeline system.

Centrifugal filter, because of its separation of water and sediment by rotating flow and centrifugation, is an ideal filter for high sand water, but it is difficult to remove impurities which are similar to water density and those with less dense than water. Therefore, sometimes it is also known as sand – stone separator. In addition, when the pump is started and stopped, the flow velocity in the system is small, the centrifugal force produced in the filter is small, the filtering effect is poor, and there will be more sand in the system. Therefore, centrifugal filter can not alone fulfill the filtration task of micro-irrigation system. It must be used in combination with screen filter or laminated filter, and the water-sand separator is used as the primary filter. This will have a better filtering effect, thus extend the cleaning period. The storage tank at the bottom of the centrifugal filter must be flushed frequently to prevent the sediment from being deposited into the system again. Centrifugal filter has large head loss, which must be taken into account when selecting and designing, see Fig. 2-10.

Fig. 2-10　Combined use of centrifugal filter and screen filter

4. Sandstone filter

Sandstone filter is also called media filter. The sand-stone is used as a filter medium to filter the water, and a basalt sand bed or a quartz sand bed is generally selected, and the particle size is determined according to the water quality, the filtering requirements and the system flow. The sandstone filter has strong filtering and storage capacity of organic impurities and inorganic impurities in water, and can supply water continuously. When the organic content of the water is high, the sandstone filter shall be selected regardless of the high or low inorganic content. The sand-stone filter has the advantages of strong filtering capacity and wide application range, and has the disadvantages of occupying relatively large space and the costing too much. It is generally used for filtrating the surface water. A single filter or more than two filter groups can be selected according to the output and filtration requirements.

The sandstone filter is mainly composed of water inlet, water outlet, filter body, filter medium sand and drainage hole, etc, as shown in Fig. 2-11. The working principle is that when the water enters the filter through the water inlet and passes through the sandstone filter bed. Because the pores in the filter medium are tortuous and small, the flow rate of the water flow is reduced, the impurities contained in the water source are blocked and deposited or adhered to the surface of the filter medium. And the filtered purified water enters the irrigation pipeline system from the water outlet. When the pressure difference between the two ends of the filter exceeds 30~50 kPa, it is indicated that the filter medium is blocked by the dirt, so it is necessary to turn back the water flow to flush. The filter valve is controlled to cause the flow of water inverse and discharges the previously blocked sewage through the outlet. In order to supply water unceasingly for the irrigation system during backwashing, more than two filters are often installed in the head hub, the working process of which is shown in Fig. 2-12.

The filtration capacity of sandstone filter is mainly determined by the properties of sandstone and grain composition. The sandstone with different grain composition has different filtration capacity. At the same time, due to the sufficient contact between sandstone and water, and the friction in backwashing, the sandstone for filter should meet the following requirements: having sufficient mechanical strength to prevent wear and breakage during backwashing; having enough chemical stability to avoid chemical reaction between the sandstone and chemical substances such as acid, alkali and so on,

Chapter Ⅱ The Main Equipment and System of the Integrated Water-Fertilizer Technology

Fig. 2-11 Sandstone filter

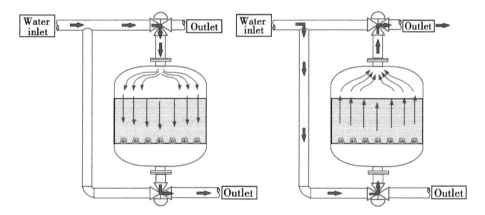

Fig. 2-12 Working state of sandstone filter

not to produce substances that cause micro-irrigation blockage, and not to produce toxic substances to animals and plants; having a certain grain composition and appropriate porosity; trying to use local material with cheap price.

5. Barrier for debris (net)

Many irrigation systems use surface water as a source of water, such as rivers, ponds, etc. These waters often contain large amounts of debris, such as dead leaves, algae, weeds and other large floats. To prevent these debris from entering sinks or pools to increase the burden of filters, a barrier is often installed at the inlet of the water pump in the reservoir as a primary purification facility for irrigated water sources, as shown in Fig. 2-13. The structure of the barrier is simple and can be designed and made according to the actual situation of the water source.

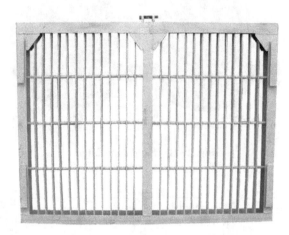

Fig. 2-13 Barrier for debris

6. Selection of filters

Filters play a very important role in micro-irrigation systems. Different types of filters have different filtering capacity to different impurities. When designing and selecting filters, they must be based on the water quality of the water source, flow and emitter requirements to meet the requirements of the system, and easy to operate. Filter selection generally has the following steps.

In the first step, the filter type is selected according to the contents of impurities in irrigation water. Surface water (rivers, lakes and ponds, etc.) generally contains more sand, stone and organic substance, so it is appropriate to select sandstone filters for primary filtration. If the volume of impurities is relatively large, it is also necessary to use barrier as the primary filter. If the sand contents are is large, it is also necessary to set up sediment ponds for primary filtration. The impurities in groundwater (well water) are mainly sand and stone, the centrifugal filter should be used as the primary filter. No matter it is sandstone filter or centrifugal filter, screen filter or laminated filter can be selected as secondary filter. For the water source with good quality we can directly select screen-screen filter or laminated filter. Table 2-2 summarizes the effec-

Chapter Ⅱ The Main Equipment and System of the Integrated Water-Fertilizer Technology

tiveness of different types of filters in removing different sewage from irrigation.

Table 2-2 Type choice of the filters

Dirt Type	Pollution degree	Standard (mg/L)	Centrifugal filter	Sandstone filter	Laminated filter	Self-cleaning screen filter	Filter choice
Soilparticle	low	≤50	A	B	—	C	Sieve
	high	>50	A	B	—	C	Sieve
Suspended solids	low	≤80	—	A	B	C	Lamination
	high	>80	—	A	B	—	Lamination
Algae	low		—	B	A	C	Lamination
	high		—	A	B	C	Lamination
Iron oxide and manganese	low	≤50	—	B	A	A	Lamination
	high	>50	—	A	B	B	Lamination

Note: Control filter refers to secondary filter. A is the first choice; B is the second choice; C is the third option.

The second step is to determine the size of the filter according to the filter capability required by the irrigation system. Generally speaking, sprinkler irrigation requires 40~80 meshes filtration, micro spray 80~100 meshes filtration, drip irrigation 100~150 meshes filtration.

The third step is to determine the filter capacity according to the system flow rate.

The fourth step is to determine the type of flushing or cleaning. If necessary, the automatic backwash type is recommended to reduce maintenance and workload. Especially when labor is in short and irrigation area is large, automatic backwash filter should be preferred.

The fifth step is to consider the price. For the different filters with the same filter effect, the selection of the filter focus mainly on the price. Generally sand medium filter is the most expensive, while the laminated filter or screen filter is relatively cheap.

Ⅲ. Control and Measurement Equipment

In order to ensure the proper operation of the irrigation-fertilization system, the head hub must be also equipped with control devices, protection devices, measurement devices, such as inlet and vent valves, reverse valves, pressure gauges and water meters.

1. Control components

The function of the control components is to control the flow direction, the flow rate and the total water supply. It is based on the irrigation scheme, distributing the planned flow to the various parts of the system. It mainly consists of various valves and special water supply components.

(1) Water supply hydrant.

The water supply hydrant refers to a water outlet of the irrigating water from the underground pipeline system to the ground, and can be divided into a movable water supply hydrant, a semi-fixed one and a fixed one according to the structure of the valves, as show in Fig. 2-14.

Fig. 2-14　Water supply hydrant

(2) Valves.

Valves are essential parts of the sprinkler irrigation system, including gate valves, butterfly valves, ball valves, globe valves, check valves, safety valves, pressure relief valves, etc, as shown in Fig. 2 - 15, Fig. 2 - 16, Fig. 2 - 17. In the same irrigation system, different valves play different roles, and different types of valves can be selected according to the actual situation.

2. Safety protection devices

During the operation of the irrigation, there will be inevitably sudden changes in pressure, pipe intake, sudden pump shutdown and other unusual conditions, which threaten the system. Therefore, safety protection devices must be installed in the relevant parts of the irrigation system to prevent the damage to irrigation equipment because of pressure change or backflow of water, ensuring the normal operation of the system. Commonly used devices include intake (vent) valve, safety valve, pressure reg-

Chapter Ⅱ The Main Equipment and System of the Integrated Water-Fertilizer Technology

Fig. 2-15 butterfly valve

Fig. 2-16 PVC ball valve

Fig. 2-17 Check valve

ulating device, reverse valve, drain valve, etc.

(1) Intake valve.

The inlet (vent) valve is a kind of safety protection device which can automatically exhaust and intake air, and can be closed automatically when the pressure water comes. The main function is to release the air in the pipe, break the vacuum of the pipeline, and some products have the function of checking the return water. The inlet (vent) valve is an important device for pipeline safety, and it is indispensable. Some non-professional design does not install the inlet (vent) valve, so it causes the tube explosion and the pipeline flat, thus the system can't work normally, shown in Fig. 2-18.

Fig. 2-18　Air Intake (vent) valve

(2) Safety valve.

The safety valve is a pressure release device which automatically opens to release the pressure when the water pressure of the pipe exceeds the set pressure to prevent the water hammer, and it is generally installed at the lower part of the pipeline. When the water column separation is not generated, the safety valve is installed at the head of the system (water outlet of the water pump) and can protect the whole sprinkler system. If the water column separation in the pipe occurred, the safety valve must be installed along one or several places along the pipeline to achieve the purpose of preventing water hammer, as shown in Fig. 2-19.

3. Flow and pressure regulating devices

When the actual flow rate and pressure in some areas of irrigation system are different from theset working pressure, it is necessary to install regulating devices to adjust the pressure and flow rate in the pipe. Especially when the natural height difference is used for self-pressure sprinkler irrigation, the pressure distribution in the pipe is not even, or the actual pressure is greater than the working pressure of the sprinkler, which leads to the uneven distribution of flow and pressure, or it is difficult to meet the re-

Chapter Ⅱ   The Main Equipment and System of the Integrated Water-Fertilizer Technology

Fig. 2-19   Safety valve

quirements and it if difficult to choose nozzles. In addition to dividing pressure zones, it is necessary to install flow and pressure regulating devices in the pipes. The flow and pressure regulating device adjust the flow rate and pressure by automatically changing the cross section of the water. In fact, it is a device that reduces the flow rate or pressure by limiting the flow rate, and it does not increase the flow rate or pressure of the system. According to this working principle, in practice, considering the investment problem, ball valve, gate valve, butterfly valve and so on are also used as regulating devices, but on the one hand, this will affect the life of the valve. On the other hand, it is difficult to adjust the flow and pressure accurately.

4. Measuring device

The main measuring devices of irrigation system are pressure gauge, flow meter, and water meter. Its function is toduly monitor the pressure and flow rate in the pipe to judge the working state of the system, and to find and eliminate the malfunction in time.

(1) Pressure gauge.

The pressure gauge is a necessary measuring device for all irrigation systems, which is an instrument for measuring the water pressure in the pipe of the system, and it can reflect whether the system is in a normal working state. When the system fails, the possible failure type can be basically judged according to the change of pressure gauge reading. The pressure gauges are often installed at the control nodes such as the head hub, the entrance of the irrigation district, the entrance of the branch pipes and so on. The actual number and specific location should be determined according to the area of the irrigation area and the complexity of the terrain. A pressure gauge must be respectively installed before and after the filter. The blocking degree of the filter is judged by the pressure difference between the two ends so that the filter can be cleaned in time to prevent blockage. Blockage reduces the cross-section of water, resulting in too small pressure and flow in the field, which affects the quality of irrigation. The pressure gauges for sprinkler irrigation should be selected to be high-quality with high sensitivity, the working pressure within the real range of the pressure gauges, large dial and easy to see. In addition to the observation, the working state of the sprinkler irrigation system is mainly shown by the pressure gauge. Therefore, it is necessary to ensure that the pressure gauge is in a normal working state and that the pressure gauge should be replaced in time after failure, as shown in Fig. 2-20.

Fig. 2-20 Pressure gauge

(2) Flow meters and water meters.

Both theflow meter and the water meter are instruments for measuring the flow rate. The difference between them is that the flow meter can directly show the change of flow in the pipe without recording the total water flow, as shown in Fig. 2-21. The water meter can not record the real-time flow because it reflects the accumulated water quantity through the pipe. In order to obtain the system flow, it needs to be calculated, and is usually installed on the head hub or main pipe. In a sprinkler irrigation system equipped with an automatic fertilizer applicator, an automatic meter should be installed because the fertilizer applicator needs to determine the amount of fertilizer applied according to the flow rate of the system.

Fig. 2-21　Flow meter

5. Automation control equipment

One of the advantages of water-saving irrigation system is that it is easy to realize automatic control. Automatic control technology can greatly improve the efficiency of irrigation. Automatic control irrigation system has the following advantages: it can timely and appropriately control irrigation water, irrigation time and irrigation period, improve water use efficiency, greatly save labor, improve work efficiency, and reduce the costs. The irrigation plan can be arranged flexibly and conveniently, and the managers do not have to go directly to the field to operate, and utilization rate of the equipment can be improved by the increase of the working time. The automatic control system of water-saving irrigation is mainly controlled by central controller, automatic valve and transmission, etc. The automation degree can be determined according to the requirements of users, economic strength, economic benefits of the crops and so on.

(1) Central controller.

The central controller is the control center of the automatic irrigation system. Managers can control the whole irrigation system byinputting the corresponding irrigation program (irrigation start time, duration time, irrigation cycle). Since the controller is expensive, the type of controller should be selected depending on the actual capacity requirements and the functions to be implemented, as shown in Fig. 2-22.

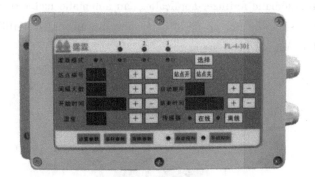

Fig. 2-22　Central controller

(2) Automatic valve.

The type of the automatic valve is various, and among them the electromagnetic valve is one of the most applied in the automatic irrigation system. It opens or closes the valve by the electric signal transmitted through the central controller, raising the metal plug to open the channel between the upstream and downstream of the valve, so that the pressure differential is formed between the upper and the lower parts of the rubber diaphragm, the valve being open, as shown in Fig. 2-23.

Fig. 2-23　Electromagnetic valve

Chapter II The Main Equipment and System of the Integrated Water-Fertilizer Technology

Section II The Fertilization Equipment in the Integrated Water-Fertilizer Technology System

The commonly used fertilization equipment in integrated water-fertilizer technology is mainly composed of the differential pressure fertilization tank, Venturi fertilizer applicator, pump fertilizer suction method, pump fertilizer injection method, self-pressure gravity fertilization method, fertilizer applicator and so on.

I . Differential Pressure Fertilizer Tank

1. Basic principles

A differential pressure fertilizer tank is connected by two thin tubes (by-pass tubes) to the main pipe, and a control valve (ball valve or gate valve) is set between the two thin tubes to produce a smaller pressure difference so that a portion of the water flows into the fertilizer tank. After the feed pipe reaches the bottom of the tank and the fertilizer is dissolved in the water tank, the soluble fertilizer is fed into the main pipe by another tube, and the fertilizer is carried to the root zone of the crop as shown in Fig. 2-24, Fig. 2-25, Fig. 2-26.

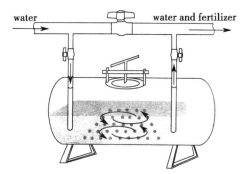

Fig. 2-24 Sketch map of the differential pressure fertilizer tank

Fertilizer tanks are made of corrosion-resistant ceramic substrates or galvanized cast iron, stainless steel or fiberglass to ensure resistance to system pressure and fertilizer corrosion. In the low pressure drip irrigation system, because of the low pressure,

Fig. 2-25　Vertical metal fertilizer tank

Fig. 2-26　Vertical plastic fertilizer tank

plastic can also be used. Solid soluble fertilizer gradually dissolved in the fertilizer tank, and liquid fertilizer can be quickly mixed with the water. As the irrigation proceeded, the fertilizer was carried away, the fertilizer solution being diluted, the nutrients getting lower and lower, and the solid fertilizer in the tank finally flowed away. The system is simple and cheap, and can reach a higher dilution degree without external power. However, the system also has some shortcomings, such as the lack of precise control of the fertilizer injection rate and the nutrient concentration. Fertilizer have to be re-fed into the fertilizer tank before each irrigation. Throttle valves increase the pressure

Chapter Ⅱ The Main Equipment and System of the Integrated Water-Fertilizer Technology

losses and the system cannot be used for automated fertilization. Fertilizer tanks are often made into the specification of 10~300 liters. In general, the small area of greenhouse uses small volume fertilizer tank, and the large area of field in rotation irrigation area uses large volume fertilizer tank.

2. Advantages and disadvantages

The advantages of differential pressure fertilizer tank are: low cost, simple operation and convenient maintenance; suitable for liquid fertilizer and water-soluble solid fertilizer; do not need additional power when fertilizing; small size occupying small area of land. The disadvantages of the tank are as follows: it is for quantitative fertilization, and the liquid concentration is uneven in the process of fertilization; it is easy to be affected by the change of water pressure; there is a certain head loss, poor mobility, which is not suitable for automatic operation; the erosion is serious and the endurance is poor; because of the small mouth of the tank, it is not convenient to add the fertilizer, especially when the irrigation area is large, the amount of fertilizer is large each time, thus the volume of the tank is limited, so it is necessary to pour fertilizer many times, which reduces the working efficiency.

3. The scope of application

The differential pressure fertilization tank is suitable for integrated water-fertilizer system including greenhouses and field planting etc. For systems with different pressure ranges, different tanks with different materials can be selected. The pressure endurance capacity of the fertilizer cans is different due to different materials.

Ⅱ. Venturi Fertilizer Applicator

1. Basic principles

When the water flows through a pipe (Venturi throat) which is smaller and then bigger, the velocity of flow increases and the pressure drops as the water flows through the narrow part, resulting in a pressure difference between the front and back, and when there is a smaller inlet in the throat, negative pressure is formed. It can extract the fertilizer solution from an open-mouth fertilizer tank through a small diameter tube. Venturi fertilizer applicator is based on this principle as shown in Fig. 2-27, Fig. 2-28.

Due to the pressure loss caused by Venturi fertilizer applicator, a small booster

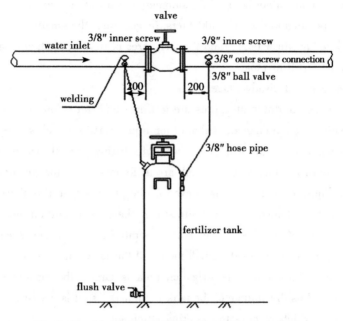

Fig. 2-27 Sketch map of Venturi fertilizer applicator

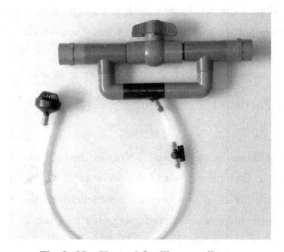

Fig. 2-28 Venturi fertilizer applicator

pump is usually installed. General manufacturers will inform the user of the product pressure loss, so we should design if we need pressure pump or not according to the relevant parameters.

The operation of Venturi applicator requires excessive pressure to ensure the necessary pressure loss, and the steady pressure at the inlet of the applicator is the guarantee of even nutrient concentration. The pressure loss is expressed as a percentage of the pressure at the inlet. The suction requires more than 20% of the loss of the inlet pressure, but the two-stage Venturi fertilizer applicator requires only 10% of the pressure. The amount of the fertilizer extracted is affected by the inlet pressure, pressure loss and straw diameter, and can be adjusted by control valves and regulators. Venturi applicator can be installed on the main road (series connection, as shown in Fig. 2-29) or as a bypass part for the pipeline (parallel installation, as shown in Fig. 2-30). In a greenhouse, a fertilizer applicator is installed as a bypass whose water flow is pressurized by an auxiliary pump.

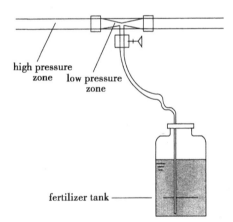

Fig. 2-29 Series connection of Venturi fertilizer applicator

The main working parameters of Venturi fertilizer applicator are as follows: first, the working pressure at the entrance (P in). The second is the pressure difference, the pressure difference (P in-P out) is often expressed as the percentage of the inlet pressure, only when this number is reduced to a certain level, the suction begins. As mentioned earlier, this number is about 1/3 of the inlet pressure, some of which are as high as 50%, and the more advanced can be less than 15%. Table 2-3 shows the relationship between the pressure difference and the amount of fertilizer absorbed. The third

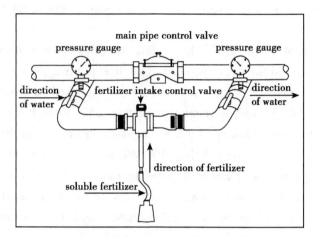

Fig. 2-30 Parallel connection of Venturi fertilizer applicator

is the amount of suction, which means the volume of liquid fertilizer sucked out in a unit of time, showing with L/h. The suction can be adjusted by some components. Fourth is the flow, the flow of water through the fertilizer applicator itself. The inlet pressure and throat size affects the flow rate. The flow range is given by the manufacturer. Each type can only work accurately within a given range.

Table 2-3 The relationship between pressure difference and amount of fertilizer absorption of Venturi fertilizer applicator

Inlet pressure P_1 (kPa)	Outlet pressure P_2 (kPa)	Pressure difference P (kPa)	Fertilizer suction flow Q_1 (L/h)	Main pipe flow Q_2 (L/h)	Total flow Q_1+Q_2 (L/h)
150	70	80	0	1 300	1 300
150	40	110	320	2 200	2 520
150	0	150	472	2 008	2 480
100	30	70	0	950	950
100	0	100	350	2 290	2 640

Venturi fertilizer applicator has obvious advantages: it does not require external energy, directly absorb the fertilizer from the open fertilizer tank, with big absorption range, simple operation, low wear rate. And it is easy to install, easy to move, suitable for automation. The nutrient concentration is even and the corrosion resistance is strong. The disadvantage is that the pressure loss is large, so the amount of fertilizer absorption is affected by the pressure fluctuation.

2. Main types

(1) Simple type.

This type of structure is simple, having only jet contraction section, no accessories. Due to the excessive head loss, it is generally not suitable for use.

(2) Improved type.

Pressure changes in the irrigation system may interfere with the normal process of fertilization or cause accidents. In order to prevent these situations, one-way valve and vacuum damage valve are added on the basis of one-stage jet pipe. When the suction pressure is too low or the inlet pressure is too low, water will flow from the main pipe into the fertilizer tank to overflow. To solve this problem, it is to install a one-way valve in front of the suction pipe or install a ball valve on the pipe. When the suction chamber of the Venturi applicator is negative, the valve core of the one-way valve is closed under suction action to prevent the outflow of water from the suction inlet, as shown in Fig. 2-31.

Fig. 2-31 Venturi fertilizer applicator with one-way valve

When the open fertilizer barrel is placed at the head of the field, the liquid in the tank may be sucked in by negative pressure at the end of the irrigation, and then run to the lowest part in the field, wasting fertilizer and possibly burning crops. So it is to install the vacuum failure valve in the pipe, no matter where the local vacuum appears in the system, the air can be replenished in time.

Some manufacturers offer various specifications ofVenturi fertilizer throat, which can be changed according to the amount of fertilizer solution required to stabilize the absorption rate of fertilizer solution at the required level.

(3) Two-stage type.

An improved two-stage structure has been developed abroad. The loss of water head during fertilization is only 12% to 15% of the inlet pressure, thus overcoming the basic defects of Venturi fertilizer applicator and making it widely used. The deficiency is that the flow decreases accordingly, as shown in Fig. 2-32.

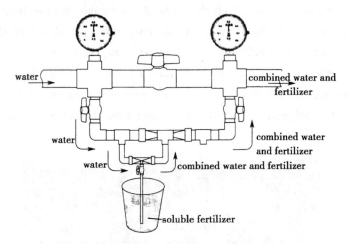

Fig. 2-32 Two-stage type of Venturi fertilizer applicator

3. Advantages and disadvantages

The advantages of Venturi fertilizer applicator are as follows: low equipment cost and low maintenance cost; the fertilization process can maintain even fertilizer concentration without external power during irrigation; the equipment is light in weight and easy to move and be used in automation system; the fertilizer tank is open when fertilizing, so it is convenient to observe the process of fertilization. The disadvantages of Venturi fertilizer applicator are as follows: the head pressure loss of the system is large; in order to compensate the head loss, the higher pressure is required in the system; the pressure fluctuation in the fertilization process is large; in order to get the steady pressure in the system, the booster pump should be equipped; the solid fertilizer can not be used directly, which should be dissolved for fertilization.

4. Scope of application

Venturi fertilizer applicator, because of its smaller flow, is mainly suitable for small area, such as greenhouse or small-scale farmland.

5. Installation methods

In most cases, Venturi fertilizer applicators are installed on the by-pass tubes (parallel installation), so that only part of the flow passes through the jet segment. Of course, the pressure drop in the main pipe must be equal to that in the jet tube. This bypass operation can use smaller (cheaper) Venturi applicator and is easier to move. When not fertilizing, the system also works. When the fertilization area is very small and the pressure was not taken into account, series connection can also be applied.

Venturi fertilizer applicators installed on by-pass tubes often use by-pass pressure-regulating valves to produce differential pressure. The head loss of the regulator is sufficient to distribute the pressure. If the fertilizer fluid flows into the main pipe after the main filter, the sucked water-fertilizer must be filtered separately. A 100-mesh nylon mesh or stainless steel mesh is often covered at the suction part, or a corrosion-resistant filter is installed at the end of the fertilizer-conveying pipe, with 120 meshes, as shown in Fig. 2-33. Some factories have installed the stainless steel mesh to their products. The end structure of the pipe should be easy to check, and can be cleaned if necessary. The fertilizer tank (or barrel) should be lower than the jet tube to prevent the fertilizer liquid flow into the the system because of self-pressure, when the fertilizer liquid is not needed. The parallel installation method can keep the constant pressure of the outlet, which is suitable for steady flow. When the inlet pressure is high, a small pressure regulator may be installed at the by-pass inlet, so that safety measures are provided at both ends.

As Venturi applicator is sensitive to pressure fluctuations during operation, pressure gauges should be installed to monitor. In general, multiple pressure gauges are installed in the head system. The pressure gauge at both ends of the control valve can measure the pressure difference between the two ends. Some of the more advanced fertilizer applicator are equipped with pressure gauges to monitor operating pressure.

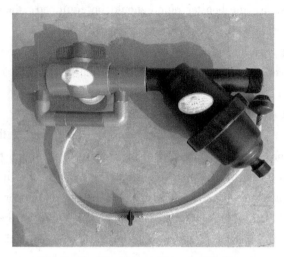

Fig. 2-33 Venturi applicator with filter

III. Gravity self-pressure fertilization method

1. Basic principles

On the occasions suitable for gravity drip irrigation or micro-sprinkler irrigation, gravity self-pressure fertilization method can be used. In hilly orchards or tea gardens in the south, mountain springs or water sources at the foot of the mountain are pumped to higher pools. Usually, an open fertilizer-mixing tank is set up near the pool and above the pool. The pool's size is $0.5 \sim 5.0 m^3$, which can be square or circular, and can be easily stirred to dissolve fertilizer. At the bottom of the fertilizer pool, pipe for the liquid to flow out should be installed and PVC ball valve should be installed at the outlet, which is connected to the outlet pipe of the water pool. A large diameter of $20 \sim 30$ cm is used in the pool. The inlet of the tube is wrapped with 100-mesh nylon screen, as shown in Fig. 2-34.

2. Scope of application

There are large hilly orchards, teagardens, economic forests and crop field in South, Southwest and Central South of China, which are very suitable for gravity self-pressure irrigation. In many mountain orchards there are pools built on the top of the mountain, and the orchards are usually irrigated with dragging pipes for spray or drip

Chapter Ⅱ   The Main Equipment and System of the Integrated Water-Fertilizer Technology

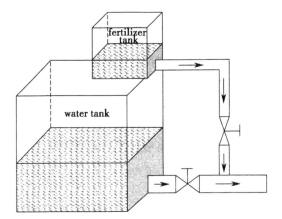

Fig. 2-34   Sketch map of self-pressure fertilizer applicator

irrigation. At this time the use of gravity self-pressure fertilization is very convenient to achieve the mixing of water and fertilizer. A considerable number of fruit farmers in citrus, lychee and longan orchards in South China use gravity self-pressure fertilization. This method is simple and convenient, and the concentration of fertilizer is even, so the it is easy for the farmers to accept. The disadvantage is that fertilizer must be delivered to the top of the mountain.

Ⅳ. Pump-suction fertilizer method

1. Basic principles

The method of pumping fertilization is to use centrifugal pump to directlysuck fertilizer solution into irrigation system, which is suitable for fertilizing the area within dozens of hectares. To prevent fertilizer solutions from flowing into the water pool, reverse valves can be installed on suction pipes. Normally, a nylon net with 100~120 mesh (stainless steel or nylon) is wrapped on the inlet of the tube to prevent impurities from entering the pipe, as shown in Fig. 2-35.

2. advantages and disadvantages

The advantages of this method arethat there is no need of additional power, simple structure, convenient operation, and an close container can be used to hold fertilizer solution. By adjusting the valve on the fertilizer tube, the fertilization speed can be controlled and the concentration can be accurately adjusted. The disadvantage is that some-

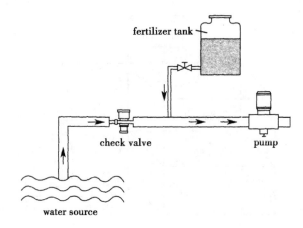

Fig. 2-35 Sketch map of pumping fertilization

one should look after the fertilizer and close the valve on the tube immediately when the fertilizer solution is almost finished, otherwise it will suck in the air and affect the operation of the pump.

V. Pump-Injection Fertilizer Method

The method of pumping fertilizer is to use a pressure pump to inject fertilizer solution into a pressurized pipe. Usually, the pressure by the pump must be greater than the water pressure of the water pipe, otherwise the fertilizer cannot be injected into the pipe. Pumping fertilizer method is the best choice for direct irrigation with deep well pump or submersible pump. Pump fertilization speed can be adjusted, fertilization concentration iseven, easy to operate, and does not consume system pressure. The disadvantage is to equip a fertilizer pump separately. For areas needing infrequent fertilization, ordinary water pumps can be used, after fertilization it should be cleaned with water, generally it will not rust. However, for area needing frequent fertilization, it is recommended to use corrosion-resistant chemical pumps.

Chapter Ⅱ The Main Equipment and System of the Integrated Water-Fertilizer Technology

Section Ⅲ Water Transfer System of the Integrated Water-Fertilizer Technology-Sprinkler Irrigation

Ⅰ. The Concept of Sprinkler Irrigation

Sprinkler irrigation uses pump pressure or natural drop to send water through a pressure pipe to the field and spray it into the air through the sprinkler head to form fine drops of water that is evenly sprayed on the farmland. It is an advanced irrigation method to provide the necessary water conditions for the normal growth of crops. Compared with the traditional surface flood irrigation method, sprinkler irrigation has obvious characteristics.

Ⅱ. The Technical Characteristics of Sprinkler Irrigation

1. The advantages of sprinkler irrigation

(1) Saving water. Because sprinkler irrigation canirrigate timely, moderately, evenly and in a planned way in small scale. It does not produce surface runoff and deep infiltration, thus increasing the effective utilization rate of water and achieving the goal of saving water. And the irrigation is even. Compared with surface irrigation, sprinkler irrigation is saves 30%~50% water than furrow or border irrigation. If on the sandy soil or slope land with strong water permeability and poor water retention ability, it can save water more than 70%.

(2) Strong adaptability. Sprinkler irrigation is suitable for all kinds of terrain and soil conditions, which does not necessarily require flat surface. Sprinkler irrigation can be applied in mountainous, hilly, sloping areas, and in the areas with high hills and hollows, which are not suitable for surface irrigation. In addition, sprinkler irrigation can be applied to a variety of crops, for all closely planted shallow root crops, such as leafy vegetables, tuber vegetables, potatoes and so on can be applied with sprinkled irigation. At the same time, the sprinkler irrigation technology is the most suitable for sandy soil with strong water permeability or subsidence soil, and for sandy soil with thin surface soil and strong permeability of subsoil.

(3) Saving labour and land. The high degree of mechanization of sprinkler irrigation and the convenience of automation with small electronic control devices can save a great deal of labor force. If the technology of spraying irrigation and fertilization is used, the effect of saving labor will be more remarkable. In addition, The application of sprinkler irrigation can also reduce the labor on digging field channels, irrigation ditch and other works. At the same time, sprinkler irrigation uses pipes to transport water, so the fixed pipe can be buried underground, thus reduced the land used for field ditches, borders, ridges and so on, saving 7%~15% of land than surface irrigation.

(4) Increasing the yield and improving the quality. First of all, sprinkler irrigation can timely and appropriately control irrigation volume, adopting the method of less irrigation and frequent irrigation, so that soil moisture can be kept within the proper range of normal growth of crops, and sprinkler irrigation can irrigate crops like rain at the same time. It can not destroy the soil mechanically, but maintain the soil aggregate structure, and effectively regulate the water, fertilizer, gas, heat and microbial conditions of the soil. Secondly, sprinkler irrigation can adjust the field microclimate, increase the humidity of ground air, adjust the temperature and temperature difference between the day and the night, and avoid the harm of dry and hot wind, high temperature and frost to crops. Its effect of increasing yield is obvious, the productivity be increased one time to twice. Secondly, sprinkler irrigation can flexibly adjust irrigation time and irrigation quantity according to crop water demand, and can timely adjust fertilization scheme according to crop growth demand and effectively improve the yield and quality of the agricultural products.

2. The disadvantages of sprinkler irrigation

(1) The spraying operation is greatly affected by wind. Droplets ejected from the sprinkler are heavily affected by the wind before sprinkling on the ground. Under the influence of the wind, the range of the nozzle in each direction and the distribution of water quantity change obviously, thus affecting the evenness and even the leakage of spray. In the windy area of irrigation season, the adverse effects of wind should be fully taken into account for the equipment selection and planning design. If it is difficult to solve the problem, other irrigation methods should be considered.

(2) Large loss of evaporation. Water droplets will cause evaporation loss before falling to the ground, especially in drought, windy and high temperature seasons. The

Chapter II The Main Equipment and System of the Integrated Water-Fertilizer Technology

loss of evaporation is greater, and its loss is related to wind speed, air temperature and air humidity.

(3) High investment in equipment. The working pressure of sprinkler irrigation system is high, and there quirement for the pressure tolerance is also high, so the investment of equipment is generally higher. This is also the main factor restricting the development of sprinkler irrigation. Another related problem is the current poor quality of sprinkler irrigation equipment, coupled with poor management, resulting in equipment damage, loss, or even the system abandoned in advance, thus the investment did not get a corresponding return. Therefore, to construct the sprinkler irrigation project, we must effectively control the equipment and construction quality, and improve the management.

(4) The high energy consumption and operation cost. The sprinkler irrigation system needs pressure equipment to provide a certain amount of pressure in order to ensure the normal operation of the sprinkler head, to meet the requirements of even irrigation. In the absence of natural water pressure, pumps must be used to increase pressure, which consumed a portion of the energy (electricity, diesel or gasoline), thus increased the operating costs. In order to solve this kind of problem, sprinkler irrigation is developing to low pressure at present. In addition, to make full use of natural water pressure where conditions exist, the operating costs can be greatly reduced.

(5) The wet surface, but insufficient in the deep layer. Compared with drip irrigation, the irrigation intensity of sprinkler irrigation is much larger, so there occurs a disadvantage that the surface layer is wetter but the deep layer is not sufficiently moist. This phenomenon is not good for deep-rooted crops, but if the sprinkler irrigation intensity is properly chosen to lessen the sprinkler irrigation intensity and extend the sprinkler time to make the water fully infiltrate into the lower layer, the problem will be greatly alleviated.

In addition, for crops that are still in the seedling stage, due to the absence of closure, using sprinkler irrigation, especially when irrigation and fertilization are combined, will cause weeds to breed on the one hand, thereby affecting the normal growth of crops. On the other hand, it also increased the waste of water and fertilizer resources. And in the high temperature season, especially in the south, when the sprinkler irrigation system is used, it is easy to form high temperature, high humidity environment during crop growth, thus cause the disease to spread and so on.

In the semi-fixed pipe sprinkler irrigation system, the main pipes are fixed, but the branch is moved and used, which greatly improves the utilization ratio of branch pipe. To reduce the amount of branch pipes, and make the investment lower than that of fixed pipe sprinkler irrigation system. This type has great potential for development in wheat areas in northern China. For the convenient movement of branch pipes, the pipes shall be light-duty pipe, such as thin-walled aluminum pipe, thin-walled galvanized steel pipe, and equipped with all kinds of quick joints and lightweight connections and water hydrant.

Hosereel spray irrigation applicator belongs to row irrigation applicator. The specification mainly is medium-sized, at the same time, there are also small-sized products. Cable reel sprinkler is also used abroad, but it is only suitable for pasture irrigation. Hose reel sprinkler has compact structure, good mobility, high production efficiency and many specifications. The control area of a single machine can reach 150 mu, and the spraying evenness is high. The water amount of the sprinkler can be adjusted in the range of several millimeters to dozens of millimeters. This type of machine is suitable for the current economic conditions and management level of our country; as long as the scale management or unified planting are formed, It can be widely used in a certain range.

Light and small sprinkler set refers to the diesel engine or electric motor set. There are two types as hand-held and hand-propelled, both belonging to the fixed-spray sprinkler. The light and small sprinkler irrigation unit has been developed to adapt to the rural power situation in China in the 1970s. After 20 years' unremitting efforts, it has become one of the leading products of sprinkler irrigation from 2 to 12 kW, with the characteristics of complete set, complete specifications and mass production. The light and small sprinkler irrigation machine adapts to the hilly areas with small and scattered water resources and the plain with water shortage, which has the advantages of less one-time investment, simple operation, convenient storage and maintenance. And the irrigation area can be large or small, suitable for drought resistance and so on.

III. Equipment and Type Selection

Sprinkler irrigation equipment, also known as sprinkler irrigationdevices, mainly

Chapter Ⅱ The Main Equipment and System of the Integrated Water-Fertilizer Technology

include sprinkler head, sprinkler irrigation pump, sprinkler pipes and accessories, sprinkler irrigation applicators, etc. The following mainly describes the the use and types of the sprinkler head and sprinkler applicators.

1. Classification, performance and selection of sprinklers

(1) Sprinklertype.

There are many types of sprinklers, usually classified according to the working pressure or structure.

① Classification by work pressure. According to the working pressure, the sprinkler can be divided into low pressure nozzle, middle pressure nozzle and high pressure nozzle, in which the working pressure of the low pressure nozzle is less than 200 kPa, the range of the spray is less than 15.5 meters, and the flow rate is less than $2.5 m^3/h$. The working pressure of the middle pressure nozzle is 200~500 kPa, the range of the nozzle is 15.5~42 meters, the flow rate is $2.5~32m^3/h$. The working pressure of the high pressure nozzle is more than 500 kPa, the range is more than 42 meters, and the flow rate is more than $32m^3/h$.

②Classificationby structure. According to the structure, the nozzle can be divided into three types: rotary nozzle, fixed nozzle and pipe with spray holes.

A. Rotary nozzle. The rotary nozzle is also called jet nozzle, which is characterized by rotating while spraying, and the water is in the form of a concentrated jet when it is ejected from the nozzle, so the range ofspray is far away, and the flow range is large, and the intensity of sprinkler irrigation is lower. It is one of the most popular forms of sprinkler used in farmland irrigation in our country at present. The disadvantage of rotary sprinkler is that when the vertical tube is not vertical, the rotating speed of the nozzle is not even, which will affect the evenness of sprinkler irrigation.

a. Vertical rocker-arm sprinkler. This is a reactive nozzle. Vertical rocker-arm sprinkler is a middle and high pressure sprinkler which can be applied to all kinds of crops including young crops. Especially, it has good stability in the walking and spraying system. In addition, it can also spray mixed liquids such as sewage or faeces.

b. Full jet nozzle. Its biggest advantage is its bumpless parts, simple structure, good spraying performance.

B. Fixed nozzle. Fixed nozzles are also known as diffuse sprinklers or scattered sprinklers. This type of sprinkler is characterized by the simultaneous spraying of water into the whole circle or part of the circle (sector), with a short range, the wetting semi-diameter being only 3 to 9 meters. And intensity of sprinkler irrigation is high, generally above 15~20 mm/h. The water distribution of most sprinklers is that the nearby intensity of irrigation is much higher than that of average irrigation, and the degree of atomization is usually higher.

C. Pipe with spray holes. This sprinkler is composed of one or several smaller diameter pipes. There are some small holes on the top of the pipe, and the diameter of the spray hole is only 1~2mm. According to the distribution of spray holes, they can be divided into two kinds: single row holes pipe and multi-row holes pipe.

a. Single-row holes pipe. The nozzle holes are arranged in a straight line, spaced at a distance of 60~150cm, and the distance between the two tubes is usually 16 m. The tube holders are in the field and can be rotated around the shaft within 90° by virtue of the action of the automatic shaker, so that both sides of the holes tube can be sprayed. The automatic shaker can be gear or turbine set driven by a water wheel or a piston-driven. Single-row holes tubes are mostly fixed, mainly used for sprinkler irrigation in nurseries and vegetable fields. This is easy to operate with high productivity, but the infrastructure investment is quite high and the holder has a certain impact on farming and other field operations. Movable single–row holes tubes are used less, usually supported by a shorter holder or other supporting thing that is easy to move, so that the investment is lower, but the labor of the mobile holes tube is larger.

b. Multi-row holes tube. There are many small holes drilled at the top of the pipes. The arrangement of the holes ensures that the field with a width of 6 m to 15 m on both sides can be evenly irrigated. The working pressure of multi-row holes tube is lower, so it is more suitable for sprinkler irrigation with natural pressure, and it does not need automatic shaker, so it is much simpler in structure than single-row holes tube.

(2) Nozzle selection.

Nozzle selection includes nozzle type, nozzle diameter and working pressure. After the nozzle is selected, the performance parameters of the nozzlesuch as flow rate, range and others are determined.

① The principle of nozzle selection. In accordance with the provisions of the National Standard of Technical Specifications for Sprinkler Irrigation Engineering, the selection principle of sprinkler nozzle is as follows: The intensity of sprinkler irrigation

Chapter Ⅱ The Main Equipment and System of the Integrated Water-Fertilizer Technology

after combination does not exceed the allowable intensity of the soil; The evenness coefficient of sprinkler irrigation after combination is no less than that specified in the specification. The atomization index shall meet the requirements of crops. It is conducive to reduce the annual cost of sprinkler irrigation projects.

②Analysis of nozzle selection. The small nozzle requires lower working pressure and the energy consumption is less, which means that the operation cost is lower. However, due to its small range, the pipes layout is required to be dense and the pipe consumption is increased. Large nozzle has long range, large pipe distance, high working pressure, high energy consumption and high operating cost. Therefore, in the primary selection of nozzle should be based on specific conditions through technical and economic analysis for further consideration.

2. Classification and selection of sprinkler applicator

Sprinkler applicator is a kind of irrigation set, which consists of sprinkler, pipe, water pump, engine and so on according to certain ways, and meets the requirements in mechanics, hydraulics, operation and so on. Being a independent system, the sprinkler applicator can move and work independently in the field. When sprinkler carries out large area irrigation, water supply system or water source should be arranged in the field. The water supply system may be open channels, and may be pressurized pipe. If the hydraulic pressure of a pressurized pipe can meet the pressure requirements of the sprinkler, the sprinkler may not be equipped with a power machine or a pump.

(1) Classification of sprinkler applicator.

In order to meet the needs of different terrains and crops, there are many forms of sprinkler. According to the spraying mode of the sprinkler head, the sprinkler can be divided into two categories: fixed spray type and movable spray type. The fixed type means that the sprinkler stops at one position for spraying. After spraying here, the sprinkler moves to the next new position according to the design requirements and then carries on the irrigation operation until all the areas have been sprinkled. Fixed-spray sprinkler includes: hand-held sprinkler, hand-lift sprinkler, hand-propelled sprinkler, tractor-mounted sprinkler, rolling sprinkler and so on.

Amovable spray sprinkler is a sprinkler that moves while it is spraying. Movable spray applicator includes coil type, center holder type, translation type and so on.

(2) Structural characteristics and application range of sprinkler.

①Hand-held sprinkler. This portable sprinkler can be moved by one person, its power using 0.37~1.5kW micro-motor or 1.5 hp air-cooled combustion engine; The pump uses micro-high-speed centrifugal pump with hand pressure pump or micro-sprinkler centrifugal pump, the pump being connected directly with the power machine; the pipe can be nylon hose or low-density polyethylene pipe; and the sprinkler uses a low-pressure sprinkler with working pressure of 150~250kPa. This portable sprinkler has simple structure, light weight, easy installation and operation, low working pressure, low energy consumption, low price, using 220V single-phase power supply, and it can make full use of small water source. It suits me. After the reform of China's economic system, it is suitable for the management mode of one household in rural areas, which can be used for small-area field crops, vegetables and cash crops under sprinkler irrigation, as well as for garden crops and green land.

②Hand-lift sprinkler. Its engine is generally 3~5 kW or 3~6 hp air-cooled diesel engine; the pump is self-priming centrifugal pumps which adapt to the characteristics of frequent movement; pump and power machine is connected directly by using coupling; pipes commonly are plastic hose or thin-walled aluminum; sprinklers are often alloy pipe withstanding some working pressure, the diameter of the pipe being generally 50mm or 65mm; the nozzle adopts the medium pressure rocker-arm type with working pressure of 300~350kPa. Fig. 2-36 is the sketch map of the hand-lift sprinkler.

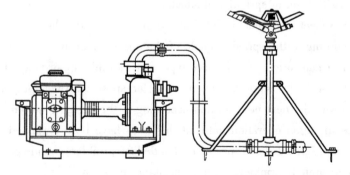

Fig. 2-36　The sketch map of the hand-lift sprinkler

The hand-lift sprinkler is compact in structure, light in weight, simple in operation and easy to maintain, does not need to leave a machine lane, has strong adapta-

Chapter Ⅱ The Main Equipment and System of the Integrated Water-Fertilizer Technology

bility and low price, but when the nozzle is installed directly at the outlet of the pump with a vertical pipe, thelarge vibration will affect the quality of sprinkler irrigation.

Hand-lift sprinkler is suitable for scattered small plots, especially in the complex terrain of mountainous and hilly areas, which can be used to irrigate food crops, vegetables, cash crops, nursery plants and fruit trees, etc.

Hand-propelled sprinkler. When the power engine and the pump are a little larger, it is more difficult to move by two people. Therefore, the pump, the power machine and the transmission mechanism are fixed on the frame with the rubber wheel, and the nozzle is installed on the branch. And it can have many nozzles for spraying. The hand-propelled sprinkler generally uses the 7.5kW motor or the 10~12 hp diesel engine; The pump is equipped with a self-priming centrifugal pump or a common centrifugal pump with a self-priming device; the pump and the power machine can be connected directly, or use the V-belt transmission; the pipes are often made up of aluminum alloy pipes connected by quick joints, using one or two branch pipes; the nozzle adopts medium or low-pressure rocker-arm sprinkler, usually equipped with 8~12 nozzles. Fig. 2-37 is the sketch map of the hand-propelled sprinkler.

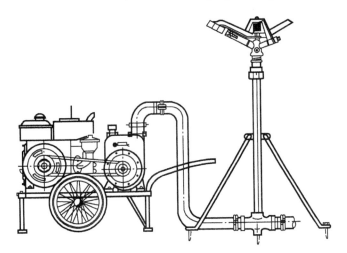

Fig. 2-37 The sketch map of the hand-propelled sprinkler

The hand-propelled sprinkler is simple in structure, low in cost of investment and operation, flexible in use and maintenance, especially when it is elected to use diesel

· 213 ·

engine as power, because 10~12 horsepower diesel engine is the largest power engine in rural areas of China at present. So one engine can have multiple uses, thus save investment. Its main disadvantage is that when used on sticky soil, spraying wet around the machine unit will cause muddy roads and difficult movements. This type of machine is suitable for a variety of crops, especially small plots in plain areas.

③Roller sprinkler. This rolling type is a more mature mechanical mobile fixed spray irrigation machine. In foreign countries, such as the United States, Germany and so on, this type is used in some parts and our country also produces this kind of machine.

In fact, the rolling sprinkler is to use machine to move the branches equipped with sprinklers. The sprinkler adopts a structure with many holders, which consists of lightweight high strength aluminum alloy tubes, large diameter steel coil type high strength aluminum alloy wheels, the central motor for drawing and the water intake hose with quick joint. Fig. 2-38 is the sketch map of the roller sprinkler.

1. Water source; 2. Pump unit; 3. Main water pipe; 4. Water supply hydrant; 5. Connecting hose;
6. Steel wheel; 7. sprinkler; 8. spraying branch; 9. drive vehicle.

Fig. 2-38 The sketch map of the roller sprinkler

The advantages of this type are simple in structure, easy in maintenance, with large control area, high production efficiency, less operators, less daily movement, convenient operation, less field project and less land use. Its main drawback is that it is affected by roller diameter and cannot irrigate high-stalk crops, and its ability to adapt to terrain slope and soil is poor.

Chapter Ⅱ The Main Equipment and System of the Integrated Water-Fertilizer Technology

Thistype is suitable for flat lands with non-cohesive soil and the slope being less than 10%. Often used in sprinkling wheat, grains, beans, vegetables, melons and other dwarf crops, especially suitable for pastures.

④Reel sprinkler. A reel sprinkler uses soft pipes to transfer water, uses sprinkler pressure water to drive the coil rotation during spraying operations. The hose (or cable) are coiling on the reel, hauling a cart fitted with a high-pressure sprinkler or cantilever trusses with several fixed nozzles to move and spray along the line of work. Fig. 2-39 is the sketch map of the reel sprinkler.

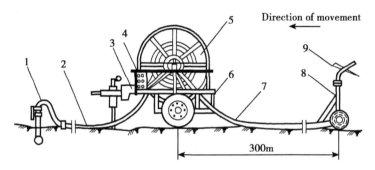

1. Water supply; 2. water pipe; 3. water power; 4. speed controller; 5. reel unit; 6. rack; 7. PE hose; 8. sprinkler cart; 9. long range sprinkler.

Fig. 2-39 The sketch map of the reel sprinkler

The sprinkler is usually composed of a sprinkler cart and a reel unit, supplying pressure water by means of a main pipe or a mobile pumping device. Reel unit includes reel, semi-flexible pipe, rack, walking wheel, motor, speed regulating device and safety mechanism. The sprinkler cart is simpler, including the nozzles and the frame. When transporting, most of the sprinklers can be loaded on the reel. Because of the large body, this sprinkler should be towed by tractor when working and transporting. Compared with other large and medium-sized sprinklers, the reel sprinkler has the following advantages and disadvantages.

The advantages are as follows: It is simple in structure, with low investment per mu; It has more specifications, with good mobility and wide range of application. The operation is simple, which may be automatically controlled with high productivity. A machine can be managed by one person (another man driving a tractor temporarily assisted work), moving one or two places a day. The sprinkler can move with the speed of 20~40m/h, and it can sprinkle 20~60 mu per day. The control area of large-scale reel sprinkler is 200~300 mu.

The disadvantages are as follows: Limited by the machine type, the pipes are with small diameter, large length, and winding layer by layer, so the head loss is large, energy consumption being much, and the operation cost being high. For example, with a diameter of 110 mm and a length of 270 m, the loss of the water supply pipe plus the water turbine is 76% to 98% of the working pressure of the nozzle. Normal pipeline loss accounts for 43%~49% of inlet pressure, that is to say, about half of the energy is lost in the pipeline. If the times of sprinkler irrigation are frequent, the applicability of this type should be carefully considered. A wider motorway (transport road, sprinkler working path) is required, which covers a large area. So, the specification of sprinkler and field planning should be considered as a whole in order to reduce the area of land as much as possible.

The reel sprinkler is suitable for plots of various sizes, shapes and terrain, suitable for irrigation of various tall and dwarf crops (such as corn, soybeans, potatoes, herbage, etc.), as well as for certain fruit trees and cash crops (such as sugar cane, tea, bananas, etc), but the soil is required not to be too sticky.

IV. The Design of Pipe Sprinkler Irrigation System

1. Technical requirements for sprinkler irrigation

(1) Timely and appropriate supply water to the crops. In order to achieve this, it is necessary to establish a reasonable irrigation system to ensure the need of water for the normal growth of crops in arid or semi-arid years. In other words, the design standard of the sprinkler irrigation project must meet the guarantee rate of not less than 85%. According to this standard, a water resource project must be matched, with reasonable volume, reliable structure, safe and convenient operation. And the specifications and dimensions of each part must ensure the implementation of the sprinkler irrigation system.

(2) Higher evenness of sprinkler irrigation. The evenness of sprinkler irrigation here refers to the combined evenness, which is associated with the water distribution of a single sprinkler, the working pressure of the nozzles, the arrangement and spacing of

the nozzles, the uniformity of the speed of the sprinkler, the inclination of the vertical tube, the slope of the ground and the wind speed and wind directions. Under the designed wind speed, the combined evenness coefficient of the fixed sprinkler irrigation system is not less than 75%, and that of the movable sprinkler is not less than 85%.

2. Arrangement of the field pipeline system

The arrangement of the pipeline system depends on the shape of the field, the slope of the ground, the direction of cultivation and planting, the wind speed and direction of the irrigation season, the combineddistance of the sprinklers, and so on. It is necessary to make a multi-scheme comparison and select the best, mainly in the following two forms.

The first one is the arrangement like the Chinese character "feng", as shown in Fig. 2-40.

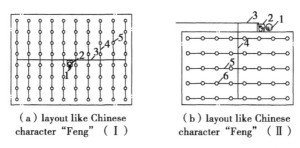

(a) layout like Chinese character "Feng" (I)

(b) layout like Chinese character "Feng" (II)

(a) 1. Well; 2. pump station; 3. main pipe; 4. branch pipe; 5. nozzles. (b) 1. water pool or well; 2. pump station; 3. main pipe; 4. sub-main pipe; 5. branch pipe; 6. nozzles.

Fig. 2-40 "Feng" type layout

The second is the arrangement like a comb, as shown in Fig. 2-41.

(1) Principles for the arrangement of field piping systems.

①It should meet the requirements of sprinkler irrigation project planning.

②The spraying branch should be consistent with direction of the cultivation and crop planting as far as possible.

③The branch pipes should be arranged paralleled with the same height. If the conditions are limited, at least the pipes should not be reverse with the slope as far as possible.

④When the wind direction is relatively constant, the branch pipe should be arranged perpendicular to the main wind direction, and the parallel arrangement with the

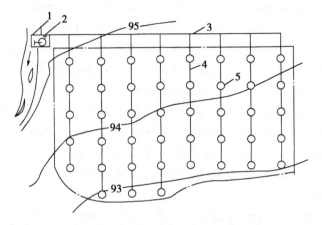

1. Channel; 2. pump station; 3. main pipe; 4. branch pipe; 5. nozzles.

Fig. 2-41　Comb type layout

main wind direction should be avoided as far as possible.

⑤The connection between the branch pipe and the upper pipe should avoid the sharp angle intersection, and the branch laying should be smooth and reduce the bending part.

In the implementation of the above principles, sometimes there will be contradictions, at this time we should analyze and compare the factors according to the specific situation, sorting out the main and secondary, and determine the layout plan according to the local conditions.

(2) The main factors affecting the layout of field piping system.

There are many factors that affect the layout of the pipeline system in the field, which often meet with the phenomenon of mutual restriction among the factors, which makes it possible to make a variety of possible layout schemes under the same conditions. In order to select a technically and economically most advantageous scheme, the main factors affecting pipeline layout are described below.

①Terrain conditions. In the up-and-down irrigation areas, branch pipes are often unable to be arranged along contour lines. At this time, the branch should be laid vertical with the contour line or hetero-tropic with it, so as to compensate for the head loss by the lowering terrain height. If the terrain slope is exactly equal to or close to the hydraulic slope of the branch pipe, it is ideal; if the terrain slope is much larger than the

Chapter II   The Main Equipment and System of the Integrated Water-Fertilizer Technology

hydraulic slope of the branch pipe, the pressure relief valve should be placed in the proper position or by reducing the pipe diameter we can solve the problem. If the upper pipe can only be laid at the lower level, the branch pipe laid on the reverse slope should not be too long.

②The shape of the land. The irregular shape of the land will bring difficulties to the arrangement of the pipeline in the field. Generally, for semi-fixed and mobile piping systems, the direction of branches in the field should be the same, and the length of most branches should be the same as possible.

③Tillage and planting direction. Some sprinkler irrigation areas are in the gentle slope zone, traditional farming and planting direction is along the slope. If the branch pipes are arranged parallel to the contour, the pipes cannot be consistent with the planting direction and tillage. When branches pipes move to use, this can cause a lot of difficulties, and even damage the crops. The spraying branch pipes should be arranged along the slope according to the direction of cultivation. Sometimes there are different farming directions in the same plot, so the direction of cultivation should be adjusted and unified by means of technical and economic analysis and scheme comparison.

④Wind direction and wind speed. Wind has a great influence on the irrigation quality. If the wind speed is very small in the irrigation area in the irrigation season, the arrangement of the branch pipe may not focus on the wind direction, but mainly to meet the other requirements. If the wind speed reaches or exceeds 2 m/s and has the main wind direction, the branch pipes should be arranged vertically to the main wind direction, so that the lateral range of the nozzle under the wind can be compensated by the number of nozzles.

⑤Water source location. Here mainly refers to plain well irrigation area; a well can control about 200 mu of land, forming a small system.

Section IV Water Transfer System of The Integrated Water-Fertilizer Technology-Micro-Irrigation

I. The Concept of Micro-Irrigation

Micro-irrigation, through a low-pressure piping system and special emitters installed on a final pipeline, is a method of evenly, precisely transferring the nutrients and water in a small amount required by the crop growth, directly to the soil surface or soil layer near the root of the crop. Compared with traditional surface irrigation and all area sprinkler irrigation, micro-irrigation is also called local irrigation because only a small amount of water is used to moisten some of the soil near the root zone of the crop.

The irrigation flow of micro-irrigation is small, the duration of one-time irrigation is longer, the irrigation cycle is short, the working pressure is lower, the irrigation quantity can be controlled more accurately, and the water and nutrient can be transferred directly to the soil near the root of the crop. Micro-irrigation can be divided into four forms, according to the different discharge modes of water flow duringirrigation, as shown in Fig. 2-42.

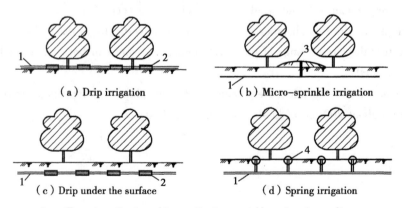

1. capillary pipe; 2. micro-dripper; 3. micro-sprinkler; 4. spring applicator.

Fig. 2-42 Types of micro-irrigation

Chapter Ⅱ The Main Equipment and System of the Integrated Water-Fertilizer Technology

Ⅱ. The Characteristics of Micro-Irrigation

1. Advantages of micro-irrigation

(1) Saving energy and reducing investment. Micro-sprinklers also belong to low-pressure irrigation, the design pressure is generally between 150~200kPa, and the flow rate of micro-sprinkler irrigation system is smaller than that of sprinkler irrigation, so the requirement for pressure facilities is much smaller than that of sprinkler irrigation and can save a lot of energy. The height difference of self-pressure irrigation is smaller than that of sprinkler irrigation. At the same time, the design working pressure is low, the flow rate of the system is small, the pipe diameter and pipe pressure are reduced, and the total investment of the system is greatly reduced.

(2) Adjusting the microclimate in the field, and easy to be automated. Because of the large atomization degree of water droplets under micro-sprinkler irrigation, the humidity of air near the ground can be increased effectively. The field temperature can be reduced effectively in hot weather, and even the micro-sprinkler can be moved to the crown of trees to prevent frost disaster and so on.

(3) The water use efficiency is high, and the effect of increasing yield is good. Micro-sprinkler irrigation also belongs to local irrigation, so the actual irrigation area is smaller than surface irrigation, and the irrigation water quantity is reduced. At the same time, micro-sprinkler irrigation has large evenness of irrigation water, does not cause local leakage loss, and irrigation quantity and irrigation depth are easy to control. According to the demand of crops in different growth period and soil water content condition, irrigation can be timely to improve the water use efficiency. The micro-sprinkler irrigation system with better management can reduce water consumption by 25%~35% compared with that of sprinkler irrigation system. Micro-sprinkler irrigation can also spray soluble fertilizer, foliar fertilizer and pesticides in the irrigation process, which has a significant effect on increasing yield, especially for some special requirements for temperature and humidity, the effect is more obvious.

(4) It is flexible and convenient to use. The sprinkler intensity of micro-sprinkler irrigation is controlled by a single nozzle and is not affected by the adjacent nozzle. The water quantity between the adjacent micro-sprinklers is not superimposed on each other. In this way, the spray diameter and sprinkler intensity can be changed by changing

the nozzles in different growth stages of fruit trees, so as to meet the water demand of fruit trees. The micro-sprinkler can be mobile and can adjust its working position at any time according to the conditions, such as in the trees, rows or plants. In some cases, micro-sprinkler irrigation system can also be transformed with drip irrigation system.

2. Disadvantages

The limitations of micro-irrigation are usually as follows: First, micro-sprinkler irrigation in the field is easily influenced by the weeds and the crop stalks, which affects the quality of spraying. Secondly, the combination of micro-sprinkler irrigation and fertilizer will result in a large number of weeds growth before the crops are ripe enough to cover the row. Thirdly, the requirement for water quality is high. The suspended substances in the water are easy to block the micro-sprinkler, so it is necessary to filter the irrigation water; then the evenness of the irrigation water is affected greatly by the wind. When the wind is greater than grade 3, the micro-spray droplets are easy to be blown away by the wind, and the evenness of irrigation water is reduced. Therefore, the installation height of the micro-sprinkler should be as low as possible to meet the requirements of irrigation, thus it can reduce the effect of wind on spraying.

III. Equipment and Types

1. Emitters

In order to meet the development of micro-irrigation, a variety of emitters have been developed abroad, its classification being as follows.

(1) Classification by the connection between emitters and tubule.

①Intertubular type. Install the emitters in the middle of the two tubules so that the emitters themselves are part of the tube. For example, both ends of a emitter is inserted into two sections of a tubule, so that the vast majority of the water flows through the cavity of the emitter to the next section, while a small portion of the water flows through the side hole of the emitter into the dripper channel, then flow out the dripper. Fig. 2-43 is intertubular emitters.

②Plug-in type. The emitters inserted directly on the wall of the tubule, such as plug-in drippers, microtubule, water inrush, orifice drips and micro-sprinklers, all belonging to the plug-in-the-tube-type emitters. Fig. 2-44 is the plug-in-the-tube-type emitters.

Chapter Ⅱ The Main Equipment and System of the Integrated Water-Fertilizer Technology

Fig. 2-43 Intertubular emitters

Fig. 2-44 The plug-in-the-tube-type emitters

(2) Classificationaccording to the output water type.

①Drip type. The characteristics of drip emitters are that the pressure flow in the tubule after energy dissipation is irrigated to the soil in the form of discontinuous droplets or fine flow. Such emitters belong to drip type as tube dripper, orifice dripper, swirl dripper, etc. Fig. 2-45 is the drip emitters.

②Spray type. The pressure water is sprayed into the soil through the holes in the emitters. According to the different spraying mode, it can be divided into two types: jet rotating type and refracting type. Fig. 2-46 is spray type.

③Spring type. The pressure flow in the tubule irrigates the soil through emitters in

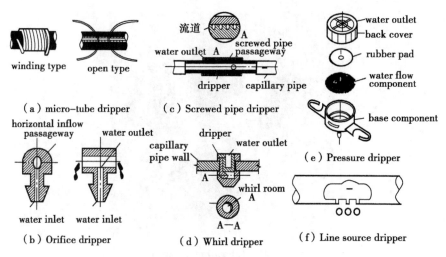

Fig. 2-45 Drip emitters

Fig. 2-46 Spray type

the form of springs. The advantage of spring irrigation is that the head is low, the orifice diameter is large, and it is not easy to clog. Fig. 2-47 is the sketch map of the spring type.

④Seepage type. Pressure water in the tubule enters the soil through many micropores or capillary in the tubule wall. There are two forms of permeable tubule, namely porous permeable tubule and edge-slit membrane tube. Edge-slit membrane tubes are seeped through capillary holes formed by seams. The capillary hole is generally 0.1~0.25mm wide, 0.7~2.5 mm in height and 150~600 mm in length.

Chapter II The Main Equipment and System of the Integrated Water-Fertilizer Technology

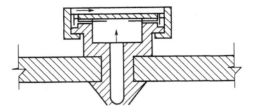

Fig. 2-47 Sketch map of the spring type

⑤Intermittent type. The pressure flow in the tubule flows intermittently and impulsively out of the emitters and infuses into the soil. Therefore, the intermittent type is called pulse irrigation. Fig. 2-48 is the sketch map of the intermittent type.

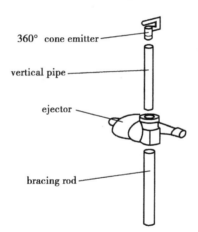

Fig. 2-48 The sketch map of the intermittent type

2. Selection of dripper and micro-sprinkler

(1) Dripper.

The so-called dripper is a device that converts the pressure flow in a tubule into a trickle or a thin flow through a channel or orifice. A dropper has a flow rate of not more than 12 L/h. It can be divided into the following types according to the ways of energy dissipation.

①Long channel dripper. Long-channel drippers adjust the volume of output water by friction and dissipation energy between the water flow and the channel wall, such as micro-irrigation dripper, inner thread tube dripper, etc, as shown in Fig. 2-49,

Fig. 2-50.

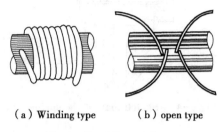

(a) Winding type   (b) open type

Fig. 2-49　Micro dripper

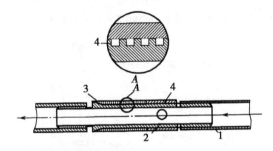

1. Capillary pipe; 2. dripper; 3. water outlet; 4. screwed thread.

Fig. 2-50　Screwed pipe dripper

②Orifice type dripper. The local head loss caused by the orifice outlet of the dripper is used to dissipate energy and adjust the flow rate as shown in Fig. 2-51.

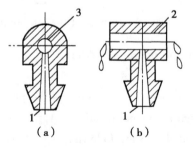

1. inlet; 2. outlet; 3. channel.

Fig. 2-51　Orifice type dripper

③Vortex dripper. Vortex-type drippers regulate the output amount of water by the eddy current formed in the vortex chamber of the emitters by the water flow into the e-

Chapter Ⅱ The Main Equipment and System of the Integrated Water-Fertilizer Technology

mitters. When the water flows into the vortex chamber, the centrifugal force produced by the flow rotation forces the water to flow towards the edge of the vortex chamber, and in the center of the vortex, a low pressure zone is generated, so that the pressure at the outlet of the center is lower, so the flow rate is adjusted, as shown in Fig. 2-52.

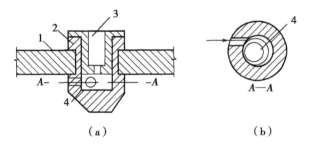

1. Capillary pipe wall; 2. Dipper; 3. Water outlet; 4. Vortex room.

Fig. 2-52 Vortex dripper

④Pressure compensated dripper. The pressure compensated dripper makes use of the effect of the water pressure on the elastic body (piece) in the dripper to change the shape of the channel (or orifice) or the size of the water section, that is to say, when the pressure decreases, the area of the cross section of the water increases; when the pressure increases, the cross-section area of the dripper is reduced, so that the output rate of the dripper automatically keeps stable, and it also has the function of self-cleaning. Fig. 2-53 shows the dripper name and code.

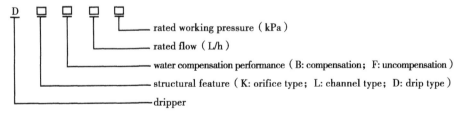

Fig. 2-53 The dripper name and code

(2) Micro nozzles.

Compared with general sprinkler irrigation, micro-sprinkler irrigation has the characteristics of small flow rate, near spraying range, small diameter of the sprinkler hole and low working pressure. The main difference between micro-sprinkler irrigation and general sprinkler irrigation is the sprinkler head (nozzle). Therefore, micro-sprinkler irrigation is often a part of micro-irrigation, but compared with drip irrigation, micro-sprinkler irrigation has larger wet area and better anti-blocking performance.

With the exception of the nozzle, the pipe network is the same as drip irrigation system or sprinkler irrigation system. There are many kinds of micro-sprinklers nozzles. According to the spraying modes, they can be divided into several types, such as refracted micro-sprinkler, rotary micro-sprinkler, centrifugal micro-sprinkler, slit-type micro-sprinkler and so on.

①Refracted micro-sprinkler. The refracted micro-nozzle mainly depends on the refraction plate at the top of the nozzle to change the direction of the pressure flow from the nozzle and then to spray it around. Under the action of air resistance, the water flow is crushed into atomized droplets and dropped to the ground for irrigation. According to the structure and shape of the refraction plate, the refraction micro-sprinkler can be divided into unidirectional refraction micro-nozzle, bi-directional refraction micro-nozzle, angle refraction micro-nozzle, plum-shaped refraction micro-nozzle and so on. The shape of the refracted micro-sprinkler is shown in Fig. 2-54. There are two kinds of spraying water: linear and full circle.

Fig. 2-54 **The sprinkling shape of the refracted micro-sprinkler**

②Rotating micro-nozzle. The rotating micro-nozzle depends on the jet of the nozzle toforce the rotating arm with the curved groove to spray the water on the ground around the nozzle. The rotary type of the rotary micro-nozzles consists of rotary arm type, rotary wheel type, rotary long arm type and so on. Among them, rotary long arm also has double arms and double nozzles, three arms and three nozzles, four arms and four nozzles and so on. Rotary micro-sprinkler is generally full-circle sprinkler irrigation.

③Centrifugal micro-nozzle. Centrifugal micro-nozzles are dispersed by centrifugal action produced by spiral channels in the nozzle. There are two kinds of sprinklers,

Chapter Ⅱ The Main Equipment and System of the Integrated Water-Fertilizer Technology

plastic and copper, and the flow rate and range can be adjusted. The range is 3 to 9 meters. Centrifugal micro-sprinkler is similar with the rotary micro-sprinkler, generally being full-circle sprinkler.

④Slit-type micro-nozzle. The slit-type micro-nozzle sprays through several outlet holes or crevice on the nozzle, whose spraying type is similar to the refracted micro-nozzle, with different spraying angles and different ranges.

Chapter III Selection of Fertilizer Suitable for the Integrated Water-Fertilizer Technology

This chapter mainly explains the principle of crop nutrient absorption, fertilizer selection and nutrient management. The emphasis is on the selection of fertilizers, which are described in detail under what conditions, and whether there is an interaction between the two fertilizers.

Section I The Principle of Nutrient Absorption by Crops

I. Essential Nutrients for Crops

There are three criteria for whether an element is the nutrient element needed by a plant during its growth (as shown in Fig. 3-1): ①The lack of an element causes the plant to be unable to complete the entire life cycle; ②The lack of an element will make the plant have strange symptoms in the process of growth, only if the element is added, these strange symptoms will disappear; ③This element has a direct nutritional effect on plant metabolism, but has no effect on growth environment.

Macro-elements include carbon, hydrogen, oxygen, nitrogen, phosphorus, and potassium, the content being several percent in plants. Carbon, hydrogen, and oxygen are also organic components, mainly from water and air. Plants are in relatively large demand for the three elements. However, the content in the soil is relatively small and can only be added by fertilization. Nitrogen, phosphorus and potassium fertilizer is the kind of fertilizer that plants need more.

Medium elements include calcium, magnesium and sulfur. The contents of these elements are more in soil than in plants, so it is not necessary to supplement them by

Chapter Ⅲ   Selection of Fertilizer Suitable for the Integrated Water-Fertilizer Technology

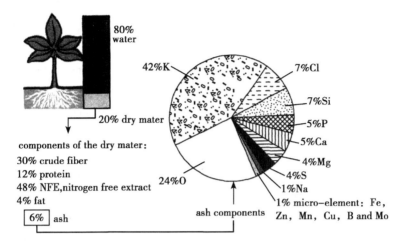

Fig. 3-1   The components of the mineral element

fertilization. However, the precipitation in southern China is relatively large, so it is necessary to supplement the contents by fertilization.

Trace elements include iron, copper, zinc, manganese, molybdenum, boron and chlorine, which are relatively few, but they are very important to plants. The content in soil can easily meet the needs of plants, but some trace elements can not meet the needs. It also needs to be supplemented by fertilization.

Some elements are called beneficial elements, which have some effect on the growth of plants, but they are not essential, or only effective under certain conditions, such as sodium, silicon, cobalt, vanadium, selenium, aluminum, iodine, chromium, arsenic, and cerium, etc. The use of these beneficial elements should be moderate, otherwise they would lead to serious damage to plants. Too little amount will influence the normal growth of plants, but too much will make plants toxic. Plant demand for beneficial elements is generally very small, so it is necessary to control the use of beneficial elements.

Ⅱ. The Important Law of Proper Fertilization

1. The theory of nutrient return (compensation)

The theory of nutrient return was proposed by J. V. Liebig, a German agricultural chemist, in the works "The Application of Chemistry in Agriculture and Physiology" in 1840. From the point of view of natural science, Libich regards the phenomenon of life on the earth as a circular process of movement, and he thinks that agriculture is the basis of material exchange between man and nature. Humans and animals absorb assimilated nutrients from the soil and atmosphere through food and return them to the soil through the plant itself and animal excrement. He said, humans grow crops on the land and take these produces away, so this is bound to make the land force gradually decreased, reducing the nutrient content of the soil. It is therefore necessary to return everything that has been taken from the soil in order to restore the ground. Otherwise it would be difficult to yield as high as it used to be, so the soil should be fertilized in order to increase the production.

The connotation of nutrient return theory (compensation) includes the following aspects: ①Plants take away some nutrients from the soil when harvest, and over time, the nutrients in the soil will be less and less. ②With the decrease of nutrients in the soil, the yield of plants becomes less and less, so it is necessary to return the nutrients to the soil in time to maintain the soil fertility. ③Fertilization is one way to keep the balance of various elements in the soil and improve the yield.

The theory of nutrient return (compensation) also has one-sidedness: ①Liebig thought that carbonate in the atmosphere is the only source of nitrogen nutrition of crops, so his estimate of soil nutrient consumption only focuses on the mineral elements such as phosphorus, potassium and so on. ②Based on the theory of mineral nutrition, Liebig over-evaluated the mineral nutrient composition and under-evaluated the role of organic nutrition. At the same time, he objected to Boussingault's view that manure mainly supplied nitrogen. He erroneously assumed that the minerals needed by plants are supplied in the form of manure. ③He opposed Boussingault's claim that legume crops can enrich soil nitrogen, and erroneously believe that plants can only absorb their living needs from the soil. Each harvest takes away the nutrients and minerals from the soil and causes the soil to be depleted. ④Everything that has been taken from the soil must be returned to restore the earth. But the actual fertilization does not need to return any nutrients taken by crops from the soil, such as high soil nutrients, or non-essential nutrient elements for crop growth, and so on.

2. The law of diminishing marginal return and the Mitscherlich's law

The law of diminishing marginal return was proposed by the 18th-century econo-

Chapter III Selection of Fertilizer Suitable for the Integrated Water-Fertilizer Technology

mists A. R. Turgot and J. Anderson, as shown in Fig. 3 – 2. The law of diminishing return is follows: the reward from a certain land increases with the increase of the amount of labor and capital invested into the land, but with the increase of the unit of labor and capital invested, the reward is gradually decreasing. In the early 20th century, on the basis of previous work, E. A. Mitscherlich et al discussed in depth the relationship between the amount of fertilizer applied and the yield (sand culture test of oat phosphate fertilizer). The experimental results show that: ①Under the premise of the relative stability of other technical conditions, the crop yield increases with the gradual increase of fertilizer, but the increase unit of the crops declines with the increase of fertilizer, that is, ($\Delta Y_1/X_1 > \Delta Y_2/X_2 > \Delta Y_3/X_3 \cdots$), which is consistent with the law of diminishing return by predecessors. ②If all the conditions are satisfied, the crop will produce some kind of maximum yield. On the contrary, if any major factors lacks, the yield will decrease accordingly.

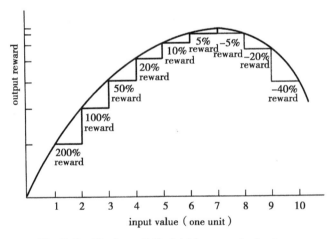

Fig. 3-2 **The law of diminishing marginal return**

The Mitscherlich's formula only shows the law of increasing the yield by increasing the fertilizer, but fails to reflect the phenomenon of crop yield reduction when the amount of fertilizer exceeds a certain range. Pfeiffer et al proposed a parabola model reflecting the relationship between fertilizer and yield. The one-dimensional quadratic equation is as follows: the equation is $y = b_0 + b_1 x + b_2 x^2$, among which b_0、b_1、b_2 are the coefficients, and they can be calculated according to the experimental data. The equation effect of this formula is as follows: when the amount of fertilizer is very low, the yield of crop increases almost in a straight line, but when the amount of fertilizer exceeds the maximum rate of fertilizer, the increase of fertilizer will not increase the yield, but on the contrary, the yield will decrease. This result has been confirmed by many experiments at home and abroad.

3. The law of equal importance

The elements needed by crops, whether macro-elements or trace elements, are indispensable. Even if the crops do not need much of this element, the lack of it will also influence the normal growth of crops, leading to some functional decrease. The law of equal importance indicates that trace elements, rare elements and macro-elements are equally important to crops.

4. The irreplaceable law

All the elements in the crop have a certain effect and can not be replaced by other elements, for example, nitrogen cannot replace the absent potassium. If an element is in the lack, the relevant chemical fertilizer containing this element must be used.

5. Law of comprehensive action of factors

Factors affecting crop growth and development include water, nutrients, light, air, variety and tillage conditions. There must be a limiting factor that plays a leading role, and the yield is also restricted by this factor to a certain extent. In order to improve the economic benefit of agriculture, when fertilizing it is necessary to combine with some agricultural technical measures, and the nutrients must also be used in coordination with each other.

III. Factors Affecting the Absorption of Nutrient Elements in Crops

Crops generally absorb nutrients through their roots in the soil. Therefore, in addition to the genetic characteristics of the crops itself, some factors affect the crop's absorption of nutrients, such as soil and other environmental factors.

The factors influencing nutrient absorption include nutrient concentration inthe medium, temperature, light intensity, soil moisture, ventilation, soil pH value, physical and chemical properties of nutrient ions, metabolic activity of roots, seedling age, and the nutrient condition in the plants during the growth period, etc.

Chapter Ⅲ Selection of Fertilizer Suitable for the Integrated Water-Fertilizer Technology

1. The nutrient concentration in the medium

Through a large number of studies, it is found that when the concentration is low, the ion absorptivity will increase with the increase of nutrient concentration, but the speed is not very fast. When the concentration is high, the ion absorption selectivity is low. The main reason is that the transpiration rate has a great effect on the absorption of ions. Therefore, in the use of fertilizer, it should be used in batches, which is conducive to the absorption of crops.

2. The types of nutrients in the medium

The nutrientinter - ions in the medium have certain antagonistic and assisting effects. The so-called antagonism of the inter-ions refers that the existence of one ion in solution can inhibit the absorption of another ion. This is embodied between cations, such as between K^+, Rb^+ and Cs^+, between Ca^{2+}, Mg^{2+} and Ba^{2+}, between NH_4^+ with H^+ and Ca^{2+}, between K^+ and Fe^{2+}; And it exists between anions, such as between Cl^-, Br^- and I^-, between $H_2PO_4^-$ and OH^-, between $H_2PO_4^-$ and Cl^-, between NO_3^- and Cl^-, between SO_4^{2-} and SeO_4^{2-}.

The so-calledassisting effect of the ions means that in the solution the existence of a certain ion is beneficial to the absorption of the roots to absorb the other ions. This is embodied between cations and anions, for example NO_3^- and SO_4^{2} are beneficial for the absorption of the cations. This is also embodied between divalent cations, trivalent anions and monovalent anions, for example Ca^{2+}, Mg^{2+} and Al^{3+} are beneficial for the absorption of K^+, Rb^+, Br^- and NH_4^+.

3. The Metabolism of Roots and the Outputs of Metabolites

The accumulation of ions and other solutes in many cases is inverse concentration gradient, which requires direct or indirect energy consumption. In the cells and tissues that do not carry out photosynthesis (including roots), the main source of energy is respiration. Therefore, all factors affecting respiration may affect the accumulation of ions.

4. The seedling age and growth stage

(1) Nutrient absorption of crop seeds. Before and after the seed germination, its absorption usually depends on the storage of nutrient, and till the trifoliate stage, it is necessary to provide nutrients through the medium.

(2) The nutrients absorbed by crops in different growth stages are different. At the

beginning of growth, the amount of nutrients absorbed by crops is relatively small, but with the passage of time, more and more nutrients are absorbed. At the later stage of maturity, the absorption of nutrients by crops becomes smaller.

5. The critical period of plant nutrition and the maximum efficiency period of nutrition

The critical period of nutrition and the maximum efficiency period of nutrient are the two most critical periods in the fertilization. The critical period of plant nutrition mainly refers to the excessive or too little nutrient elements, which causes the elements in the plant unbalanced, and has obvious adverse effects on the normal growth of the crops. It occurs in the early stage of the crop, such as the critical period of the phosphorus nutrition appears in the seedling stage. The maximum efficiency period of the plant nutrition mainly refers to the period in which the absorbed nutrient can exert the greatest effect in the plant body, in the most vigorous period of crop growth. During this period, due to the rapid growth of the crops, the nutrient absorbed capacity is also very large. To achieve the purpose of increasing the yield, the need for nutrients must be met during this period. In the process of managing fertilizer and water, it is necessary to pay attention to the critical period of fertilization, and to face the continuity of plant absorption, adopting the method of combining base fertilizer, top dressing and seed fertilizer.

IV. Misunderstanding of Fertilization

At present, there are manyimproper ways, such as the methods of fertilization, the time of fertilization and the amount of fertilizer, and there are many misconceptions, so that the fertilizer will cause the crop to die. The main errors are as follows:

1. Preferring nitrogen fertilizer, but neglecting the phosphorus and the potassium fertilizer.

The focused nitrogen fertilization will cause the stem and leaf to be long, the tissue soft, the disease resistance reduced. It will influence the normal growth and development of the later stage, the transformation of the nutrient substances. The yield is reduced, and the quality is reduced. Therefore, the nitrogen fertilization need proper amount and right time, and the fertilizer should be combined with the potassium fertiliz-

Chapter III Selection of Fertilizer Suitable for the Integrated Water-Fertilizer Technology

er and the phosphate fertilizer to ensure the normal growth of the crops to improve the yield.

2. Neglecting the application of trace elements.

Only paying attention to the macro-elements, but ignoring the trace elements will lead to the fact that the trace elements can not keep up with, but also the absorption of the macro-elements will be affected, resulting in unnecessary waste. Neglect of trace elements will lead to plant deformities, the fall of flowers and fruits, the decline of crop yield and quality, etc. Therefore, according to the characteristics of crops, macro-elements should be applied together with the trace elements, so as to ensure the normal growth and development of crops.

3. The more fertilization, the better.

Although sometimes excessive amount of fertilizer can increase the yield of crops, but because of the high cost, the actual benefits are not high. Sometimes overfertilization can lead to underdevelopment of the crops' reproductive organs, resulting in reduced yields, which can only be counterproductive. Therefore, according to the characteristics of crops, soil fertility and planting density, we should find the best fertilization scheme to avoid waste and give full play to the effect of fertilizer, so as to increase the economic benefit.

4. Fertilizing after some symptoms because of lacking fertilizer.

After the fertilization, it will take some time for crops to absorb and use it. If the crop symptoms are found, then fertilize the crops, it will not only prolong the time of lacking fertilizer, but also lead to yield reduction. Therefore, the time of fertilizer should be based on the general characteristics of the crops, light, temperature, water, and the fertilization method and other factors.

5. Surface fertilization.

Fertilizer is easy to volatilize and lose or it is difficult to reach the root of the crop, which makes it difficult for crops to absorb, resulting in a very low utilization rate. Therefore, when fertilizing, we must determine the position of fertilization, ensure that the roots to successfully absorb and ensure the effects of fertilization.

Section II   Selection of Fertilizers

I. Characteristics and Selection of Chemical Fertilizer

There are many kinds of fertilizers, generally divided into two categories: chemical fertilizers and organic fertilizers. Chemical fertilizer, also known as mineral fertilizer or inorganic fertilizer, refers to fertilizer which is synthesized by chemical method or produced by certain minerals, some of which are industrial by-products. Compared with organic fertilizer, chemical fertilizer has many advantages and disadvantages.

The components of the chemical fertilizer are inorganic substances, and the nutrient are concentrated, such as the main nutrient components of the nitrogen fertilizer are nitrogen, the main components of the phosphorus fertilizer are the phosphorus, and the main components of the potassium fertilizer are potassium, and the like. The compound fertilizer contains several nutrient elements, but the main nutritional components are two or three of the nitrogen, the phosphorus and the potassium. The nutrition contents of the chemical fertilizer are higher than that of the organic ones, for example, ammonium acid carbonate contains N 17%, ammonium sulfate contains N 21%, and urea contains N 46%. But the organic fertilizers contain less N, such as night soils contain N 0.5% ~ 0.8%, phosphate 0.2% ~ 0.4%, potassium 0.2% ~ 0.3%, and average nitrogen content of manure is N 0.55%, phosphorus 0.22%, potassium 0.55%. The nitrogen content of 1kg urea is equivalent to 60 to 70kg of human fecal urine (night soil) and 100kg of manure. In addition, the transportation and storage of chemical fertilizers are convenient, so if managed properly, there would be little nutrient loss. But the management processes of organic fertilizer are more complex, such as the accumulation, heap rot, storage, etc. If it is not properly managed, it is easy to lose nutrients. Therefore, the chemical fertilizer has the advantages of high nutrient content, low rate of application, remarkable fertilizer efficiency and it is convenient for transportation, storage and application.

The common chemical fertilizer is mostly water-soluble, and can be easily dis-

Chapter III Selection of Fertilizer Suitable for the Integrated Water-Fertilizer Technology

solved in water after being applied to the soil, and can be quickly absorbed by the roots of the crops. And the fertilizer effect is fast, such as the ammonium sulfate and the urea etc. The fertilizer can be effective in a few days after the fertilization. Because the nutrient content of the chemical fertilizer is high, the fertilizer efficiency is fast and timely, and the yield increasing effect on the crops is very obvious. According to the results of the National Chemical Fertilizer Test Network, the grain yield can be increased by 5~10kg per kilogram of chemical fertilizer. According to the Food and Agriculture Organization of the United Nations (FAO), the contribution of chemical fertilizer to crop production is about 40%~60%. The organic fertilizer has many kinds of nutrients, but its fertilizer efficiency is slow, stable and lasting, and it has a long time of aftereffect, which is superior to chemical fertilizer in improving and fertilizing the land and soil. Crop growth requires an appropriate proportion of nutrients, but the chemical fertilizer has a single nutrient so it can be combined in different proportions. On one hand, it can meet the requirements of various crops for different nutrients. On the other hand, the effect of combined fertilization is better than that of single application, which is called the interaction effect of fertilizer. The application of chemical fertilizer has strong pertinence. Different kinds and characteristics of fertilizers can be applied according to soil and different crops. And according to different growth period of crop, different fertilization methods would be applied, such as topdressing different fertilizer at different stage, spraying on the leaf, and so on. Especially in modern formula fertilization technology, it is easy to operate with chemical fertilizers.

However, chemical fertilizer has some shortcomings. At present, the common chemical fertilizer in our country is still mainly unit fertilizer or single-substance fertilizer, such as nitrogen fertilizer, phosphate fertilizer, potassium fertilizer, etc. This kind of fertilizer is single in nutrient type and high in content, and if it is not applied properly, it is easy to cause imbalance in the proportion of nutrients absorbed by crops. Most of the commonly used chemical fertilizers are water soluble, easy to lose, volatilize and fixed by the soil when applied. So the fertilizer efficiency is not long-lasting, and the use efficiency is low. For example, the ammonia volatilization is easily caused by surface application of ammonium nitrogen fertilizer in alkaline soil, and nitrate nitrogen fertilizer is easily lost with water in rainy areas or under irrigation conditions. And the water-soluable phosphatic fertilizer is easily lost and lead to the fixation when applied to soil, thus reducing the fertilizer efficiency. Long-term large-scale application of some chemical fertilizer will cause soil consolidation, soil structure deterioration, increase of soil salt content, secondary salinization and physical and chemical deterioration of soil properties to a certain extent. In addition, another important difference between chemical fertilizer and organic fertilizer is that it does not contain microorganisms, which is disadvantageous to the transformation of soil nutrients and the improvement of soil biological properties. Therefore, chemical fertilizer is not as good as organic fertilizer in improving soil. Chemical fertilizer is a high energy-consuming product, with high production cost and use cost, so chemical fertilizer alone will increase the cost of agricultural production. But the organic fertilizer can be accumulated and used on the spot, which is beneficial to reducing the agricultural production cost. In addition, from the effect of fertilization on the environment, the chemical fertilizer is more likely to cause environmental pollution compared with the organic fertilizer.

II. Types of Fertilizers Used for Irrigating Fertilization

Irrigating fertilization techniques have strict requirements, especially for equipment, fertilizers and management methods. The channel of drip irrigation emitters is relatively small, so only water-soluble solid fertilizer or liquid fertilizer can be used in fertilization to prevent the blockage of the channel. The larger emitter is the sprinkler head, and it spurts like rain, so foliar fertilizer can be sprayed. The choice of fertilizer in the use of sprinkler fertilization is not very demanding.

Because fertilizers such as urea, ammonium bicarbonate, ammonium chloride, potassium sulfate and potassium chloride are relatively high in purity and relatively few with impurities, so they do not subside when dissolved with water. And they reached the national standards and industry standards, which can be used as top fertilizer. When phosphorus is lacking, potassium dihydrogen phosphate soluble fertilizer can be used as top-dressing fertilizer. When adding trace elements with topdressing, it can not be used together with phosphorus, for it is easy to form insoluble precipitate, blocking the dripper or sprinkler.

Special attention should be paid to the following aspects: first, the fertilizer containing phosphate ions is precipitated when mixed with fertilizers containing metal ions such as calcium, magnesium, iron and zinc. Second, when the fertilizers containing

Chapter Ⅲ　Selection of Fertilizer Suitable for the Integrated Water–Fertilizer Technology

calcium ions mixed with fertilizers containing sulfate ions, precipitation will occur. Third, it is better to use immediately after mixed. Four, for the fertilizer that will precipitate after mixing, separate injection method should be adopted to solve the problem.

All kinds of special fertilizer for drip irrigation can be used directly, such as the Japanese commercial fertilizer produced in Qingdao, Shandong Province, have a formula of 16 : 16 : 16, 19 : 7 : 19, 16 : 6 : 21, 20 : 9 : 11, etc. All kinds of water soluble fertilizer or biogas liquid after filtration can be used directly, as shown in Fig. 3-3.

Fig. 3-3　Main soluble fertilizer

When using the organic fertilizer in micro-irrigation system, it is necessary to solve two problems: first, the organic fertilizer must be liquid. Second, the organic fertilizer must be filtered several times. The general organic fertilizer is suitable for micro-irrigation and fertilization, especially if the organic fertilizer is easy to rot and the residue is small. However, it is difficult for the organic fertilizer containing cellulose and lignin to flow through the micro-irrigation system. Some organic fertilizer is not suitable for the the micro-irrigation system for the fertilizer has much residue after the rot process. A micro-irrigation system may be used to fertilize as long as there is no residue that blocks the system after the decay.

Ⅲ. The Interaction of Various Factors among Fertilizers

1. Reaction during fertilizer mixing

In order to prevent the accumulation of residues that block the micro-irrigation system, more than two storage tanks should be used in the system, one containing calcium, magnesium and trace elements, and the other containing sulfate and phosphate. Thus it can effectively avoid fertilizer mixing, and can irrigate and fertilize safely and effectively.

2. Temperature changes during fertilizer dissolution

Different fertilizers cause thermal reactions when they are dissolved. For example, phosphoric acid will release a large amount of heat when dissolved, causing the temperature of water to rise, and urea will absorb the heat around it and decrease the temperature of water when dissolved. These reactions are of great guiding significance for the mixture of nutrients in the field. In order to prevent salting-out, for example, fertilizers should be dissolved in a reasonable order when the weather and temperature are low, and the heat generated during dissolution should be used to prevent salting-out. The solubility of each liter of fertilizer varies at different temperatures, as shown in Table 3-1.

Table 3-1　Solubility of chemical fertilizer at different temperatures　(g)

Compound molecular formula	0℃	10℃	20℃	30℃
$CO(NH_2)_2$	680	850	1 060	1 330
NH_4NO_3	1 183	1 580	1 950	2 420
$(NH_4)_2SO_4$	706	730	750	780
$Ca(NO_3)_2$	1 020	1 240	1 294	1 620
KNO_3	130	210	320	460
K_2SO_4	70	90	110	130
KCl	280	310	340	370
K_2HPO_4	1 328	1 488	1 600	1 790
KH_2PO_4	142	178	225	274
$(NH_4)_2HPO_4$	429	628	692	748
$NH_4H_2PO_4$	227	295	374	464
$MgCl_2$	528	540	546	568
$MgSO_4$	260	308	356	405

Chapter Ⅲ Selection of Fertilizer Suitable for the Integrated Water-Fertilizer Technology

Section Ⅲ Nutrient Management

Ⅰ. Soil nutrient test

Soil testing is a necessary means of determining fertilizer requirements for plants grown in the soil. The soil analysis should clarify whether the content of a certain nutrient element in the soil is sufficient or deficient for the crop to be planted. The soil itself contains a variety of nutrients, and nutrients remain in the soil by the prior chemical fertilizer or an organic fertilizer. But only a small portion of the nutrients in the soil can be absorbed by the plant, i. e. , to be effective to the plant. The nitrogen exists mainly in the organic matters, and only the nitrate nitrogen and the ammonium nitrogen can be absorbed and utilized by the plant only by the decomposition of the microorganism. Only a small part of the phosphorus in the soil is quick-acting phosphorus, but the soil phosphorus pool will release them to maintain the phosphorus concentration of the soil. Only the exchangeable potassium in the soil and the potassium in the soluble liquid can be used by the plant, but as the effective potassium is continuously absorbed, the dynamic equilibrium between it and the fixed potassium is broken and the potassium is converted to the soil solution. The total content of the nutrient elements in the soil measured cannot show their effectiveness on the plants. Now it has been found a method of extracting the potential effective nutrients, which is widely used in soil analysis laboratories and the analysis of data can be reliably used to estimate the availability of nutrients.

The extraction methods of different elements and different soils are also different. Some methods use weak acid or weak base as extractant and some use ion exchange resin to simulate the absorption of nutrients by roots. Cation availability (such as potassium ions) is usually the exchangeable part of the extraction. The results of crop response to nutrients in field must be used to correct the analytical data before diagnosis with analytical data.

When determining the nutrient requirement of a crop, it is necessary to subtract the effective nutrient content contained in the soil from the total crop demand for nutrients from. In addition, the water-soluble nutrients used in irrigation and fertilization, especially phosphorus fertilizer, will react in the soil and the effectivity will be reduced, which must be taken into account when fertilizing. For example, the amount of phosphate fertilizer is usually larger than the actual needs of plants, so as to meet the absorption of plants.

The tests on the soil and growth medium should include two other parameters: conductivity (EC) and pH. The conductivity of water extract in soil or growth medium can show the content of soluble salt. The nutrients which were not absorbed or lost by plants after fertilization and irrigation would result in salt accumulation. The increase of salt concentration would increase the osmotic pressure of the roots environment, thus decrease the absorption of water and nutrients by root system, resulting in the reduction of yield. Some excess ions can be toxic to plants and have side effects on soil structure.

The pH value of soil and growth extracts reflects the acidity and alkalinity of soil and growth medium. Most plants grow best when pH is close to neutral. Some fertilizers can be acidified, such as ammonium compounds that increase acidity by oxidation to nitrate nitrogen. In the medium with weak buffering effect, such as sandy and coarse soil, acidification is more obvious than that in fine soil. When irrigation water contains excessive sodium ions, the soil will be alkaline.

II. Tests of Plant Nutrients

1. Methods of plant testing

(1) Methods of plant testing.

According to the test method, there are chemical analysis, biochemical method, enzymatic method, physical method and so on.

Chemical analysis is the most commonly used and most effective method for plant testing. According to the different analytical techniques, it is divided into routine analysis and tissue rapid measurement. Dry samples are often used in plant routine analysis, and the tissue test refers to the analysis of the concentration of active ions in fresh tissue juices or extracts. The former method is the main technique for evaluating plant nutrition. The latter is simple and rapid and is suitable for direct application in the field.

Biochemical method is one for determining the nutritional condition of plants by some chemical substance, such as the determination of aspartic acid in rice leaf sheaths

Chapter III Selection of Fertilizer Suitable for the Integrated Water-Fertilizer Technology

or leaves, or the use of starch-iodine inverse ratio as a nutritional diagnostic method for nitrogen.

Enzymatic method: the activity of some enzymes in crop is closely related to the amount of some nutrient elements. According to the change of the enzyme, the abundance and shortage of certain nutrient elements can be judged.

Physical methods: such as leaf color diagnosis, blade color ⇔ Chlorophyll ⇔ Nitrogen.

(2) Measuring position.

Generally speaking, there is a certain relationship between the nutrient concentration of different parts of the plant and that of the whole plant. However, the nutrient concentration of different parts of the same organ is also very different, and the difference between different nutrients is not the same. The nitrogen concentration of the leaf and root is more sensitive to the change of nitrogen supply. Therefore, it can be used as a sensitive indicator.

2. Determination of indexes in plant testing

(1) The representation method of Diagnostic Indexes.

①Critical value

②Nutrient ratio: because of the interaction of nutrient elements, the change of the concentration of one element often leads to the change of other elements. Therefore, it is better to use the ratio of nutrients as a diagnostic index than the critical value of a single element to reflect the relationship between abundance and shortage of nutrients.

③Relative yield

④DRIS method: a comprehensive system of diagnosis and fertilization recommendation. Based on the principle of nutrient balance, the DRIS method is used to study the soil by using leaf diagnosis technique, considering the balance of nutrient elements and the factors affecting crop yield, researching on the relationship between plant, soil and environmental conditions, and their relationship with yield.

Index method: first, a large number of leaf analysis was carried out to record the yield results and various parameters that might affect the yield. The materials were divided into two groups: high yield group (B) and low yield group (A). The results of leaf analysis were shown by N%, P%, K%, N/P, N/K, K/P, NP, NK, PK and so on. Calculating the average value of each form, the standard deviation (SD), coefficient of variation (CV), variance and the variance ratio (S_A/S_B) of the two forms. The form of keeping the highest variance ratio is chosen as the representation of diagnosis. The diagnosis of N, P, K is shown by N/P, N/K, K/P.

When applied, N, P, K index of can be obtained according to the following formula:

N index = + [f (N/P) + f (N/K)] /2
P index = − [f (N/P) + f (K/P)] /2
K index = + [f (N/P) − f (N/K)] /2

When the measured N/P > N/P, the formula is

F (N/P) = 100× [N/P (measured) /N/P (Standard) −1] ×10/CV

When the measured N/P is less than the standard N/P, the formula is

F (N/P) = 100 × [1−N/P (Standard) /N/P (instance)] ×10/CV

f (N/K), f (K/P) and so on

The algebraic sum of three exponents is 0. The higher the negative index value is, the greater the nutrient need intensity is, and the higher the positive index is, the smaller the nutrient need intensity is.

If a wheat index is N=−13, P=−31, K=44, the fertilization need is P>N>K.

Evaluation: this method has a higher accuracy on a variety of crops, not affected by sampling time, location, plant age and variety, superior to the critical value method, but it only points out the degree of crop demand for a certain nutrient, but does not determine the amount of fertilization.

(2) Method of determining diagnostic indicators.

The diagnostic index shall be determined only if a lot of test data is obtained from the production test. The following methods are generally used.

①Field investigation and diagnosis. A representative plot is selected in a region, and the chemical diagnosis is carried out before the sowing period or growing period, and various data are collected according to local experience. We can find and the law of the variation of the yield and the nutrient and the like under different conditions, and divide them into different grades as the diagnostic standard.

②Field calibration. In other words, the research on the division of nutrient sufficiency or deficiency index. Using the field multi-point tests, the curve between the nutrient determination value and the relative yield is found, generally dividing into four

Chapter III Selection of Fertilizer Suitable for the Integrated Water-Fertilizer Technology

level "high, medium, low, very low". The advantage of field verification is that the local natural conditions can be fully reflected, and the factors affecting the supply of nutrients are shown in the grading index. The accuracy of the obtained index is high, but the time and the repetition are required.

The field trials are divided into short-term and long-term. The short-term test is generally a year, with fertilization and non-fertilization of a certain nutrient, requiring the repetition of multiple test points. According to the relative yield, the level is graded. The overlapping effect of the fertilizer cannot be reflected due to the short service life. As a result, it is only possible to decide whether to apply fertilizer and not to determine the amount of fertilizer applied.

Long-term field experiments can provide data for the nutrient use, and the same type and amount of fertilizer can be used in a single point for many years, thus the correlation data between the test value and the yield can be obtained. It serves as a basis for the amount of fertilizer to be applied.

③Contrast method. Under the same conditions of varieties and soil types, the normal and abnormal robust plants were selected and the nutrient contents of soil were measured. The index is determined by the comparison of the two.

The diagnostic index should be determined by nutrition diagnosis, field production and fertilizer test, and the rule of diagnosis should be found out many times. Any diagnostic index is obtained under certain production conditions. Indicators in the field can only be used for reference, not hard-copy. When quoting, it must be used after the inspection of the production.

(3) Problems needing attention in the Application of Diagnostic Indexes.

①The variety of crops and the characteristics of the variety. Different crops, different varieties of the same crop and different growth stages of the same variety have different requirements for nutrients and critical concentration, and the application indexes shall be taken into account. In some case, the nutrient content is high, but the amount of growth or the yield is not necessarily high.

②The interaction between the nutrient elements. Antagonism such as Ca^{2+} and Mg^{2+}; Assisting action such as Ca^{2+}, Mg^{2+} and Al^{3+} can promote the absorption of the K^+ and NH_4^+, etc.

Therefore, when the diagnostic index of some kind of element is applied, it is not only to know the relative quantity of the nutrient, but also to know the relation of the related elements.

③The technical conditions of diagnosis require consistent; the sampling analysis should be consistent with the formulation of indicators, and should be comparable, otherwise the indicators will have no practical value. Indicators should be constantly revised with changes in production levels and technical measures.

In conclusion, the application of the diagnostic index should be applied when fertilization, but the application of the index should not be in isolation. The application of the index must be used flexibly according to the specific conditions, so that the diagnostic index is more practical and the diagnosis technology is gradually improved.

III. Formulation of Fertilization Programs

The fertilization programs must specify the amount of fertilizer, the type of fertilizer, and the period during which fertilizer is to be used. The fertilizer application is influenced by the comprehensive factors such as plant yield, soil fertilizer supply, fertilizer utilization rate, local climate, soil conditions and cultivation techniques, etc. There are many methods to determine the amount of fertilizer applied, such as nutrient balance method, field test method and so on. Taking nutrient balance method as an example to illustrate.

1. Determination of the amount of fertilizer to be applied

(1) The total nutrient requirement of the planned plant yield. Soil fertility is the basis for determining high and low yields. There are many ways to determine how high a plant's planned yield should be based on local factors, not blindly too high or too low. The usual approach is to build on the average yield of local plants in the first three years and to increase production by another 10% to 15% as planned. According to the planned yield, calculate the total amount of nitrogen, phosphorus and potassium required according to the following formula.

Nutrients required for planned plant production (kg) - (planned yield/100) × nutrients required for 100kg production

(2) The amount of fertilizer supplied by soil. The amount of fertilizer supplied by the soil refers to the nutrients absorbed from the soil when the plant reaches a certain level of yield (excluding the amount of fertilizer applied). There are many ways to get

Chapter III Selection of Fertilizer Suitable for the Integrated Water-Fertilizer Technology

this value. Generally speaking, the amount of fertilizer supplied by the soil is expressed by the total amount of nutrients in the total harvest in the soil-free area, and different places should conduct multi-site experiments on different plants according to the type of soil. After obtaining the local reliable data, estimate the soil fertilizer supply with the following formula: soil fertilizer supply = the measured value of soil nutrient (mg/kg) ×0.15 ×correction factor.

(3) Fertilizer utilization rate. Fertilizer utilization rate refers to the percentage of nutrients absorbed by plants from the total amount of fertilizer applied. It is an important symbol of rational fertilization, and it is also an important parameter needed to calculate the amount of fertilizer applied. It can be obtained by field experiment and indoor chemical analysis as follows: fertilizer utilization ratio (%) = [(absorption amount of the elements above the ground in the fertilization area plus (-) absorption amount of the elements above the ground in the fertilization-free area) /the total amount of the elements in the fertilizer applied] ×100.

If we know the total amount of nutrients needed to achieve the planned yield, the amount of fertilizer supplied by the soil, the efficiency of the fertilizer to be applied, and the content of a certain nutrient in the fertilizer, the planned fertilization amount can be estimated according to the following formula: planned fertilization amount (kg) plus (total nutrient required for planned yield plus the fertilizer amount supplied by the soil) / (the available nutrient content in fertilizer times fertilizer utilization rate).

2. Determination on the fertilization period

It is one of the most important bases for rational fertilization to master the nutrient characteristics of plants. Different plant species have different nutritional characteristics, and even the same plant has different character at different growth stages. So only to understand the characteristics of plant demand for nutritional conditions at different growth stages, the nutrient conditions can be adjusted effectively by applying fertilization to improve the yield, improve the quality and protect the environment. Plants go through many different stages of growth and development throughout their lives. In these stages, apart from the early stage of seed nutrition and the stage at which the roots stop absorbing nutrients at the later stage, there are many stages of growth and development. During the whole period the plant absorbs nutrients from the soil or medium through other organs, such as roots or leaves, and absorbs nutrients from the environment, which is called the vegetative period of plants. The species, quantity and proportion of nutrient elements absorbed from the environment in different growing stages of plants have different requirements, which is called the stage vegetative period of plants. Although there are stages and critical periods in plant nutrient requirements, we must not forget the continuity of nutrient absorption. In addition to the critical stage of nutrition and the maximum efficiency period, it is necessary to provide adequate nutrients for any plant at all stages of growth.

3. Determination on the fertilization links

The plant has nutrition stages and parts of nutrition stages, and the fertilizer is applied according to the seedling condition during the vegetative period, so fertilization cannot be completed only one time. For most annual or perennial plants, fertilization shall include base fertilizer, seed manure and topdressing for 3 periods (or links). Each fertilization period (or link) plays a different role.

(1) Base fertilizer. The mass also often call it bottom fertilizer (underground), which is a fertilizer that is used to combine the soil with the soil before sowing (or planting). The effect is two fold, on one hand it can improve the fertilizer and the the soil, on the other hand, it can supply the nutrient required for the whole growth and development. A multi-purpose organic fertilizer is used, and a part of the chemical fertilizer is used as the base fertilizer. The application of base fertilizer shall be applied in accordance with the principle of fertile soil, rich fertilizer and mixing soil and fertilizer.

(2) Seed fertilizer. It is a fertilizer applied near or mixed with the seeds at the time of seeding (or field planting). Its effect is to create a good nutrition condition and environment for the germination of the seeds and the growth of the seedlings. Therefore, the seed fertilizer is generally multi-purpose decomposed organic fertilizer or quick-acting chemical fertilizer and bacterial fertilizer, etc. At the same time, in order to avoid the bad effect when the seeds are close to the fertilizer, we should select a fertilizer which is less corrosive or less toxic to the seed or the root system. Any of the following fertilizers shall not be used as seed fertilizer, for they have excessive concentration, excessive acid or alkaline, strong moisture absorption, high temperature during dissolution and even have toxic components. For example, ammonium bicarbonate, ammonium nitrate, ammonium chloride, and calcium superphosphate made with local method are not suitable for seed fertilizer.

Chapter Ⅲ   Selection of Fertilizer Suitable for the Integrated Water-Fertilizer Technology

(3) Topdressing. It is the fertilizer applied during plant growth and development. Its function is to replenish the nutrients needed in the process of growth in time in order to promote the further development, improve the yield and improve the quality. Generally we use fast-acting chemical fertilizers as topdressing.

4. The determination on fertilization methods

Fertilization methods are as follows: ① Spreading fertilization. Spreading is a method of applying base and topdressing, by spreading fertilizer evenly on the surface and then toppling it into the soil. All plants with large amount of fertilizer or dense planting such as wheat, rice, vegetables, etc. , and plants with wide distribution of root system can be applied with the method of spreading fertilizer. ②Strip fertilization. It is also a method of base fertilizer and topdressing, that to say, furrow application of fertilizer followed by soil mulching. In general, wheat, cotton and ridge-planted sweet potato are more often applied with strips when there is less fertilizer. And for wheat, for example, the fertilizer can be put into the soil with a fertilizer applicator or planter before closing the line. ③ Holes fertilization. Holes fertilization is the application of fertilizer in the sowing hole before sowing, and then cover it. It is characterized by concentrated fertilization, small amount of fertilizer, good effect on increasing yield, so it is mostly applied to fruit trees and woods. ④Layered fertilization. It is to apply fertilizer to different levels of soil in different proportions. ⑤Fertilization with water. A method by which fertilizer is dissolved in irrigated water and applied to the soil during irrigation (especially sprinkler irrigation). This method is mostly used for topdressing. ⑥External topdressing. The fertilizer is mixed into a certain concentration solution and sprayed on the foliage of plants for absorption. ⑦Circular and radial fertilization. Ring-shaped fertilization is often used in orchards to fertilize, digging a ring-shaped ditch on the vertical surface around the crown, 30cm deep and 60cm wide, covered with soil after fertilization. The next year refertilization can be carried out on the outside of the ditch in the first year in order to expand the range of fertilization year by year. Radial fertilization is located at a certain distance from the trees, taking the trunk as the center, digging 4 to 8 radial straight grooves, 50cm wide and 50cm deep. The furrow length is the same as that of the tree crown. Fertilizer is applied in the ditch, and the next year, the furrows are excavated at the other positions.

Chapter IV Application of the Integrated Water-Fertilizer Technology to Agricultural Production

When popularizing and applying integrated water-fertilizer technology in crop cultivation of agricultural production, it is necessary to observe the objective laws, use scientific methods, make reasonable planning and meticulous layout, take appropriate measures according to local conditions, and perform with the correct standard, so as to increase the crop yield and harvest.

Section I Planning and Design of the Integrated Water-Fertilizer

I. Summary of Intelligent Irrigation System of Integrated Water-Fertilizer

Intelligent irrigation system based onthe Internet of things is a kind of modern agricultural equipment with wide application potential, which involves the key technologies such as sensor technology, automatic control technology, data analysis and processing technology, network and wireless communication technology. In this system, the soil moisture data are provided timely by soil moisture monitoring station, and the sensing data are analyzed and processed according to the actual conditions of the model area, such as irrigation area, geographical conditions, the distribution of different crops, the laying of pipe network, etc. According to the sensing data, the irrigation standard is set, and then the intelligent irrigation of integrated water and fertilizer is realized by different ways such as automatic, regular or manual irrigation. The Central Station Administrator can use computers or intelligent mobile terminal equipment, log in the system to monitor the crop growth in the model area, and remote control the irrigation equipment (such as fixed sprinkler, etc.).

The intelligent irrigation system based on Internet of things can realize precise and intelligent irrigation in the model area, which can improve the utilization rate of water, alleviate the problem of increasing shortage of water. Thus it can increase crop yield, reduce crop costs, save human resources, and optimize the management structure.

II. The Overall Design Scheme of Intelligent Irrigation System of Integrated Water-Fertilizer

1. The overall objective of intelligent irrigation system of integrated water-fertilizer

The intelligent irrigation systemcan timely collect the soil moisture content and display it on the management interface in the form of dynamic graph. According to the arrangement of irrigation pipes in the model area and the installation position of the fixed sprinkler, the corresponding irrigation mode (including automatic mode, manual mode, timing mode, etc.) is set up in advance. Then the intelligent control is realized through the analysis and processing of the timely collected soil water content and historical data. The system can record the time of each irrigation, the period of irrigation and the change of soil moisture content in each region. It has the function of comparing historical curves, and can input the fertilizer distribution and growth potential of the crops in each region, information on spraying pesticide and crop yield, etc. The system can assign user rights through administrator system, and open different functions to different users, including data inquiry, remote viewing, parameter setting, device control and product information input and so on.

2. Intelligent irrigation system structure with integrated water-fertilizer

The system is equipped with soil moisture monitoring stations and remote control system, intelligent gateway and cameras, to realize the sensing data collection and irrigation equipment control in the model area. The model area communicates with the data platform through 2G/3G network and optical fiber, and the data platform mainly realizes the functions of environmental data acquisition, threshold alarm, historical data recording, remote control, the condition of the control equipment and so on. The data platform further realizes the data transmission with the remote terminal through the Internet, and the remote terminal realizes the remote monitoring of the model area by the users as shown in Fig. 4-1.

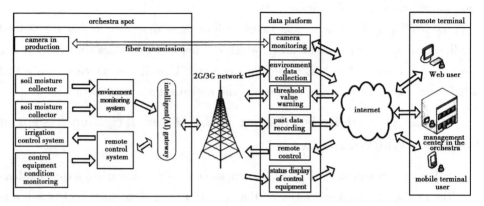

Fig. 4-1 Structure chart of intelligent water-fertilizer integration

According to the arrangement of irrigation equipment and irrigation pipeline and the division of area, the core controller node is arranged; a small local area network is formed through ZigBee network, and the equipment location is realized by GPRS (General Packet Radio Service). Then it connects to the base station of the 2G/3G network through the embedded intelligent gateway, and transfers the data to the server. Camera video is transmitted to the server via optical fiber; the server transmits data to the remote terminals over the Internet (Fig. 4-2).

3. The composition of intelligent irrigation system with integrated water and fertilizer

Intelligent irrigation system can be divided into six subsystems: crop growth environment monitoring system, remote control system, video monitoring system, communication system, server, and users management system.

(1) Crop growth environment monitoring system. Crop growth environment monitoring system is mainly soil moisture monitoring system (soil moisture monitoring system). The soil moisture monitoring system is equipped with a variety of soil moisture sensors to collect the soil moisture content based on the square of the model area, topography and the types of crops planted, and analyzes and processed the collected data. And through the embedded intelligent gateway the data are sent to the server. According to the actual demand of planting crops, users in the model area realize intelligent irrigation based on the soil moisture (soil moisture) parameters collect-

Chapter Ⅳ Application of the Integrated Water-Fertilizer Technology to Agricultural Production

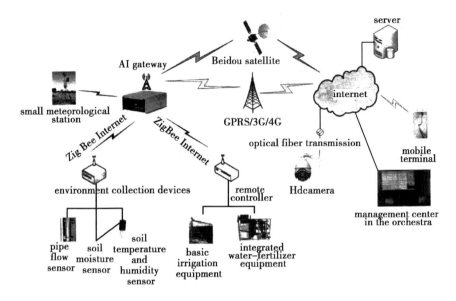

Fig. 4-2 Map of the intelligent water-fertilizer integration

ed. Transmission of data over a wireless network to meet network communication within the range of distance, the user can adjust the location of the collector as needed.

(2) Remote control system. The remote control system realizes the remote control of fixed sprinkler and integrated water-fertilizer infrastructure. The threshold value of sprinkler is set in advance to generate automatic control instructions according to the soil moisture data collected, so as to realize the automatic irrigation function. The remote control of sprinkler can also be realized by different modes such as manual or timing. In addition, the system can detect the open and close state of sprinkler timely.

(3) Video monitoring system. The video monitoring system realizes the visual monitoring of the key parts of the model area. According to the layout of the area, a high-definition camera is installed, which is generally installed in the crop planting area and near the fixed sprinkler. Video data is transmitted to the monitoring interface via fiber optic, allowing managers to view crop growth status and irrigation effects through due video.

(4) Communication system. If the area is wide and the terrain is complex, the wired communication is difficult. ZigBee network can realize the functions of ad hoc net-

work, multi-hop, near identification and so on. The reliability of the network is good, when a node in the field has problems, the ZigBee network can realize the communication in the area by using ZigBee network. The rest of the nodes will automatically find other optimal paths without affecting the communication lines of the system.

The data transmitted by the ZigBee communication module is collected in the central node to pack and compress the data, and then sent to the server through the embedded intelligent gateway.

(5) Server. The server is a computer that manages data resources and provides service to users. It has high security, stability and processing ability. It also provides database management service and Web service for intelligent irrigation system.

(6) User management system. Users can log into the user management system through a Web browser through personal computers and handheld mobile devices. Different users need to be assigned with different permissions, and the system will open different functions to it. For example, the senior administrator is generally the main person in charge of the model zone, has the authority to view the information, compare the historical data, configure the system parameters, control the equipment and so on. General administrators are plant managers, purchaser and sales personnel, etc., have access to view data information, control equipment, record crop fertilizer information and in-out warehouse management; visitors are consumers and government personnel, with the ability to view the growth information, growth condition and other authority. The user management system is installed in the management center of the garden, including the operation platform of the user management system for the due view of crop growth in the area.

4. The function of intelligent irrigation system with integrated water-fertilizer

Intelligent irrigation system can have the following functions: environmental data display, analysis and processing, intelligent irrigation, crop growth record, product information management, and so on.

(1) Environmental data display and analysis and processing. The first is the display of the environmental data for reference. The data information of each soil moisture point can be displayed on the interface of the system, and the data can be updated timely. The data display type contains timely data and historical data, and we can view the current soil moisture content and that of any time (for example, soil moisture

Chapter IV Application of the Integrated Water-Fertilizer Technology to Agricultural Production

data of the area every month or the day); The method of data display includes list display and graph display, which can be compared according to the data of different planting areas of the same crop or different time periods of the same area, and it is shown in the form of curves, column charts, and so on. Second is the environmental data analysis and processing. According to the soil water content collected and the specific demand for soil moisture content in the actual growth process of crops, the threshold value of water content was set to open the irrigation valve. According to the demand of different crops for soil moisture content, we set irrigation time, irrigation period and so on.

(2) Intelligent irrigation. This system can realize three irrigation control methods. First, irrigation should be scheduled and periodic according to the conditions. According to the situation of crop planting in different areas, scheduled and periodic irrigation was carried out. Second, multi-parameter setting irrigation. The system automatically opens the solenoid valve in the corresponding area and irrigates the different crops by setting the upper and lower limits of the multi-parameter suitable for their growth, when the timely parameter values exceed the set threshold, and the system will automatically open the solenoid valve in the corresponding area to irrigate the region to stabilize the parameter value within the set value. Third, manual controlling of the remote irrigation. Administrators can manually carry out remote irrigation through the management system.

(3) Crop growth record. The information such as the environmental data, the irrigation condition, the fertilizer distribution information, the crop growth potential and the yield of each area are recorded through the database.

(4) Produce information management. The information such as the distribution of the fertilizer, the growth condition, the spraying of the pesticide, the quality and output of the produce, in-and-out of the produce, the inventory of the warehouse and the grade classification of the crop products are recorded by the administrator of the park.

5. Characteristics ofintegrated water-fertilizer intelligent irrigation system

The system adopts the design idea of expansibility, and emphasizes on the stability and reliability of the system in the design. The whole system consists of a plurality of groups of gateways and a ZigBee self-organizing network unit, each gateway serves as a network center of a ZigBee local area network. The network comprises a plurality of nodes, each node composed of a soil moisture sensor instrument or a remote control device controller, which are respectively connected to the soil moisture sensors and the fixed sprinkle. The system can conveniently and rapidly form an intelligent irrigation system according to the requirements of the user. The user only needs to increase the number of equipment at all levels, so that the capacity expansion of the whole system can be realized, and the original systemic structure does not need to be changed.

6. Design of intelligent irrigation system with integrated water and fertilizer

(1) System layout. Because the communication sub-module of this system adopts Zigbee wireless local area network, which has the characteristics of flexible structure, self-organizing network and close recognition, it is relatively flexible to arrange the control nodes of soil moisture sensors. According to the variety of crops planted and the different soil moisture demand of various crops, soil moisture sensors are installed. Depending on the irrigation pipes laid in the garden, the location of the fixed sprinkle, the time intervals of the crops and the sub-area water supply, remote controller equipment needs to be equipped (each remote controller equipment includes the core controller, wireless communication module, several controller extended modules and the installation accessories). Each set of controller equipment is installed next to the fixed sprinkler according to the principle of proximity to realize the remote intelligent control function of irrigation in the area; In addition, the open-close state signal and video signal of the fixed sprinkler are detected automatically by the control equipment, and the open-close state is grasped timely by long-distance inspection.

In the implementation of the project, soil moisture monitoring stations areset up according to the specific conditions of the model area (including geographical location, geographical environment, crop distribution, regional division, etc.). The remote control equipment needs to be installed next to the control cabinet of irrigation equipment in the later stage, and the remote control of sprinkler including integrated water-fertilizer infrastructure can be realized by means of wire.

(2) Network layout. Soil moisture monitoring equipment and remote controller equipment are built-in ZigBee module and GPRS module respectively, as the nodes of communication network. The embedded intelligent gateway is the central node of the ZigBee network in a certain area. It forms a small local area network, realizes the net-

work communication in the corresponding area of the campus, and realizes the data transmission with the server through the 2G/3G network.

The system adopts wireless communication mode, including ZigBee network transmission and GPRS module positioning. Due to the flat terrain, with no tall buildings or other shelter, so it is suitable for wireless transmission.

Section II Equipment Installation and Adjustment of the Integrated Water-Fertilizer Technology

The equipment installation of the integrated water-fertilizer technology mainly includes the installation of the head hub equipment, the installation of pipe network equipment and the installation of micro-irrigation equipment.

I. Installation and Adjustment of Head Equipment

1. Installation of negative pressure variable frequency water supply equipment

The installation position of the negative-pressure frequency-conversion water supply equipment shallmeet the requirements of the control cabinet for the environment. The cabinet shall have enough space or access channels in front of and behind the cabinet. The power line diameter of the control cabinet, the capacity of the low-voltage cabinet in front of the control cabinet shall have a certain storage. The various test control instruments or equipment shall be installed in the place with smooth system and stable pressure, and it should not have obvious adverse effect on the control instrument or equipment. If installed at a high temperature (higher than 45℃) or in a corrosive place, a specific explanation shall be made when signing the order. If the installation environment is found to be out of conformity when installed, contact the original supplier in time for replacement.

When the water pump is installed, no leakage of the inlet pipe should be paid attention to, drainage ditch should be set up on the ground, and necessary maintenance facilities should be prepared. As to the pump installation dimensions, please refer to all kinds of pump installation instructions.

2. Installation of centrifugal self-priming pump

(1) Installation and use method.

The first step is to builda pump house and a pool, the house covering an area of more than 3m ×5m, and installed an anti-theft door, the pool being 2m ×3m. The second step is to install the ZW horizontal centrifugal self-priming pump, the inlet being connected to the bottom of the inlet pool, and the outlet being connected with the filter. External installing the water gauge, pressure gauge and exhaust valve (the exhaust valve is installed outside the wall of the outlet pipe, the exhaust valve will overflow when the pump starts and stops, thus it can keep the pump room from being wet by the water overflow). Third, install the fertilizer pipe, connect the valve at the three-link suction pipe, then attach the filter, the filter and the direction of the flow should be consistent with the wire hose and bottom valve. The fourth step is to prepare about three fertilizer buckets, each with a capacity of about 200 liters. The fertilizer can be sucked by means of the different suction tube. Thus, the fertilizer can be mixed in the pipe by inhaling the fertilizers which are not mixed at the same time. The fifth step is to confirm fertilization concentration. According to the diameter of the inlet and outlet pipe, we choose the diameter of fertilizer suction pipe to maintain the fertilization concentration being 5%~7%. Usually for the pump with 4-inch water inlet pipe and 3-inch outlet water pipe, 1-inch fertilizer pipe should be installed, and thus it ensured the final fertilization concentration about 5%. The amount of fertilizer intake always varies with the flow rate of the pump, and maintains a relatively stable concentration. The amount of irrigation in the field is large, that is, the flow rate is large, and the rate of fertilizer suction is also accelerated, on the contrary, the rate of fertilizer suction will slow down, so the concentration should be remained relatively stable all the time.

(2) Notes for precautions.

The fertilizer-suction filter and the water outlet filter should be kept clear when fertilizing. In case of blockage, we should timely clean them. During the fertilization, when the fertilizer solution in the fertilizer bucket is to be sucked dry, the sucking valve shall be closed in time to prevent the air from entering the pump body to generate cavitation.

Chapter IV Application of the Integrated Water-Fertilizer Technology to Agricultural Production

3. Installation of submersible pumps

(1) Installation method.

Remove the upper outlet joint of the pump and connect the check valve with the flange. The check valve's arrow points to the direction of the water flow. The pipe extends vertically out of the pool surface, turns into the pump room through the bend, connects with the filter in the pump room; open a fertilization opening with the diameter of 20mm in front of the filter, connect the fertilizer pump, and install a pressure gauges before and after the pump. Water pump in the the pool needs to be cushioned about 0.2 meter high from the bottom to prevent sludge accumulation, which will influence the heat dissipation.

(2) Fertilization method.

The first step is to turn on the motor, making water supply normal and the pressure stable. The second step is to turn on the fertilizer pump, adjust the pressure, start to inject fertilizer, and take care of the change of pressure and amount of fertilizer at any time. The pressure of the fertilizer tube is slightly higher than the pressure of the outlet pipe to ensure that the fertilizer liquid can be injected out of the water pipe. But the pressure should not be too high, so as not to cause backflow. After the fertilizer has been used over, go on running the water pump for about 15 minutes and send the remaining fertilizer solution from the pipe network to the roots of the crops.

4. Integration of micro-storage and fertilizer in mountainous areas

The integrated technology of micro-fertilization in mountainous area is to make use of the difference of natural terrain height in mountain area to obtain water pressure, and to carry out micro-irrigation on relatively low fields, that is to say, "micro-pool" and "micro-drip irrigation" are combined to form "micro-storage and micro-irrigation". This approach does not require electricity power and pump power, and is suitable for irrigation of crop plantations in mountainous, semi-mountainous and hilly areas. Filters and exhaust valves are directly installed at the outlet of the water pool, and then connected to the pipe network to deliver water and fertilizer to the roots of the plants.

The use of integrated water-fertilizer technology in mountain area can save more than 15 labor per mu per year, and increase crop yield by more than 15%. This method can make all the fertilizer into the soil tillage layer, thus reduced the waste of fertilizer and surface evaporation, and saved of fertilizer by more than 15%.

The head equipment of integrated water-fertilizer technology applied to the mountainous area is mainly composed of source water pool, sand sink pool, pipe, sediment reservoir (pool), main valve, filter and exhaust valve, as shown in Fig. 4-3. This type of head equipment is simple, safe and reliable. If the filter performance is good, there is little need for care in the fertilization process.

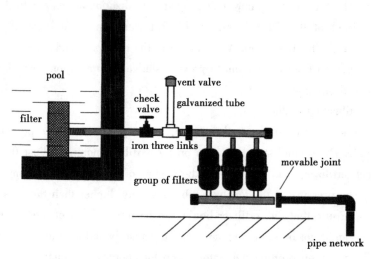

Fig. 4-3 Sketch map of the head hub of the Integrated water-fertilizer in mountainous areas

The source water pool and the sediment pool play a primary role in filtering the water source and storing water, keeping out the sediments, branches and leaves. The aqueduct pipe is the pipe which leads the water from the water source to the reservoir, which is suitable to be buried under the surface of the ground for 0.3~0.4 meter to prevent freezing crack and man-made damage. If the pipeline exceeds 1 km and there is a rolling slope on the way, an exhaust valve should be installed at the height of the fluctuation to prevent air resistance.

The difference between the reservoir and the irrigated land should be 10~15 meters, and the size of the pool should be determined according to the size of the water source and the irrigation area, generally, 50~120m³ being the best choice. The construction quality of the sediment pool is high, better to have reinforced concrete structure, and the pool body should be buried deeply, with the surface part being not more than 1/3 of the whole body. When building the pool, cleaning valve, outlet valve and overflow exit should be pre-installed. Special attention should be paid to the installation of the valves when the reservoir was being built, so that they can be connected with the pool. The pool should not be drilled after the construction in case it is easy to leak. The top cover is added as the maintenance gate to ensure safety.

The filter is installed at the outlet valve, and it is best to install 2 sets of filters for easy cleaning. The fertilizer pool can also be replaced with fertilizer bucket, with the volume of being $1\sim2m^3$, connected to the outlet pipe.

In practice, there are two fertilization methods suitable for mountainous integrated water-fertilizer technology. First, direct fertilization. When the amount of fertilizer and water supply are determined, the soluble fertilizer can be added according to the pool volume and the concentration of fertilizer. The fertilizer needs to be stirred in a small bucket in advance until it is fully dissolved, the residue being filtered out, and then the fertilizer is poured into the reservoir and evenly mixed. The valves are opened during irrigation, and the fertilizer liquid flows directly to the root of the crops. After fertilization, fill the pipe with clear water for about 15 minutes, rinse away the emitters and the remaining fertilizer in the pipe, and prevent the fertilizer crystals from blocking the drippers. Second, pressure fertilization. Place the fertilizer bucket at a slightly higher level than the highest level of the reservoir. And the fertilizer pipe is connected with the outlet pipe, and the fertilizer opening is placed in front of the filter so that the fertilizer liquid passes through the filter to remove impurities. Open the main valve of the water outlet during irrigation, and then open the fertilizer valve. The saturated fertilizer solution and irrigation water are mixed evenly in the pipeline and then transported to the working area through the various pipe networks. After the fertilization is completed, water is used to flush the pipe.

Ⅱ. Installation and Test of the Network Equipment

1. Pipe networks in the flat field

In the construction of integrated water-fertilizer facilities, besides choosing the proper head equipment, a reasonable layout and economical and practical water supply network are also needed. In recent years, the plastic pipe industry develops rapidly, with mature quality day by day, and the plastic pipe replaces galvanized pipe with the advantage of low price and high quality. At present, plastic pipe is used in the construction of irrigation pipe network, among which PVC (PVC) and polyethylene (PE) pipes are widely used, among which PVC pipe needs to be glued with special glue, and PE pipe needs hot melt connection.

(1) Ditch excavation.

The first step of laying pipe network is to digditch. Generally the ditch is 0.4 meter wide and 0.6 meter deep. It is U-shaped. The trench should be straight and consistent in depth. The turning point should be dealt with at 90°and 135°. The slope of the ditch is inverted trapezoid, that is the upper is wider and the lower is narrower, preventing soil from collapsing and causing double work. Mechanical construction is needed in larger field, and manual work is required in smaller field as shown in Fig. 4-4.

Fig. 4-4 Ditch excavation

(2) PVC pipe installation.

PVC pipes are equipped with PVC fittings. The pipes and fittings are bonded with special glue. This glue can dissolve the PVC pipe and the surface of the fittings, and after joining, the materials permeate each other and become together after 72 hours. Therefore, when coating glue, you should pay attention to the amount of glue. The glue should not be too much, for too much glue will deposit on the bottom of the pipe, and part of the pipe wall will be dissolved soften, which will reduce the pipe stress. In the face of extreme pressure such as water hammer, this part is the most vulnerable to break, resulting in higher maintenance costs, and also affecting agricultural production.

(3) PE pipe installation.

PE pipeline is connected by hot-meltway, and there are two kinds of hot-melt way: butt hot-melt type and socket hot-melt type, generally large-diameter pipes (above DN 100mm) are connected by butt hot－melt connection, and there are customized hot-melt machines. Specific operation can be carried out according to the instructions for the use of the machine. For the pipes with DN less than 80 mm, hot melt connections can be carried out by means of socket. The advantage is that the hot-melt machine is light and portable, and the disadvantage is that the operation requires more than two people. After insertion, the hot melt part of the pipe is easily overheated to shrink, thus the water flow will be affected.

2. Pipe networks in the mountainous area

PE pipe network is suitable for mountainous irrigation network, and the general installation direction is the same as that of the pipe network in the flat ground. But the laying method is different from that on the flat ground. The main pipe is laid down from the reservoir along the slope. The pressure dissipation tank is installed every 20~30 meters and the floating ball valve is installed in the tank. Buried 0.3 m below the surface, the branch pipes would be perpendicularto the slope and exposed on the ground, and valves should be installed, protected by valve well.

All the branch pipes should be laid according to the design requirements, and the key is that the laying of the drip irrigation pipes needs to be along the contour line, and the water out is more even.

III. Installation and Adjustment of the Micro-Irrigation Equipment

1. Installation and adjustment of micro-sprinkler irrigation

The micro-sprinkler system comprises a water source, a water supply pump, a control valve, a filter, a fertilizer valve, a fertilizer tank, a water pipe, and micro-spray heads etc. Here, the micro-sprinkler installation in the greenhouse is taken as an example.

The selection and installation of material: hanging pipe, branch pipe and main pipe should be PVC tube, with the diameter of 4~5mm, 8~20mm, and 32mm respectively, and the wall thickness of 2mm. The spacing of micro-nozzle is 2.8~3 meters; the working pressure is about 0.18 MPa, and the flow rate ofa single-phase water pump is 8~12 liters per hour. It is required that the pipe has good anti-clogging performance, and the range diameter of the micro-nozzle is 3.5~4 meters. The spray atomization should be even, and the distance between the two branches of the pipe is

2.6 meters. The expansion bolt is fixed in the position of two meters to the ground in the length direction of the green house, and the branch pipes are fixed. Connect the micro-sprinkler, the pipe and the elbow, and install the micro-sprinkler upside-down.

(1) Installation step.

①Tools preparation. Hacksaw, strip rolling, perforator gloves, etc.

②Installation way. In the greenhouse, micro - sprinklers are usually mounted upside-down as shown in Fig. 4-5, which not only does not occupy the field, but also facilitates the field operations. According to the effect of field experiment and practical application, the optimum spacing of micro-sprinklers is 2.5~2.6m, and the length of lower-hanging is 1.8 ~ 2m above the ground. Generally, G-type micro-sprinkler is chosen, and the micro-sprinkler G-type bridge should be oriented in one direction. In this way, the spray of water droplets can complement each other and improve the evenness.

Fig. 4-5 Upside-down micro-sprinkler

③Installation of anti-drop device. During installation, an anti-drop device can be installed so that the micro-sprinkler can stop the dropping of the remaining water in the tube when the spray is stopped, so as not to influence the growth of the crop. It can also not be installed, and the trick is to adjust the border position and branch installation position when installing the micro-sprinkler. The nozzle is installed directly above the border, thus the remaining water drops in the border ditch.

④Sprinkler installed at the two ends. At the end of the greenhouse, two nozzles

Chapter IV Application of the Integrated Water-Fertilizer Technology to Agricultural Production

are installed at the same time, with a height difference of 10cm. One of them has a flow rate of 40 liters per hour. Its function is to make both ends of the greenhouse moist more even.

⑤ Pre-installation of the nozzle. Cut the tube (cut evenly at predetermined length), install the sprinkler, install the weight, finished product.

⑥ Black tube fixed. Spread the black tube longitudinally along the direction of the greenhouse, and adjust the twisted part so that the black tube is smoothly laid on the ground. Punch holes at the predetermined distance, then install the sprinkler, starting at the end of the greenhouse, reserving 2 meters. Bind the black tube with the sprinkler to fix it. Attention! Wire-like binding should not be used, because it is easy to pierce the plastic film or rust.

(2) Installation type.

① Sprinklepaths (channels) selection. The paths number of nozzles installed in a greenhouse should be determined according to the width of the greenhouse.

8-meter wide greenhouse should have two channels, and the nozzles flow of 70 liters per hour are chosen. The double channels' model is LF-GWPS6000 and has 6-meter spray amplitude. The distance between the two black tubes is about 4 meters, and the distance between the nozzles is 2.5~2.6m, cross arranged.

6-meter wide greenhouse should have a single channel with the nozzle flow being 120 liters per hour. The single channel, model LFGWP8000, 8-meter spray amplitude, spacing 2.5~2.8m. Both ends of the greenhouse are installed with two nozzles, with a height difference of 10cm, one of which is 70 L/h.

② Sprinkle pipe selection. Pipes are usually black low-density (high-pressure) polyethylene tubes, orcalled black tubes for short. This kind of pipe is anti-aging, which can adapt to the harsh field climate, and the new material pipe can be used in the field for more than 10 years.

③ Diameter selection. According to the length of singlechannel irrigation, the diameter of the pipe can be determined, that is, the outer diameter 16-mm black tube can be used when the channel length is within 30 meters; the outer diameter 20-mm black tube can be used when the channel length is 30~50 meters; the outer diameter 25-mm black tube is used when the channel is 50~70 meters, and the outer diameter 32-mm black tube is used when the channel is 70~90 meters. Generally, the length is not more than 100 meter, so it can save the cost. If the length is more than 100 meters, it is suggested to open three links for water to flow through it.

2. Installation andadjustment of drip irrigation equipment

The biological characteristics of crops are different, so the plant spacing and row spacing are also different (usually 15~40cm). In order to achieve the goal of even irrigation, the drip hole distance and the specification are required to be the same. The distance between the drip holes is usually 15cm, 20cm, 30cm, 40cm, and the commonly used ones are 20cm and 30cm. In the process of choosing drip irrigation facilities, it is necessary to consider the products with the same evenness of water output at the head and the end, and the error limit is less than 10% at the head and the end of the drip irrigation system. In the process of design and construction, it is necessary to choose the suitable drip irrigation belt according to the actual situation, and according to the flow rate and other technical parameters of the drip irrigation belt, we determine the best laying length of a single drip irrigation belt.

(1) Installation of drip irrigation equipment.

①Selection of emitters. In greenhouse, drip belt of 16mm ×200mm or 300mm is usually used in greenhouse cultivation, and the wall thickness can be selected as 0.2mm, 0.4mm, 0.6mm according to farmers' investment demand. The holes face up and spread flat under the film on the surface.

②Number of drip irrigation belts. According to the crop planting requirements and investment will, the number of belts laid per border can be determined, usually at least one, two belts are the best.

③Installation of drip irrigation belt. The head pipe of the greenhouse is 25 inches, with a total switch in the house, and a by-pass valve per border bed. During rainy season, the soil moisture in the middle of the greenhouse is different from that on the edge, and the irrigation volume can be adjusted by the by-pass valves.

When laying drip irrigation belt, first pull outthe belt from the below, and is controlled by one person, another person pull the drip irrigation belt. When the belt is slightly longer than the border surface, cut it off and fold the end of it to prevent foreign body entering. The head part is connected with the by-pass or by-pass valve, requiring the belt to be trimmed neat with scissors. If there is a drip outlet nearby, cut off the dripper. Put the drip irrigation belt smoothly into the mouth of the bypass valve, and

Chapter IV Application of the Integrated Water-Fertilizer Technology to Agricultural Production

then tighten the nut. The tail of belt is folded and tied with fine rope to make it easy to rinse (it can also be clogged with tape plug, but it is affected by water pressure, sediment and so on in the process of use, so it is not easy to unscrew and rinse. It's convenient to tie the belt with rope.).

Connect the black pipe to the main pipe, install the ball valve at the tee outlet, and install the valve well or valve boxfor protection. After the installation of the whole pipe network, test the water pressure, flush out the debris left in the pipeline during the construction process, adjust the defects, then close the water, finally plug on the drip belt, and plug the 25-inch black pipe.

(2) Equipment use technology.

①Test the drip irrigation belt with water. When the drip belt is pressed for the flow of the water, the normal output of water is dripping. If there are other holes, the water will spray, and the sound of the impact of the water column will be heard under the film, so it is necessary to inspect all parts of the belt to check for insect bites or other mechanical holes, and repair them in time. Before laying drip irrigation belt, be sure to kill the underground pests or overwintering pests in the border bed.

②Irrigation time. During the first irrigation, because the soil particles are loose, the water droplets are easy to flow down the gap of the soil directly into the ditch, and the horizontal wetting can not be realized on the border surface. Therefore, it is necessary to irrigate water for a short time, many times, so that the soil of the border can form capillary tubes and make the water moist horizontally.

The water quantity of fruit crops should be controlled properly in the growth stage to prevent the excessive growth of branches and leaves from affecting thefruiting. After the fruits are borne, the drip irrigation time should be determined according to drip flow, soil moisture, fertilizer interval and so on. Generally, drip irrigation for 3 hours to 4 hours when the soil was dry, but when the soil moisture was in the medium and it is only for the purpose of fertilization, it was more suitable for water and fertilizer to be irrigated for about 1 hour.

③Cleaning the filter. After each irrigation is completed, the filter needs to be cleaned. After irrigation for every 3 times, especially after water-fertilizer irrigation, it is necessary to open the drip irrigation plug and rinse the impurities left in the pipe wall. After harvest, rinse the belt once, collect them and set aside. If it is in the greenhouse, just remove the drip irrigation belt and hang it on the arch pipe of the greenhouse, and spread and use it under the film the next time.

Section III Operation and Maintenance of the Integrated Water-Fertilizer system

I. Operation of Integrated Water-fertilizer System

The operation of integrated water-fertilizer system includes preparation, irrigation operation, fertilizer operation, rotation irrigation replacement and end operation.

1. Preparation

The preparation work before operation is mainly to check whether the system is installed in place according to the design, to check whether the main equipment anddevices of the system are normal, and to repair the damaged or leaky pipe segments and accessories.

(1) Check the pump and motor.

Check whether the voltageand frequency marked on the pump and the motor is consistent with the voltage, whether the grounding of the motor shell is reliable, and whether the motor is leaking oil.

(2) Check the filters.

Check whether the filter position meets the design requirements, whether there is damage, and whether it needs to be washed. Before the first use of the dielectric filter, the tank is filled with water and a pack of chlorine balls is put into the tank. After 30 minutes, the filter will be washed twice according to the normal method. We can also stir the medium in advance to make its particles loose and the contact surface unfolded. Then fully clean all parts of the filter and fasten all screws. When the centrifugal filter is washed, open the cover and rinse the sand out. When the mesh filter is manually cleaned, pull the handle, loose the screw, open the cover, take out the filter, brush the dirt on the screen with a soft brush and wash it clean with water. The lamination filter should be checked and the deformed laminate replaced.

(3) Check fertilizer tanks or fertilizer injection pumps.

Chapter IV Application of the Integrated Water-Fertilizer Technology to Agricultural Production

Check the parts of the fertilizer tank or fertilizer injection pump andsee if the connection to the system is correct or not; remove the accumulated dirt from the tank to prevent it into the piping system.

(4) Check other parts.

Check if the vertical pipes at the end are broken or the plugs are missing. The former need to be repaired, the latter need to be supplemented. Check that all valves and pressure regulators are open and closed normally; check the pipe network system and its connecting microtubule, and repair the defect in time. Check that the intake and exhaust valve is in good condition and open. Close the drain bottom valve on the main & branch pipes.

(5) Check the electronic control cabinet.

Check the installation position of the electronic control cabinet. The electric control cabinet should prevent sunlight and be installed separately inan isolation unit to keep the room dry. Check the wiring and safe of the electronic control cabinet to meet the requirements, to see whether there is grounding protection.

2. Irrigationoperation

The integrated water-fertilizer system includes unit system and combined system. The combined system requires rotation irrigation. The different complexity of the system, different irrigation crops and soil conditions will affect the irrigation operation.

(1) Water trial operation of pipeline.

When used for the first time in the irrigation season, the pipes must be washed with water. The drain valve should be opened before filling; all control valves should be closed; the control valve on the outlet pipe of the pump shall be opened slowly after the pump is in normal operation, and then the pipe shall be flushed one by one from upstream to downstream. It is necessary to observe whether the exhaust device is working properly when the water is filling. The drain valve should be closed slowly after the pipe is flushed.

(2) Startup ofthe water pump.

It is necessary to ensure that the engine starts under no or light load. Before starting the pump, first close the main valve and open all exhaust valves on the pipes to be filled with water to let out the gas, and then start the pump to fill the pipe slowly. After startup, observe and listen to find the abnormal sound of the equipment. When it is confirmed that the start-up is normal, slowly open the filter and control valve which control the rotation irrigation group, and begin to irrigate.

(3) Observation of pressure gauges and flow gauges.

It is observed whether the difference of pressure gauge reading before and after the filter is within the prescribed range. If the pressure difference reading reaches 7-meter water column, it indicates that the blockage in the filter is serious and the pump should be stopped and the pipes should be flushed.

(4) Flushing pipe.

When the newly installed pipeline (especially the drip irrigation pipe) is used for the first time, it is necessary to firstopen the plug at the end of the pipeline, fully discharge water to flush the pipeline system at all levels, and after washing the impurities gathered when installing, close the end plug. Then you can start using it.

(5) Inspecting in the field.

Check whether the pipe joint and pipe is leaking or not in the irrigation area, and to see if the emitter is normal.

3. Fertilizer application

(1) Differential pressure fertilizer tank.

①The operation of differential pressure fertilizer tank. The operation order of the differential pressure fertilizer tank is as follows. The first step is to calculate the amount of fertilizer according to the specific area of each irrigation area or the number of crop plants (if trees). Weigh or measure the fertilizer amount in each irrigation area. The second step is to connect the bypass pipe to the main pipe with two pipes with one valve respectively. For easy movement, a quick joint can be used on each pipe. In the third step, the liquid fertilizer should be poured directly into the fertilizer tank. If solid fertilizer is used, it should be dissolved separately and injected into the fertilizer tank through the filter. Some users put solid fertilizer directly into the fertilizer tank to dissolve it during irrigation, in which case smaller tank is OK, but more than five times as much water is needed to ensure that all fertilizers are used up. The fourth step, after finishing injecting the fertilizer solution, fasten the lid of the tank. The fifth step, check that the inlet and outlet valves of the bypass pipe are closed and the control valve is open, and then open the main pipe valve. The sixth step, open the bypass inlet and outlet valve, then slowly close the control valve, and pay attention to observe the pres-

Chapter IV Application of the Integrated Water-Fertilizer Technology to Agricultural Production

sure gauge to get the required pressure difference (1~3 meter water pressure). In the seventh step, for the users with good economic conditions, the time required for fertilizing can be measured by conductivity meter, or the fertilizing time can be estimated by Amos Teitch's economic formula. Close the inlet valve after fertilizing. The eighth step, to fertilize the next tank, some of the water in the tank must be discharged. A 1/2-inch exhaust valve or 1/2-inch ball valve should be installed at the inlet of the fertilizer tank. Before opening the drain switch at the bottom of the tank, open the exhaust valve or ball valve, otherwise the water will not be discharged.

② Method of fertilizing time in the pressure differential tank. The pressure difference fertilizer tank is applied according to the quantity of fertilizer. At the beginning of the fertilizing, the concentration of fertilizer flowing out is high. With the fertilizing going on, the fertilizer in the tank becomes less and less, and the concentration becomes thinner and thinner. Amos Teich summarizes the law of the decreasing concentration of the solution in the tank, that is, 90 percent of the fertilizer has entered the irrigation system after the four times the volume of the tank flows through the tank (but the fertilizer should be completely dissolved at the beginning). The amount of water flowing in the tank can be measured by the flow rate at the entrance to the tank. The time of irrigation and fertilizer depends on the volume of the fertilizer tank and its outflow rate:

$$T = 4V/Q$$

In the formula, T is the fertilizing time (hour), V is the volume of fertilizer tank (Liter); Q is the rate of outflow liquid (L/h), 4 is that it takes 480-liter water to bring 120-liter fertilizer solution into the irrigation system.

Because the volume of the fertilizer tank is fixed, the flow rate of the bypass pipe must be increased when the fertilizer rate needs to be accelerated. At this point, the control valve should be closed more tightly. The Amos Teich formula is an approximate formula obtained when the fertilizer is completely dissolved. In the field, solid fertilizer is often used (the amount of fertilizer does not exceed 1/3 of the tank), in this case the fertilizer is slowly dissolved. Some researchers compared the application time of the same amount of potassium chloride and potassium phosphate fertilizer into the fertilizer tank under the same pressure and flow rate under the condition of complete dissolution and solid state. To judge the application of irrigation water by monitoring the change of electrical conductivity of irrigation water at the dripper. The time of fertilization indicates that the fertilizer is completed when the conductivity in water reaches stability. 50kg solid potassium nitrate or potassium chloride (or dissolved) was poured into the fertilizer tank, the volume of the tank was 220 liters, the amount of water flowing into the tank per hour was 1 600 liters, the flow rate in the main pipe was 37.5m^3 per hour, the pressure difference through the fertilizer tank was 0.18kg/cm^2, and the temperature of irrigation water was 30℃. The results showed that the time of fertilization was basically the same when the flow rate, pressure and dosage were the same, whether the fertilizer was solid or dissolved in the fertilizer tank. The two fertilizers were finished in about 40 minutes. About 10 minutes after the beginning of fertilization, the maximum concentration of the dripper is reached, which is related to the area of the irrigation area (about 150 mu when fertilizing). The larger the area, the farther the fertilizer will go and the longer it will take to fertilize. Because the speed of fertilization is related to the flow rate through the fertilizer tank, when rapid fertilization is needed, the pressure difference at both ends of the tank can be increased, and on the contrary, the pressure difference can be reduced. Where there are conditions, the following methods can be used to determine the time of fertilization.

  a. EC method (conductivity method). Most of the fertilizer is inorganic salt (except urea), so after dissolved in water, the conductivity of the solution increases. The fertilization time of each tank of fertilizer can be known by monitoring the change of conductivity of effluent. Pour a simple fertilizer or compound fertilizer into a tank to a volume of about 1/3, weigh it and record the inlet pressure (in the case of a pressure gauge) or mark the tightening position of the throttle valve (no pressure gauge at the inlet). The EC value of effluent was measured by conductivity meter, and the starting time of fertilization was recorded. The EC value is measured every 3 minutes during the application process until the EC value at the outlet is equal to the EC value at the inlet, which indicates there is no fertilizer in the tank. Record the end of the time. The time difference between the beginning and the ending is the fertilization time.

  b. Reagent method. It is judged by the white precipitation of potassium ion and ammonium ion with 2% sodium tetraphenylborate, which is similar to EC method. Potassium nitrate, potassium chloride, ammonium nitrate and other fertilizers containing potassium or ammonium can be used as experimental fertilizers. Record the time when the fertilization began. Every time, 3 ml of fertilizer solution is taken with a 50 ml beaker, 1 drop of sodium tetracenyl boron solution is dripped into it, shake it well. The solution turns white and precipitates at the beginning of fertilization, and then there is no reaction when the concentration becomes thinner and thinner. The time between the beginning and end is the time of fertilization.

Urea is the most commonly used nitrogen fertilizer in irrigation and fertilizer. However, neither of the above two methods can detect thefertilization time of urea. By measuring the fertilization time of the same amount of potassium chloride, the fertilization time of urea was deduced according to the solubility. If at normal temperature, the solubility of potassium chloride is 34.7g per 100g water and urea is 100g per 100g water. When the fertilization time of potassium chloride is 30 minutes, because the solubility of urea is higher than that of potassium chloride, the fertilization time of urea of equal weight should also be 30 minutes. Or the urea and potassium fertilizer were added into the tank in the proportion of 1 : 9, and the application time of urea was known by monitoring the conductivity. Because the solubility of potassium fertilizer is smaller than that of urea, so if the increase of conductivity can not be monitored, it indicates that the fertilization of the urea has finished.

c. Water flow method. According to Amos Teich formula $T = 4V/Q$, when the fertilizer is liquid fertilizer or solid fertilizer with good solubility, such as urea, the time required for one fertilizer can be calculated. Therefore, a flow-meter can be installed at the end of the outlet of the fertilizer tank. When the recorded volume is about 4 times as much as the pressure difference fertilizer tank, it is shown that the fertilizer in the tank has been almost used. The time during this period is the time of fertilization.

It is of great significance to know the time of fertilization for the application of pressure difference fertilizer tank. When applying the next tank of fertilizer, the water in the tank must be released at least 1/2 to 2/3, otherwise no fertilizer can be added into the tank. If you do not know the timing of each tank of fertilizer, it may occur that the we may stop fertilization before the fertilizer is used up, thus the remaining fertilizer solution will be discharged and the fertilizer is wasted. Or if the fertilizer has already been used up but the irrigation is still going on, it will waste water or electricity, which increased the labor, especially when the irrigation is simply for the purpose of fertilization. Especially during the rainy season or when the soil does not need irrigation but only fertilizer, it is needed to speed up the rate of fertilization.

In a field survey, the researchers found that there were some problems in the use of fertilizertanks. In the large irrigation area, some fertilizer tanks are too small, so fertilizer tanks with the volume of more than 300L should be equipped with in order to facilitate users. Some fertilizer tanks are not installed with inlet and exhaust valves, resulting in difficult operation. Some inlet and outlet pipe diameters of the fertilizer tanks are too small to control the rate of fertilization. Generally for more than 200-liter fertilizer tank, 32-mm steel wire hose should be used. From the point of view of the convenience of pouring fertilizer, the horizontal fertilizer tank is superior to the vertical fertilizer tank. Usually a bag of fertilizer is 50kg, so it is difficult to pour them into the vertical tank with the height up to the waist.

(2) Venturifertilizer applicator.

Although the Venturi fertilizer applicator can be applied in a proportional manner, and constant concentration supply is maintained throughout the fertilization process, but the amount of fertilizer should be calculated when making fertilization plan. For instance how much fertilizer is required for a rotation irrigation area should be calculated in advance. If the liquid fertilizer is used, the desired volume of liquid fertilizer is added to the tank (or bucket). If the solid fertilizer is used, the fertilizer is firstly dissolved into mother liquor, and then is added into the fertilizer tank, or directly making the mother liquid in a fertilizer tank. After the fertilizer is applied to a rotation irrigation area, then the next irrigation area will be arranged.

When continuous fertilization is needed, the amount of fertilizer is calculated for each irrigation area. On the premise of constant fertilization rate, each irrigation area can be quantitatively determined by recording the fertilization time or observing the scale on the inner wall of the tank. For the fertilizer applicator with auxiliary pressurized pump, a timer is installed to control the work time of the pressurized pump on the premise of the amount of fertilizer applied in the irrigation area (the volume of fertilizer mother liquid). In the automatic irrigation system, the fertilization time of different irrigation areas can be controlled by the controller. When the whole fertilization can be completed on the same day, the pipeline can be washed after completing the fertilization, otherwise the fertilized pipeline must be washed on the same day. The washing time and requirement is the same with the bypass tank.

(3) Gravity self-pressing fertilization method.

Due to the small water pressure (usually less than 3 meters) and the blocking effect of the filter in the conventional filtration method (such as laminated filter or screen filter), the irrigation and fertilizer process can not be carried out. Zhang Chenglin et al. used the following methods to solve the filtration problem in the gravity drip irrigation system. A section of 1~1.5 m long PVC pipe with a diameter of 90 mm or 110 mm is connected at the outlet of the tank. Drill some round holes with a diameter of 30 mm and 40 mm on the pipe. The more the number of holes, the better. A 120-mesh nylon net is sewn into the shape of the pipe size, with one end opening, which is directly wrapped on the pipe, and the opening end will be fastened. With this method the water flow area is greatly increased, although the nylon mesh is also blocking water, but due to the increase of the water flow area, the total flow rate is also increased. The same method is used to solve the filtration problem in the fertilizer ponds. When the nylon mesh becomes dirty, replace a new net or wash it before using it. After several years of application, the effect is very good. Because the cost of nylon net is low, easy to buy, it is easier for the user to accept and adopt.

(4) Pump absorption method.

According to the area of therotation irrigation area, the amount of fertilizer is calculated, and then the fertilizer is poured into the fertilization pool. Turn on the pump and pour in the water to dissolve the fertilizer. Turn on the switch at the fertilizer mouth and the fertilizer is inhaled into the main pipe. Usually, the PVC pipe of 50~70mm is used in large irrigation area, which is convenient to adjust the rate of fertilization. Some farmers choose pipes with too small diameter (25mm or 32mm), when accelerated fertilization is needed, they cannot meet the needs for the too small diameter. For larger irrigation areas (such as more than 500 mu), you can mark on the fertilizer pond or bucket. Dissolve the fertilizer at one time and distribute them to each irrigation area by scale. Suppose an irrigation area required a scale unit of fertilizer, when the fertilizer solution reaches the scale, immediately turn off the fertilizer switch and continue to irrigate and flush the pipe. After washing, open the switch for the next irrigation area. When the fertilizer reached the second scale, it means that the fertilization is finished in the second irrigation area and continue in turn. Using this method to fertilize large irrigation area can improve work efficiency and reduce labor intensity.

◘ Modern Practical Integrated Water-Fertilizer Technology on Agriculture (Chinese-English)

In some irrigation areas with well's water in the north, the water temperature is low and the fertilizer dissolves slowly. Some fertilizers dissolve slowly even at higher temperatures (such as potassium sulfate). At this time, the installation of the mixing equipment in the tank can significantly accelerate the dissolution of fertilizer. The general mixing equipment is composed of speed reducer (power being 1.5~3.0kW), impeller and fixed support. The impeller is usually made of stainless steel 304.

(5) Pumping and injecting fertilizer method.

Usually there are pesticide dispensers in the south of China. Many farmers use them as pumps to inject fertilizer. The specific measures are as follows. A fertilizer pool with brick cement structure is built on the outside of the pump room, generally 3~4m^3. Usually it is 1 meter high, 2 meters long and wide, with no leakage as the quality requirement. It is best to install a drainage valve at the bottom of the pool to facilitate cleaning and removing impurities from the pool. The inside of the fertilizer pool had better be marked with paint, 0.5 meters for a grid. Install a fertilizer suction pump to inject dissolved fertilizer into the water pipe. Vortex self-priming pump is usually used to draw water, and the lift range must be higher than the maximum limit designed for irrigation system. The usual parameters are as follows. The power is 220V or 380V; the supply power is 0.75~1.1kW; the lift range is 50 meters and the flow rate is 3~5m^3 per hour. This kind of fertilizer has the characteristics of showing whether their is fertilizer left or not, and it is convenient to adjust fertilization rate. And it is suitable for irrigation system such as hour hand sprinkler, sprinkler belt, coil sprinkler, drip irrigation and so on. It overcame all the shortcomings of the differential pressure fertilizer tank. Especially in the case of groundwater use, because of the low water temperature (9~10℃) and the slow dissolution of fertilizer, the fertilizer can be put in the pool and heated up in advance and dissolved by automatic stirring. Usually, the motor power of the deceleration mixer is 1.5 kW. The mixing device is made of non-rusty material into an inverted T-shape.

(6) Mobile irrigation fertilization machine.

Mobile irrigation fertilization machine is developed for lands without power supply. It is mainly composed of gasoline pump, fertilizer tank, filter and trolley. It can be directly connected with irrigation and fertilizer pipeline in the field, and the movement is convenient and rapid. When users need to irrigate and fertilize the field, they

Chapter IV Application of the Integrated Water-Fertilizer Technology to Agricultural Production

can use tractors to pull the irrigation machine into the field, connect it to the pipeline in the field, and take turns to irrigate and fertilize different fields. Mobile irrigation and fertilization machine can replace the fixed first system of pump room, the cost is low, easy to popularize, and can meet the requirements of small area irrigation and fertilization system. At present, the specifications of the main pipe of the mobile irrigation and fertilization machine is 2 inches and 3 inch respectively. (1 inch is about 3.3cm). Each mobile machine can be responsible for an area of 50 mu.

①Operating rules of mobile irrigation and fertilization machine are as follows. In the first step, the level of engine oil and gasoline should be checked before each use of the machine. If there is not enough oil, add enough oil before use; gently lift one end of the gasoline engine when adding oil, tilt it not too much, and the oil should not be too much. Check that all connectors are proper; check that the channel water level is at the safe level. In the second step, the water chamber of the pump should be filled with pre-water before starting, otherwise it will damage the sealing ring of the pump, and there will also be the phenomenon that the water can not be pumped. The third step is to turn on the fuel switch and turn off the choke valve pull rod when starting the fertilization machine. Place the engine switch on, move the air valve lever slightly to the left, and then shake the starting handle. When the mobile fertilization machine is working, slowly open the choke valve and place the air valve at the desired speed. In the fourth step, after the mobile fertilization machine starts smoothly, observe the reading number of the pressure gauge, and when the water is pumped up, the pressure meter shows the usual lift range, it is necessary to slowly open the main valve of the system for irrigation to prevent the heavy pressured water flushes off the outlet joint at once. In the fifth step, when the water is normal, the pressure gauge reading on the left is between 0.08~0.20 MPa (8~20m) and the reading on the right is 0~0.05 MPa (0~5 m). In the sixth step, when the filter works normally, the pressure gauge reading difference between the two sides of the filter is 0.01~0.03 MPa, which indicates that the filter is clean at this time. When the pressure gauge reading difference between the two sides of the filter exceeds 0.04 MPa, the filter must be cleaned as soon as possible. In the seventh step, when the system is operating, managers should go to the relevant irrigation areas to inspect to see if the operation is normal. When such phenomena occur as broken pipe, disconnected pipe, clogged pipe, damaged emitter and leakage, they should be dealt with in time. If you can't handle it, you should immediately notify the relevant technical personnel to assist. In the eighth step, when the system stops, the choke valve rod will be moved to the left and to the bottom, turn off the engine switch, fuel switch, and then turn off the ball valve switch.

②Fertilization operation. In the first step, pour the calculated fertilizer into the fertilizer tank, stir and dissolve with water before turning on the fertilizer switch. The second step, before fertilizing, turn on the switch to start water irrigation. When all the emitters in the field spray water normally, turn on the fertilizer switch and start fertilizing. It is appropriate to control the fertilization time at 30~60 minutes, the slower the better (the specific situation can be adjusted according to the dry and wet conditions in the field). The speed of fertilization can be controlled by the switch of fertilizer pool. The third step, after the fertilization, the irrigation system cannot be closed immediately, but go on irrigating 10~30 minutes with clean water, thus the fertilizer liquid in the pipe will be completely discharged. If fertilizing in rainy weather, the pipes can be washed when the weather is fine. Otherwise, the emitter will be blocked by algae, moss, microbes, etc. (this measure is very important and it is the key to the success of drip irrigation).

4. Replacement of rotation irrigation group

According to the integrated irrigation and fertilizer system, when observing the water meterto find the quantity reached the required irrigation amount, replace to the next irrigation plot. Don't open all the irrigation valves at the same time. First open the valve of the next irrigation group, then close the valve of the first irrigation group to irrigate. Repeat the operation steps mentioned above.

5. End of irrigation

After allfields are irrigated and fertilized, turn off the pump switch in the irrigation system and then turn off the switches in the field. Comprehensively inspect the filter, the fertilization tank, the pipeline and so on to meet the standard for next normal operation. Note that after the irrigation in winter, open the drain valve on the main branch pipe and drain the water as much as possible to avoid impurities in the pipe.

II. Maintenance of Integrated Water-Fertilizer System

In order to maintain the normal operation of the integrated water-fertilizer system

Chapter IV Application of the Integrated Water-Fertilizer Technology to Agricultural Production

and improve its service life, the key is to use it correctly and maintain it well.

1. Water source works

There are many forms of water source, such as underground water, river canal water, pond reservoir water and so on. It is the primary task of water source engineering management to keep these water sourceworks in good condition, to operate reliably and to meet the designed water supply.

The pumping station, reservoir and other projects are often repaired and maintained, and the annual repair should be carried out every year in the non-irrigation season to keep the project in good condition. Sediment and other dirt deposited in the reservoir should be removed and washed regularly. Algae in the still water of open reservoir are easy to propagate, and alum should be put into the pool regularly in irrigation season to prevent algae from breeding.

At the end of the irrigation season, all the water should be excluded from the pipes and valves and wells should be blocked.

2. Fertilization system

In the maintenance of fertilization system, close the water pump, open the fertilizer injection entrance connected to the main pipeline and open the water inlet driving the fertilization system, thus eliminate the pressure.

(1) Fertilizer injection pump.

First wash the fertilizer tank of the fertilizer injection pump with clean water, open the lid to dry, then rinse the fertilizer injection pump with clean water, then decompose the pump, take out thedriving piston, oil the parts with the lubricating oil as the normal lubrication maintenance. Finally, dry the parts and reassemble them.

(2) Fertilizer tank.

First carefully clean the residual liquid in the tank and dry it, then remove the hose from the tank and wash it with clean water. The hose should be stored in the tank. Apply antirust solution to the top cover and handle thread every year. If the metal coating on the surface of the tank is damaged, immediately get rid of the rust and repaint it. Be careful not to lose any connecting part.

(3) Maintenance of mobile irrigation and fertilization machine.

The use of mobile irrigation-fertilization machine should be managed by special personnel as far as possible, and the management personnel should be carefully responsible. All operations should be carried out in strict accordance with the technical operation rules. The idling of the power machine should be strictly prohibited, and the suction pump must be immersed in the water when the system is turned on. Managers should regularly check and maintain the system to keep tidy and clean. Rain is strictly prohibited. Regularly change engine oil (half a year), check or replace spark plugs (1 year); timely clean the filter core manually, and strictly forbid to open the filter under pressure; when ploughing the land we need to move the ground pipes. They should be gently put up and down, and should not be dragged.

3. Field equipment

(1) Drainage bottom valve.

Before thecoming of winter, in order to prevent the pipeline from freezing in winter, the drainage bottom valve located on the main branch pipeline in the field is opened, and the water in the pipeline is discharged as far as possible. The valve is not closed in winter.

(2) Field valve.

Openthe manual switch of each valve.

(3) Drip irrigation pipe.

Straighten each drip irrigation pipe in the field so as not to twist it. If recollecting them in winter, we should also be careful not to distort them.

4. Prevention of blockage of drip irrigation system

(1) Irrigation water and fertilizer solution are first filtered or precipitated.

Before the irrigation water or fertilizer solution enters the irrigation system, they should pass through the filter or sedimentation tank, and then enter the water pipeline after the filter.

(2) The capacity of water transmission would be properly improved.

According to the test, the waterblockage is reduced to very small when the flow rate ranges from 4 to 8 L/h, but considering that the larger the flow rate is, the higher the cost is, so the optimal flow rate is about 4 L/h.

(3) Drip irrigation pipes should be regularly washed.

After using the drip pipe system for 5 times, the end plug of the drip pipe should be opened for flushing, thus the impurities accumulated in the pipe can be washed out of the drip irrigation system.

(4) The water quality should be tested in advance.

Before determining the use of drip irrigation system, it is best totest the water quality. If the water contains more iron, hydrogen sulfide, tannin, it is not suitable for drip irrigation.

(5) Fertilizers that are completely soluble in water should be used.

Only fertilizer that is completely soluble in water can be fertilized by drip irrigation. Do not apply general phosphate fertilizer through drip irrigation system, because phosphorus reacts with calcium in water to form precipitation and clog the emitter. We'd better not to mix several different fertilizers to avoid precipitation due to chemical reactions.

5. Maintenance of small components

Water-fertilizer integration system is a set of precision irrigation device. Many components are plastic, so in the process of use, we should pay attention to the close cooperation of each step of operation, and should not twist hard the knob and switch. When opening each container, note that some of the small parts should be restored as they were and should not be lost.

The service life of integrated water-fertilizer system is directly related to the maintenance level of the system. The better the maintenance is, the longer the service life is, and the longer the benefit is.

Chapter V Application of the Integrated Water−Fertilizer Technology to Fruits

Irrigation and fertilizer are two important management measures of orchards. Traditionally, irrigation and fertilizer are carried out separately, which is undoubtedly costly. Irrigating and fertilizing at the same time is the best measure. The root system of fruit trees absorbs water and fertilizer at the same time, which will greatly improve the utilization of fertilizer and the growth of fruit trees. The technology of simultaneous management of water and fertilizer is called integrated water−fertilizer management technology. Especially after using pipeline to irrigate and fertilize (drip irrigation or micro−sprinkler irrigation is the most suitable for orchard), the labor of irrigation and fertilization can be greatly saved. Tuopuyunong integrated water−fertilizer technology is a scientific, economical and efficient technology.

Section I Application of the Integrated Water−Fertilizer Technology to Apples

Apple is an important fruit tree in China, with a large cultivation area, mainly distributed around Bohai Bay and the plateau area of Northwest China. At present, apple planting technology in China is developing towards mechanization, standardization and simplicity. Apple is a fruit tree with multiple fertilization and frequent irrigation. Water and fertilizer management is closely related to yield and quality. At present, there are some problems in apple production, such as labor shortage, labor price rising annually, fertilizer cost increasing, excessive fertilizer, unbalanced fertilizer, soil salinization, acidizing, hardening soil and worse root microecological environment, etc. Fertilization and irrigation are two important field performances. How

Chapter V Application of the Integrated Water-Fertilizer Technology to Fruits

should it be reasonable and scientific? The majority of farmers are eager for theoretical and technical guidance. In particular, well-illustrated popular science books, which is simple and easy to understand, are more popular with fruit farmers.

I. The Origin of Cultivated Apples

Apple is one of the most commonly cultivated deciduous fruit trees in the world, which was cultivated more than 5 000 years ago. However, there is still no conclusion about the origin of the cultivated apples. Many scholars in many countries have ever studied it, but there exist a lot of difficulties to solve this problem, mainly due to the long history of apple cultivation, wide geographical distribution, complicated varieties, easy hybridization between species and varieties,

Most scholars believe that the cultivation of apples originated from multi-centers. According to the distribution of apple in the world, cultivated apples may originate from three species: forest apple, Caucasian apple and Sieversii apple. According to expert Bonomalinc's idea, it is generally believed that the species to which apples are cultivated do not actually exist at all, because no such wild type has been found in nature so far.

Bonomalinnc has done a lot of research on the origin of cultivated apples. He conducted extensive investigation, collection and comparative study on wild species of the genus apple, especially on the morphology, biological characteristics, distribution of some wild primitive species that may be cultivated apples, and the comparative study of the characters of cultivated varieties, he points that Sieversii apple is the real ancestor of cultivated apple.

Forest apple is a widely distributed species, west to Bay of Biscay Spain, east to the Volga River Basin, north to Scandinavia, south to the Mediterranean Sea and the Black Sea, but its type is poor and quite simple. Its fruit pulp is loose, sour, occasionally bitter, and the color is mostly green or yellow, locally called sour apple. In the typical forest apples, there is no sweet or sour-sweet fruit which is typical of the modern cultivated apple varieties.

Like forest apples, Caucasian apples are simple in type andthe fruits are sour. Although a large number of investigations have been carried out, no sweet type has been found, and the color of the fruit is also green. Only some fruits with red color is occasionally seen on the southward hillside at high altitude.

Sieversii apple is different from the above two, and its morphology and biological types are very diverse. As far as the fruit is concerned, the diameter of the fruit can vary from 1.5~7cm; the weight of a single fruit is 6~50g or heavier; the shape is round, oblate, long, etc., and the edges are diverse; the color is green, light yellow to bright red, and some have rust and fruit powder; The pulp color is white, milky yellow, pink; the fruit flavor is also very diverse, with every kinds from too-sour-to-eat type to sour-sweet type; mature stage is also early maturing, late ripening. According to this, Bonomalinc believes that sieversii apples are the ancestors of cultivated apple in the world so far, whose center is distributed in the Kazakh Mountains of Central Asia and the Tianshan Mountains of China.

According to Bonomalinc, the most important difference between Sieversii apple and other wild apple species is that it has the decisive traits that determine the cultivation of apple varieties, that is, it has sweet and large fruit. It is the large and sweet fruit traits that determines today's cultivated varieties.

According to his research, Bonomalinc believed that the cultivated apples in the world were spread from Central Asia to Europe. He believed that the oldest orchards in the world were produced in Central Asia and Western Asia. At that time, people did not know grafting but propagate with root tillers, and Sieversii apples were prone to root tillers. The characteristics of excellent plants were preserved by root tiller planting. In addition, Sieversii apples can also be propagated with seeds. Through the continuous selection and cultivation of wild fruit forests in ancient times, some fine types have been continuously improved. These fine types gradually spread from Central Asia to neighboring countries, such as Iran, Afghanistan, Albania, Greece and Turkey, and later passed through the Caucasus to Europe.

About the diversity of Sieversii apple, scholars in China have seen a similar situation when investigating the wild fruit forest in Xinyuan, Xinjiang province.

II. The Growth Habits of Apples

Apple is a kind of deciduousarbor fruit tree, which buds in spring every year and falls into dormancy in winter. The growth potential of apple shoots also changes with the evolution of its age. At young age, the plant grows vigorously, the new shoots are

Chapter V Application of the Integrated Water-Fertilizer Technology to Fruits

strong, and the annual growth is more than 120cm. When the fruit is in full productive age, the growth potential becomes significantly weaker, and the annual growth of new shoots is generally 30~50cm. At the end of the fruiting stage, the shoot growth is weaker, and the annual growth is often less than 20cm. After entering the aging period, the growth of the outer shoots is very small, mostly in 5cm.

In addition to the exuberant growth and the stout new shoots, the crown branches of the young apple trees are not open and the vertical growth is also strong. Because of this characteristic of young trees, for most varieties, special attention should be paid to the angle of opening branches, which can weaken the growth of branches and improve the ventilation and light conditions in the crown at the same time. It is beneficial to the future fruiting of young trees.

Some apple varieties have large branch angles, and the fruiting branch is easy to droop, so it is not necessary to stretch out the main branches for such varieties at young age.

After most varieties enter the fruiting stage, due to the burden of the fruits, the branches gradually stretch out.

Another characteristic of apple growth is the hierarchical distribution of branches, which is formed during its long-term systematic development, which is a manifestation of adaptation to external conditions. Due to the continuous growth of apple plants year by year, in order to solve the problem of light exposure of the lower branches, in the process of its systematic development, it is formed that the new shoots are sprouted into long branches at the top, and the following buds germinate into short branches, leafage branches or enter into a hidden state that do not germinate, so that the branches of the first year and the branches this year naturally become hierarchically distributed.

The internal reason of the hierarchical distribution of apple branches also lies in the heterogeneity of buds on the branches and the advantage of the top growth of the branches. Because of the different budding time and the nutritional status of the tree, the bud at the top of the branch is fully developed, while the bud at the base is thin. And after germinating in spring, the water and nutrients flow first to the top buds, while the lower buds get less water and nutrients, so the lower buds grow weakly and cannot grow into long branches. In this way after several years year, the result is hierarchical layer.

Whether the layers are obvious or not is closely related to the branching power of the apple varieties. Usually, the varieties with strong branch power, such as red jade, golden crown, Qin crown and so on, have no obvious layers, while the varieties with weak branching power, such as Guoguang, green banana, etc., are more obvious in layers.

The distribution degree of apple branches change with the plant age. In the young stage, the branches grow healthily, and the stratification is obvious, and after entering the fruiting stage, the growth potential of the branches decreases and the stratification gradually becomes less obvious. By the time of aging, the branches almost stop growing, so the hierarchical distribution of branches cannot be easily seen.

Most apple varieties show secondary growth of new shoots in the growth cycle, that is, the new shoots are divided into spring shoots and autumn shoots in that year, and the growth length of new shoots is short in northern China. According to the observation of the Fruit Trees Institute of the Chinese Academy of Agricultural Sciences, in Xingcheng area of Liaoning Province, the peak of new shoot growth is mostly in late May regardless early, middle or late maturing varieties, and special varieties appear in early June. There is only one growth peak in early maturing varieties and most middle maturing varieties, and two growth peaks in late maturing varieties, most of which appear in the first ten days of August. The spring shoot growth stop at the begging of July and begin to grow twice at the end of July.

According to the observation of orchard in Xuzhou City, Jiangsu Province, in the ancient path area of the Yellow River, the new shoots begin to grow in the first ten days of April, and become vigorous in the first ten days of May. There is another growth peak in early June, and then the growth slows down gradually. Most of the autumn shoots begin to grow in early July, reach the peak in the middle ten days of the month, and stop growing in the middle ten days of August. The vigorous trees can continue to grow three times until the end of September. Due to its late growth, the filling degree of the top bud is generally poor.

In the ancient path area of the Yellow River, the shoot growth of young apple trees is generally 90~120cm, and the difference of shoot length in spring and autumn is very small. Some varieties, such as Fuhong, Golden Crown, Hua Guan, Zhu Guang and so on, often form a large number of axillary flower buds or top flower buds on autumn shoots. Under the principle of ensuring the normal growth and good shaping of main

Chapter V Application of the Integrated Water-Fertilizer Technology to Fruits

branches, attention should be paid to the utilization of these axillary flower buds when pruning.

III. The Various Nutrients Needed by Apple and Its Role on the Tree Physiology

Fruit tree is a kind of greenplant, whose important difference from other living organisms is that fruit tree can use the inanimate substance (inorganic matter) of nature as its own nutrition, that is to say, the nutrition of plant comes from inorganic matter. But it should be noted that many studies have shown that plants also have the ability to directly utilize nitrogen in amino acids, nucleic acids, and other organic compounds. The process of plant absorption of inorganic substances in soil is closely inseparable from the activity of soil microorganisms. 90% of the plants on the earth are in a state of symbiosis with soil microorganisms.

Green plants have organs to carry out photosynthesis and can use solar energy to transform mineral elements into organic forms. The contents of plant cells such as protoplasm and cell nucleus are made up of carbon, nitrogen, phosphorus and sulfur. The normal activity of plant cells requires other mineral elements, such as potassium, calcium, magnesium, iron, manganese, zinc, copper and so on.

Although nitrogen and other ash account for a small proportion in plants, they play an important role in plant life. The assimilation of leaves and the formation of chlorophyll require not only light, temperature and moisture, but also various mineral elements. Therefore, in order to regulate the nutritional conditions of plants, we must understand the role of various nutritional elements in the physiological activities.

1. Nitrogen

Nitrogen plays an important role in the fact that it is the most important component of nitrogen-containing substances (amino acids) in plants, and amino acids are the material basis of proteins and nucleic acid molecules (the contents of the nucleus). In addition, The formation of chlorophyll, which plays an extremely important role in photosynthesis, can not be separated from nitrogen. Therefore, nitrogen is an indispensable element for plants to carry out their whole life activities. If nitrogen is insufficient, the synthesis of nitrogen organic matter is blocked, which leads to the stop of growth, smaller leaves and lighter color. Under the insufficient supply of nitrogen, fruit tree can only use nitrogen-containing organic compounds stored in the roots, stems and branches to ensure the growth of some new shoots, but due to the sharp decrease of nitrogen content, the fruit rate is low and the young fruits falling in large numbers. If sufficient nitrogen supply can be obtained in the stage of vigorous growth of fruit trees, a large number of nitrogen and organic substances can be synthesized, the growth process can be promoted and accelerated. However, if there is too much nitrogen supply, the new shoots can not stop growing for a long time, and lead to the poor maturity of branches to decrease the cold resistance.

The supply of nitrogen has a significant effect on the size of the Gala apple. The increase of nitrogen can increase the nitrogen of the leaves and increase the fruit and yield. To increase the size of the fruit, the nitrogen content of the leaves should not be less than 2.0%, but if it exceeds 2.5%, the storage capacity of the fruit will be affected.

Compared with other fruit trees, theamount of nitrogen in apples is not high. The requirement of pure nitrogen in adult orchards is 30~100kg per hectare. Many reports indicate that on fertile soils apples can grow and harvest normally without nitrogen fertilizer (Atkinson, 1980). The recommended nitrogen application rate for apple is 75~150kg per hectare of pure nitrogen.

The physiological effect of nitrogen on fruit tree plants is closely related to the growth stage of plants and the supply of water and other nutrients, especially phosphorus and potassium.

The content of nitrogen in different organs of fruit treeis different. The content of nitrogen in leaves, fruits and fruiting branches is the highest, and the content in the vegetative branches is very little, while the content of nitrogen in the trunk, perennial branches and roots is the least. Many studies have shown that the content of nitrogen in different organs of apple plants (percentage of its dry weight) is 0.40%~0.80% in fruits, 2.30% in leaves, 0.54% in vegetative branches and 0.88% in fruiting branches. The trunk and perennial branches are 0.49%, and roots are 0.32%.

The nitrogen content and other nutrients content of fruit trees change with the increase of plant organ age, different growth stages and different seasons in the annual cycle. The young and tender organs contains most nitrogen. At the beginning of spring growth, the nitrogen content in leaves and shoots is the highest, and the source of nitrogen is the nitrogen compounds stored in plants in the first year. With the change of growth stage, the relative nitrogen content in leaves, shoots and fruits decrease gradually, but the total content increases sharply. Before leaves fall in autumn, some nitrogen and other nutrients (phosphorus, potassium) are transferred from leaves to branches, trunks and roots for storage and reuse in the coming year. Nitrogen is a very active element, which can also be transferred from the old tissue to the new tissue.

2. Potassium

Although potassium is not a part of the main organic matter of plants, it also plays an important role in the process of plant life. Potassium exists ina state of ion, which is the catalyst of enzymes in plant metabolism and the synthesis of proteins and carbohydrates. Because it exists as an ion, it is an active buffer for cell solutions and various organic acids. Potassium is also an osmotic agent, which plays a particularly important role in the movement of water in plants; it can freely enter the xylem and phloem of plants and play an important role in the assimilation of carbonic acid and the transport of carbohydrates. In the case of insufficient potassium supply the synthesis process of plants was blocked, the amount of carbohydrates, especially sucrose and starch were obviously reduced, and the output of plastic substances from the leaves decreased, while carbohydrates were consumed due to respiration.

The absorption of nitrogen and the synthesis of protein in plants both need potassium for its smooth operation.

Under the condition of insufficient potassium supply, apple fruit became smaller, darker in color, lighter in flavor, lower in cold resistance, andits buds and flowers are susceptible to frost damage in spring. Potassium plays a key role in root growth and stomatal opening and closure in leaves. The content of potassium should not be less than 1.4%.

The content of potassium in leaves and fruits was the most, accounting for 1.6% of its dry weight in leaves and 1.2% in fruits.

3. Phosphorus

Phosphorus also plays a very important role in plant life activities and is a component of nuclear proteins and various nucleic acids, and nuclear proteins play an important role in the construction ofcell nucleus and other organic compounds (phospholipids, coenzymes, inositol). In addition, phosphorus is also involved in carbohydrate metabolism, and it can promote many fermentation processes. In the absence of phosphorus, starch can not be converted into sugar. It has been reported that phosphorus can also improve the oxidation reduction reaction of the plants.

The solubility of phosphorus is low in most soils. Therefore, the utilizability of phosphorus in soil solution is very small.

When phosphorus supply is insufficient, the growth of new shoots and leaves of plants will be weakened. Due to the weakening of meristem activity, the accumulation of soluble nitrogen and anthocyanin will be decreased, which cause the thick green leaves to have purple and red spots before leaves fall in autumn. If the supply of phosphorus is insufficient, the budding and fruiting will be seriously affected.

The content of phosphorus in leaves is the highest, accounting for 0.45% of its dry weight, followed by that in the fruiting branches, accounting for 0.28% of its dry weight, and the phosphorus content in fruits accounts for 0.09% ~ 0.20% of its dry weight.

The content of phosphorus in plants is the same as nitrogen, with the increasing age of plant and its different organ, and the process of phenological period in the annual cycle.

4. Magnesium

Magnesium is acomponent of chlorophyll, and magnesium plays an important role in the assimilation of plant carbonic acid. With the aging of leaves and the destruction of chlorophyll, magnesium flows to seeds and stored with phosphorus in inositol. A small amount of magnesium is involved in the synthesis of pectin.

If the supply of magnesium in plants is insufficient, the growth stops, the color of leaves becomes lighter, brown spots appear, and the color is normal only along the veins of leaves. If magnesium is seriously insufficient, it will lead to early shedding of leaves, leaving only a small number of leaves at the top of new shoots.

5. Calcium

Calcium plays an important role in the physiological balance of nutrient solutions in plants, when there are too many cationic substances, such as monovalent hydrogen ion, potassium ion, ammonium ion, sodium ion, divalent magnesium ion, and trivalent iron ion and aluminum ion, calcium can play a balancing role. The existence of calcium ion in nutrient solution is particularly important for root growth, especially in a-

Chapter V Application of the Integrated Water-Fertilizer Technology to Fruits

cidic soil.

The sufficient supply of calcium is a very important condition for plants to absorb and utilize ammonium nitrogen. If calcium supply is insufficient, plants can not fully absorb ammonium nitrogen.

The biggest difference between calcium and nitrogen, phosphorus and potassium is that it can not be reused by plants, because calcium is mainly accumulated in theaging parts or parts lacking of vitality. The content of calcium in the fruit is very small. In recent years, scholars from all over the world have pointed out that the bitter pit disease of some apple varieties is mainly caused by the lack of calcium supply. Practice has proved that regular spraying of calcium chloride during the growing season can reduce the incidence of this disease.

6. Iron

Iron content in plants is very small, although iron is not acomponent of chlorophyll, iron is essential in the normal process of chlorophyll formation in plants.

In the absence of iron, it often causes the plant to stop growing andthe leaves become yellow. Because iron cannot be reused in plants, iron cannot be transferred from the original green leaves to yellow leaves.

7. Sulfur

Sulfur is a component of the amino acids that make up proteins. Plants can only absorb sulfur in an oxidized state from sulfate. Insufficient sulfur supply is rarely found in plants, because there is often a lot of sulfur in the soil, and because of the continuous application of ammonium sulfate and superphosphate, sulfur in the soil can be constantly supplemented.

8. Boron

Boron is inactivated in plants and, like calcium and iron, cannot be reused by plants. Plant demand for boron is very small, and plant flowers contain more boron. Once boron is insufficient, flowers cannot be fertilized normally. In the case of boron deficiency, root development is hindered and suberification spot disease often occurs in fruits.

9. Manganese

Manganese, like iron, has some effect on the formation of chlorophyll. In addition, it has some influence on the oxidation reduction reaction process of plants.

Because plants need little boron and manganese, they are usually called trace elements. In addition to boron and manganese, such trace elements are important to apples as zinc, copper, aluminum and so on. In spring, some new shoots of leaves with small and hard texture often appear at the apex of apple plants, and this is caused by zinc deficiency.

IV. The Period and Method of Fertilization

1. Application of base fertilizer in autumn

Base fertilizer is the most important fertilizer applied to fruit trees. The nutrients needed for fruit tree growth mainly come from base fertilizer, which is usually applied before soil freezing in late autumn and early winter, so it is also called autumn fertilizer.

Organic fertilizers, such as stable fertilizer, compost, etc., are used for base fertilizers. The root system cut off at the end of autumn can be recovered very well before the plant begins to grow in the spring next year, and after a winter decomposition the fertilizer applied into the soil would be well absorbed and utilized by the root system of the fruit tree.

When applying base fertilizer, some phosphate fertilizer or potassium fertilizer can be added at the same time.

Mineral phosphate fertilizer and potassium fertilizer have poor activity in the soil and basically stay in the soil layer. In order to benefit the absorption and utilization of fruit tree roots, they must be applied to the deep soil layer of 20~30cm. Therefore, it is beneficial to mix phosphorus and potassium fertilizer when applying base fertilizer in autumn.

The application method of base fertilizer is generally based on annular trench, that is, a circular ditch with a depth of 40cm and a width of 30~40cm is excavated along the edge of the crown, in which the base fertilizer is applied and then buried with soil. The position of fertilizer should expand outward with the expansion of tree crown every year.

In addition to the application of circular trenches (ditch), a trench can also be excavated on both sides of the crown. The depth and width of the trenches can be the same as those of the circular trenches, and the position of the trenches should be

Chapter V Application of the Integrated Water-Fertilizer Technology to Fruits

rotated every other year.

In addition, 5 radioactive trenches can be excavated under the crown of the tree, 1 meter away from the trunk, and then the base fertilizer can be applied into the ditch and buried in the soil.

The practice has proved that the utility of base fertilizer is maintained for a long time, so the base fertilizer can be applied every two years to solve the problem of insufficient source of base fertilizer in orchard.

2. Topdressing during the growing season

The purpose of topdressing is mainly to supplement the deficiency of base fertilizer. Most of fertilization are applied in time according to the needs of growth and development in the growing season of the fruit trees. The source of topdressing are generally nitrogen fertilizer, such as ammonium sulfate, ammonium nitrate, urea and so on.

Because nitrogen fertilizer is easy to be lost in soil, the amount of nitrogen should not be too mucheach time. For fruiting trees with 5~6 years old, generally each plant can be applied 100~150 g, and 250~500 g per plant for the tree that begins to bear a large number of fruits at the age of more than 10 years. The nitrogen content of urea is high, so the amount of urea applied can be less.

In addition to ammonium sulfate, ammonium nitrate and urea, mature human feces and urine can also be used as topdressing. Usually a serving of human feces and urine can be mixed with five portions of water.

Poultry dung (chicken dung, pigeon dung, etc.) can also be used as topdressing for apples after dissolving with water. The general concentration of topdressing can be of 1kg poultry manure mixed with 10kg water.

Topdressing is usually applied in radioactive trenches, that is, dig several shallow trenches near the outside of the crown using hoes, and then evenly put the chemical fertilizers into them, or apply human manure, poultry manure into them, and then cover them with soil. Topdressing is best when the soil is wet after rain, and the effect is better if it can be combined with irrigation after topdressing.

According to the objective law of apple plant growth within one year, topdressing should be carried outfor many times, and the effect is better.

In proper management, the following topdressing is usually required each year:

(1) Topdressing before anthesis (two weeks before anthesis). It is suitable for

varieties with smaller flowers and buds but fruiting the next year.

(2) Topdressing after anthesis (after flowers falling). Fruit trees need a lot of nutrition when flowering and fruit setting, and after anthesis it is the period for the new shoots to grow exuberantly, so they need to consume a lot of nutrition. At this time, topdressing fertilizer has a good effect on improving fruit setting rate, reducing fruit drop and ensuring the vigorous growth of new shoots.

(3) Topdressing after physiological fruit drop. The topdressing was about a month after the falling of flowers. In the ancient path area of the Yellow River, topdressing is in late May, when the fruit setting has finished finally, young fruits are expanding rapidly, flower buds are approaching the eve of differentiation. This fertilization is to ensure the normal growth of young fruits, promote flower bud differentiation, and lay the foundation for high yield the following year, which is extremely important.

Of the above three topdressing, the second and third must be applied under any circumstances. When topdressing, except nitrogen fertilizer, human feces, chicken manure and sheep manure can be used for topdressing after diluted with water.

(4) Topdressing in autumn. In addition to the above three topdressing, autumn topdressing in August, mid September and late September has more obvious effect on plant growth, high yield and good quality in the successive years. At this time, the middle-late maturing and late maturing varieties of apple are in the stage of fruit expansion before harvest. Topdressing is beneficial to their growth, at the same time, it can promote the assimilation of leaves, enhance root activity and increase the accumulation of storage substances in autumn. And it creates a good basis for the overwintering and the blooming, fruiting and growing in the next year.

Topdressing in the growing season should also take into account the age of the tree and the fruiting condition of the year. During the period of topdressing in young apple orchards, emphasis should be placed on the exuberant growth period of early spring and new shoots. For orchards that have already entered a large number of fruiting periods, the period of topdressing in "Danian" (productive fruiting year) should focus on the time after anthesis and before the physiological fruit drop, the purpose of which is to promote the growth of new shoots and increase the nutritional area of leaves in order to meet the nutrient accumulation of the yield in that year. It can also form enough flower buds to ensure the yield of the following year. In the "Xiaonian" (less fruiting year),

Chapter V Application of the Integrated Water-Fertilizer Technology to Fruits

topdressing should be placed on the early spring and in the period of exuberant growth of new shoots, in order to prolong the period of exuberant growth of new shoots and avoid excessive flower bud formation and Daxiaonian phenomenon.

(5) Topdressing of foliage spray. In the apple growth season, in addition to the application of soil topdressing, the foliage spraying method can be used for topdressing out of the root, that is, the solution of the mineral fertilizer is sprayed on the leaf surface of the apple tree by a sprayer, and the mineral is absorbed and utilized through the sponge tissue cells of the leaf back, which can also play the role of topdressing.

Leaf topdressing is usually sprayed in combination with pesticide. All apple trees after foliage spray show such traits as dark green leaves, good fruit coloring and good branch maturity.

Among the fertilizers used as leaf topdressing, urea is the best. Urea is easily absorbed by leaves. When spraying urea, the effect is the best when the petals fall off, and it is not suitable to spray before fruit falls in June, otherwise it is easy to cause Daxiaonian.

Urea enters the leaf through the stomata on the back of the leaf or directly through the cutin of the leaf; within the first few hours after spraying, urea is absorbed the fastest. Because of its stomata, the effect of urea absorption on the back of the leaf is stronger than that on the surface of the leaf. Young leaves absorb faster than old leaves. The time for leaf to absorb urea can last two days and nights.

In addition to ureaas leaf topdressing, other mineral fertilizers can also be used as leaf topdressing, usually sprayed with the following concentrations:

Urea 0.5%

Ammonium sulfate 0.3%

Ammonium nitrate 0.3%

Superphosphate 1.0% (This concentration is prone to drug damage to some varieties, especially weak trees)

Potassium chloride 0.5%

Nitrogen application (spraying branches) by foliage spray in early spring is beneficial to the future growth, fruit setting rate and early fruit development. The leaf color of the plants sprayed with nitrogen in early spring is thick green, the total nitrogen content increases from 1.4%~1.8% to 2.6%, which is increased by 30%~40%; chlorophyll is increases by 5%~10%, and the dry weight of leaves increases by 15%~25%, photosynthetic intensity increases by 20% ~ 35%, and leaf area increases by 15%~30%.

V. Period of Irrigation in Production

The vast number of fruit farmers in apple industry in China have accumulated rich experience in apple irrigation according to the soil, climate and the requirements of apple tree growth, flowering, fruit and dormancy in the long-term cultivation of apple. It has played an important role in the development of apple production. According to the different irrigation period, it is generally divided into winter–spring irrigation and growth period irrigation.

Winter-spring irrigation is carried out before soil freezing and early spring bud germination in late autumn and early winter. In Northeast China, North China, Northwest China and the ancient paths area of Yellow River, precipitation is less in winter and spring, air humidity is low, and cold wind is often blowing, evaporation is large. Therefore, it is very necessary to carry out winter and spring irrigation.

Irrigation at the end of autumnand the beginning of winter, called "frozen water" or "overwintering water" by people, has a good effect on the overwintering of apple trees, and can significantly reduce the dried-up and freezing phenomenon of the first year branches or young trees because of the drought and cold wind during overwintering. It can also reduce the adverse effects of early spring drought on apple trees. Early spring is a drought-prone season in these areas, and irrigation is beneficial to the early germination and growth of apple plants. The practice has proved that winter and spring irrigation is an important factor to ensure the high and stable yield of apples.

According to the experiment at the vegetable andpotato Irrigation experimental station in Moldavia, the former Soviet Union, for the same apple variety, if the fields were not irrigated, the output was 9 795kg per hectare and if it was irrigated three times during the growing season, the yield was 17 100kg per hectare. If irrigated in the three growth period and one time in winter, the yield was 20 490kg, so it can be clearly seen that winter and spring irrigation plays an important role in increasing the yield per unit area.

The specific periods and functions of irrigation in apple areas in China are as fol-

Chapter V Application of the Integrated Water-Fertilizer Technology to Fruits

lows:

(1) Before the germination in early spring. At this time, irrigation can promote the early germination of buds and reduce the adverse effects of spring drought on fruit trees.

(2) After germination or before flowering. At this time, irrigation can make the flowers open normally, which is beneficial to fertilizing and setting fruit.

(3) After anthesis. At this time, the young fruit began to expand, the new shoots were in the period of exuberant growth, and the plants needed more moisture. Timely irrigation could improve the fruit setting rate, reduce the fruit drop, and promote the vigorous growth of young fruits and new shoots.

(4) After physiological fruit drop (about 1 month after anthesis). At this time, irrigation can accelerate fruit growth and benefit flower bud differentiation.

(5) 2 or 3 weeks before harvest. At this time, irrigation is beneficial to the final growth of fruit, can improve the yield and quality per unit area, and create good water conditions for the growth and development of plants.

(6) After the fallen leaves and before the frozen time. At this time, it should be fully irrigated in order to improve the overwintering ability of apple plants, reduce the freezing damage in cold regions, and lay a good foundation for the high yield of the following year.

In the apple orchards, irrigation is usually carried out in combination with fertilizer or topdressing. The number of times of irrigation depends on the specific circumstances.

The amount of irrigation water is usually based on the sufficient water infiltration in soil layer of the root system. The irrigation of early autumn and early spring should make the soil of the whole root layer of apple plant wet with water.

According to the materials of the Symposium on High Yield and Stable Yield of Apples held by Soviet scholars in 1955, drought in summer, especially insufficient water supply in flower bud differentiation stage, is one of the main causes of Daxiaonian. During this period, a large amount of water was supplied to the growth of young fruits, so the auxiliary shoot of fruits could not differentiate flower bud, which led to the following Xiaonian. Therefore, irrigation in the stage of flower bud differentiation is an important measure to ensure high yield and stable yield.

VI. Irrigation Method of Integrated Water-Fertilizer

1. Sprinkler irrigation method

Sprinkler irrigation is to send the pressured water by water pump or natural drop to the field by pipes, and then spray into the air through the sprinklers to form small water droplets, which fall evenly on the tree like natural rain, to achieve irrigation purposes. Sprinkler irrigation is an economical and effective irrigation method, the whole orchard irrigation is more uniform, and can be used in sloping land or complicated orchards or sandy, gravel soil. In addition, the fixed sprinkler irrigation system can also be used to spray medicine, to apply fertilizer and prevent frost. It can also be used to spray water for cooling in particularly hot weather. Compared with surface irrigation, sprinkler irrigation can generally save 30%~50%.

The disadvantage of sprinkler irrigation is that thecost of purchasing and installing the machine is high, which increases the energy cost; moreover, sprinkler irrigation above the tree increases the humidity of orchard, which is easy to cause diseases and insects; furthermore, sprinkler irrigation is affected by wind.

2. Drip irrigation method

Drip irrigation is a kind of irrigation method which uses plastic pipe to send water to the root of the tree through the orifices or drippers of the 10 cm capillary pipes in diameter. It is the most effective way of water-saving irrigation in arid and water-scarce areas. Drip irrigation is better than sprinkler irrigation in water resources utilization, and its utilization rate can reach 95%. Drip irrigation can also be combined with fertilizer, so that the fertilizer efficiency can be more than doubled, the labor can be saved, the pollution can be reduced, the soil structure can be maintained, and the yield can be increased.

Drip irrigation can make the whole or part of the soil of fruit trees reach or approach to the maximum water holding capacity in the field. It can increase the growth of new shoots, prolong the flowering period and pollination period, which is beneficial to the high yield and stable yield, increase the fruits and improve the quality.

The disadvantage of drip irrigation isthat the cost is high. In addition, due to the sediment of impurities and minerals, it is easy to block the capillary and drippers.

Drip irrigation system consists of control system, main pipe, side pipe, valve and

Chapter V Application of the Integrated Water-Fertilizer Technology to Fruits

dropper. The control system has filters, flow meter, control valves, chemical drugs, fertilizer injector and automatic controller pumps. The quality of these components must be well guaranteed, otherwise the whole system will not work well and often will not get the desired results.

Section II Application of the Integrated Water-Fertilizer Technology to Pear Trees

I. The Growth Characteristics of Pear Trees

1. Growth and fruiting habits

(1) Distribution and growth of root system.

The root distribution of pear tree was deeper and wider, the vertical distribution was 0.2~0.4 times of tree height, the horizontal distribution was about 2 times of the crown width, and some to 5 times. Root distribution is affected by tree species, varieties, stocks, soil, groundwater level and cultivation management. According to Yu Dejun's investigation in Dingxing, Hebei Province, the root distribution of Yali pear varies with topography, soil quality, soil layer and groundwater level. Therefore, deep plough of soil and reduction of groundwater level can induce root growth vertically and transversely, and make root growth develop well.

The activity and growth of pear root system are closely related to tree age, tree potential, soil physical and chemical properties, water and temperature. Generally, when the soil temperature is 0.5℃ before germination, it begin to germinate, and the growth rate is better at 15~25℃. It stops when the temperature is higher than 30℃ or below 0℃. Under suitable conditions, it can grow for the whole year without obvious dormancy period.

There are 2 or 3 peaks in the annual growth of pear roots. In Wuhan, there are three growth peaks in the roots of young trees: the first growth peak is from late March to late April, the longest duration and the largest root growth. The second growth peak is from mid-May to late July, which lasts for a long time and the amount of root growth is larger, and the third growth peak is from the first or middle ten days of October to the first ten days of November, which lasts for a short time and the growth is also small. Due to the influence of flowering and fruiting, there are only two growth peaks for the fruiting trees in general: the first growth peak is after the new shoot growth stops, and the leaf area forms till the high temperature comes (that is, from late May to the first ten days of July). Due to the sufficient supply of assimilation nutrients in the shoot, and the underground soil temperature is suitable for root growth, the root growth is the fastest in this period, and then the growth is gradually slow. The second growth peak is after the fruits harvest and the soil temperature is not less than 20℃. At this time, due to the suitable soil temperature, and more nutrients are accumulated, the root system grow rapidly (from September to October). After that, the root system grow slowly and enter the relative dormant period after leaves fall. If the fruiting are too much, the trees are very weak, the management is extensive, the disease and insect pests are more serious, and the roots are affected by drought and flood, these trees will have no obvious growth peak in one year.

(2) Shoot growth.

Pear bud belongs tothe late maturing bud, which is not easy to germinate in the year of formation, so except for special varieties or trees with strong potential (especially young trees), there is usually only one shoot from the buds in the previous year. The germination rate is high, but the branch formation is low. Except for the base blind node, almost all the obvious buds can germinate and grow, but the number of growing into branches is not large. Therefore, most of the branches of pear stop growing earlier, the contradiction between branches and fruits for nutrients is small, flower buds are easier to form, and the fruit setting rate is also higher. The growth potential of each branch is quite different; the growth of the first branch is very strong, and the second and third branches are weakened in turn, so the top branches are very strong. It is easy to see that the upper part of the crown is strong, the lower part is weak and the main branch is strong and the lateral branch is weak.

The new shoots begin to grow from the germination, and its growth intensity varies with tree species, varieties, tree age, tree potential and nutritional status. Under the same conditions, it mainly depends on the time of differentiation and the growth degree of the young stems in the buds. According to the study of Laiyang Agricultural College, the differentiation of leaf bud begin from the time of budding to dormancy, forming 3~7

Chapter V Application of the Integrated Water-Fertilizer Technology to Fruits

leaf primordium; after dormancy it differentiates 3~10 leaf primordium, so the short branch has 3~7 leaves, and the medium and long branches do not exceed 14 leaves. Only a few new shoots with strong growth and bud differentiation can have more than 14 leaves because they continue to differentiate after growth.

The one-year branches with flower buds are fruiting branches, and those without flower buds are growing branches. According to the development characteristics and length, the growing branches can be divided into three kinds: short branches (below 5cm), medium branches (5~30cm) and long branches (above 30cm). According to the length, fruit branches can be divided into short fruit branches (below 5cm), medium fruit branches (5~15cm) and long fruit branches (above 15cm).

In Guangdong, Guangxi, Fujian, Yunnan provinces and other subtropical areas, the shoots of pear trees grow more frequently, due to high temperature, more rain water, long summer and short winter. The summer is generally cooler than that in the middle areas of the Yangtze River due to the regulation of the sea breeze. For example, the pear in Huiyang, Guangdong Province, can send out new shoots three times a year. Except for warm South China, the autumn shoots of pears enter the period of low temperature dormancy before they are mature. Therefore, the autumn shoots is disadvantageous to the growth and fruiting of pear trees.

(3) Leaf growth.

The leaves grow with the growth of the new shoots, and the first leaf at the base is the smallest and gradually increases from bottom to top. On the leaves on the short branches, one or two leaves on the top are the largest; the leaves on the middle or the long branches vary in size according to the nutritional status and external conditions during the growth period. On the long branches with bud differentiation, the first leaf from the base is small, and it becomes bigger gradually from bottom to top. When the maximum leaf appears, there are 3 smaller leaves, then it gradually becomes bigger and then smaller gradually. The first leaves from the base to the large are called the first round leaves, which is the leaf that differentiates within the bud, and the above leaf is the one that differentiates outside the bud, which is called the second round leaf. From germination to spread, it takes about 10 days, and it takes 16 to 28 days from spreading leaves to stopping growth. Where the leaf is small and the tree is weak, the growth period of the single leaf is short, on the contrary, it is longer.

◘ Modern Practical Integrated Water-Fertilizer Technology on Agriculture (Chinese-English)

The leaf area of different types of new shoots is different; the leaf of the long shoot is the largest, the middle shoot is the second, and the short shoot is the smallest. However, as far as the shoots' per unit length having the leaf area is concerned, the condition is on the contrary. The short shoot per unit has the largest leaf area, the middle shoot is the second, and the long shoot is the smallest, so it is the short and middle shoot that will accumulate the most nutrient, which is beneficial to flower bud differentiation and fruits' big size. However, without a certain number of long shoots, there will be no basis for the production of medium and short shoots, which is not conducive to the growth and fruiting of pear trees. At the same time, the leaf area of the long shoots is large, and the synthesized nutrients are more than that in the medium and short shoots. So in addition to supplying their own needs, there exists some surplus for others' use. Therefore, the reasonable proportion of all kinds of branches is the biological basis of high yield and stable yield.

(4) Flower bud differentiation.

Pear is a kind of species which is easy to form flower bud, which can not only form topflower bud from top bud, but also form axillary flower bud from side bud. Some varieties have large number of axillary flower buds and the fruit setting is reliable. The morphological differentiation of pear flower bud can be divided into five stages: flower bud differentiation stage, flower primordium and calyx stage, Corolla stage, stamen stage and pistil stage. The flower bud differentiation of pear usually begins not long after the shoot growth stops, or when the bud growth point is active after rapid growth and the tree conditions in and out are suitable, it begins to differentiate. If the new shoots stop growing prematurely, the leaves become small and few, malnourished, or if it stops growing too late, and the climate is not good, it can not form flower buds. According to Sato's study, for most early maturing varieties of Japanese pear, the flower buds begin to differentiate in mid-June, medium-maturing varieties in late June and late-maturing varieties from late June to mid-July. The temperature in the Yangtze River Basin of China is higher than that of Japan. Therefore, the flower bud differentiation of Japanese pears in Wuhan and other places is slightly earlier than that in Japan. Changshiro begins to differentiate in mid-May and Mingyue begins to differentiate in mid-June. Generally, the floral organ is basically formed by October, after that the temperature decreases, the tree gradually enters dormancy period, and the flower bud

Chapter V Application of the Integrated Water-Fertilizer Technology to Fruits

differentiation is suspended. In the early spring, the temperature rises and it continues to differentiate until the ovules in the pistil are well developed in the first and middle ten days of April, and the rest of the flowers are also growing until blooming.

(5) Flowering andbeing pollinated and fertilized.

①Flowering. After dormancy in winter, when the average temperature is above 0℃, the organs in the floral organ grow slowly and develop slowly. With the increase of air temperature, the growth of the flower bud gradually accelerates, and the volume of the flower bud also expands and germinates. The blooming period of flower buds in the Yangtze River Basin is generally in the middle of March. After the flower bud opens, it experiences budding and inflorescence separation, finally the petal extends and blossoms.

The flower bud of pear is a mixed bud, and there is a young shoot in the bud. After germination the flower bud develops from the young shoot to the fruiting new shoot, and a cymose inflorescence is formed at the top of the bud. Generally, there are 5~8 flowers per inflorescence, but there are great differences among different varieties, some with less than 5 flowers, some with more than 10 flowers. The leaf axils of the new fruiting shoots can have auxiliary shoots, and the number, strength of the auxiliary shoots are related with the characteristics and nutritional status of the variety. Generally speaking, most have 1 or 2 auxiliary shoots, but some have nothing or some have as many as 3 shoots.

Most pear varieties are those of flowering first and then spreading leaves. A few varieties are that flowers and leaves open at the same time or leaves spread first before flowering, the former such as Kant, the latter such as Jingbai pear. The order of flowering, as far as a single inflorescence is concerned, the base flower blossoms first, the central flower blossoms later, the first blooming flower develops well, which is easy to be pollinated and fertilized, and the fruit setting is reliable. The anther and pollen of the same flower grows in the order of first outer and then inner, and grows centripetally.

The time of blooming early or late, andthe length of flowering of the pear trees vary with variety, climate change and soil management. Generally speaking, the flowering period of Qiuzi pear and white pear is earlier, the flowering period of Xiyang pear is late, and the sand pear is in the middle. Even for the same variety with different ages, the flowering period is also different. But the flowering period of different varieties and ages will become relatively consistent sooner or later, which can be divided into three types: early, middle and late, and this can be used as a reference for the selection of pollinating trees.

Early flower types include Chilipear, Yali pear, Cangxi pear, Jinchuan Sydney, Ergongbai pear, Jinshui No. 1 pear, yellow honey pear, Kant pear and so on.

Middle flower types include Dangshan pear, Xingao pear, Xiangnan pear, Huanghua pear, Changshiro pear, 20th century pear, Jinshui No. 2 pear and so on.

Late flower types include Sanji pear, Jiangdao pear, Taibai pear, Crown pear, Fuli Pear, Bali pear, Rimianhong pear and so on.

According to the observation of Zhejiang Agricultural University in Hangzhou, no matter which type it is, the initial flowering time is short, about 2 days, the full flowering period is longer, about 4 days, as for the final flowering period, there are great differences among the varieties, some varieties 3~4 days, some as long as 10 days. In a word, the flowering period of each type is only 5 to 14 days from the beginning to the end, and the majority is 5 to 6 days, and it takes only 1 to 3 days to change from the beginning bloom to the full bloom. The time and the length of flowering of the same variety in the same place are mainly affected by temperature.

In general, pear trees bloom in spring, but there are also special circumstances, such as secondary flowering and late flowers.

Autumn blooming is often called secondary flowering. Secondary flowering is that the flower buds differentiated bloom in autumn of the same year. This phenomenon is often due to the early falling leaves caused by natural disasters such as drought or diseases or insects during the flower bud differentiation period, and the transpiration is greatly reduced, while the water and inorganic salt absorbed by the root system are still transported upward and the concentration of cell liquid is decreased. The process of flower bud development increases after the qualitative change stage, and promotes the flower bud to germinate ahead of time and bloom in autumn. Secondary flowering can also be fertilized to fruit, but because it is late and the temperature is not enough, it can not ripen normally or the fruit is too small, and there is no value. Therefore, it is necessary to strengthen the supply of fertilizer and water and the control of diseases and insect pests in summer and autumn, to prevent early leaves' falling and secondary flowering.

After normal flowering in spring, part of the top of the accessory shoot blos-

Chapter V Application of the Integrated Water-Fertilizer Technology to Fruits

soms. The flowering stage of this flower is 10~20 days later than that on the new shoot, so it is called late flower. The late flowers bloom at the top of the accessory bud that has differentiated before dormancy, and continue to differentiate from dormancy to germination. Due to the short differentiation time, the floral organ is developing imperfectly and the morphological variation is great, such as more petals, fewer stamens, calyx being like leaf and so on. Generally speaking, it can not bear fruits.

②Being pollinated andfertilized. Pear is a fruit tree with cross-flower pollination, and self-pollination will cause no fruiting or low setting rate. Therefore, it is necessary to match the pollination tree in production. When selecting pollination tree, it is appropriate to select strong affinity, large pollen quantity and high pollen germination rate. The flowering stage of the pollination tree should be the same as the main variety. The first batch of flowers on a tree or the first to second flowers in the same inflorescence had sufficient nutrients, good differentiation, the highest fruit setting rate, and the fastest fruit development with the largest type and the best quality; the flowers blooming at full flowering stage are listed second; the last blooming flowers are generally the worst.

(6) Fruit setting, fruit drop and fruit development.

①Fruit setting, flowers falling and fruit falling. Pear is a species with high fruit setting rate. When the tree is strong and can be fertilized and properly managed, it can generally meet the requirements of high yield. However, there are great differences in fruit setting ability among different varieties. According to such standard as that it is strong to set more than 3 fruits on 1 inflorescence, 2 are medium, 1 is weak, the varieties with high fruit setting rate are Huali No. 2 pear, Fengshui pear, 20th century pear, Yali pear, etc. The varieties with medium fruit setting rate are Xiangnan pear, Dangshan crisp pear, Changshiro pear and so on. Varieties with low fruit setting rate are Cangxi pear, etc. The production practice shows that the varieties with high fruit setting rate often inhibit the new fruit in the year of a large number of fruits and influence the differentiation of flower bud, which results in the phenomenon of Diaxiaonian or fruiting every other year, because of excessive fruit setting and heavy load. Varieties with low fruit setting rate often fail to meet the requirements of high yield because of their low fruit setting.

During the yearly cycle, pears has three peaks of physiological fruit drop. In addition, there is a slight fruit drop before the fruit maturity. The fruit of pear is larger and the handle is crispy. If there is no wind-proof facilities, the fruit yield will be reduced by the heavy wind with 5~6 degrees.

②Fruitdevelopment. Pear fruit is developed from the fertilized flower, and the receptacle develops into pulp, the ovary develops into the core, and the embryo develops into seed. There is a close relationship between the three in the process of fruit development. According to the speed of fruit development, it can be divided into three periods:

Rapid fruit expansion period: this stage begins from the expansion after ovary being fertilized to the the emergence of embryos in young seeds. Endosperm cells proliferate and occupy the largest space in the seed coat. Some of the cells in the receptacle and fruit core split rapidly, the number of cells increases, and the fruit increases rapidly, and the longitudinal diameter increases faster than the transverse diameter, so most of the young fruits are oval.

Slow growth period: from the appearance of the embryo to the maturity of the embryo. During this period, the embryo develops rapidly and absorbs the endosperm and gradually occupies the space of all endosperm in seed coat. During the rapid increase of embryo, the volume of receptacle and fruit core increases slowly. Stone cells in the pulp begin to occur and reach their inherent number and size.

Rapid fruit growth period: During this period, the embryo occupies all the space in the seed coat until the fruit ripens. Due to the increase of pulp cell volume and cell gap volume, the volume and weight of fruit increase rapidly, but the volume of seed increases little or even doesn't increase, but the seed coat changes from white to brown and enters the mature stage of seed.

The size of the fruit depends on the number of cells and the size the cells. If the number of cells is large and the volume is big, the fruit size is big; if the volume is small and the number is big, the fruit size is medium.

All active plant organs havenycterohemeral (day and night) periodicity in growth rate, and the main factors that affect the growth of plants are temperature, water status in the plant and light. There are usually two growth peaks in a day, one before the noon and the other in the evening. The variation of fruit growth mainly follows the law of contraction in the day and expansion at night. Pear fruit contracts in the sunny day, due to high temperature and dryness; at night, the fruit will expand, and the accumulation of

Chapter V Application of the Integrated Water-Fertilizer Technology to Fruits

photosynthetic products in the fruit is mainly before the midnight, and the increase of the fruit after the midnight is mainly from the water absorption. Therefore, in high temperature period in summer, the fruit should be picked on sunny days in the early morning, the fruit is large with sufficient water, crisp and cool. On rainy days, due to the saturated state of air humidity, the fruit does not contract at all, and even because of rainfall fruit expansion becomes extraordinary even the fruit cracking occurs, especially when the fruit osmotic pressure is high and the rainstorm or a large amount of irrigation will promote the fruit cracking. Bagging can prevent fruit cracking, because the effect of bagging fruit on fruit expansion due to rainfall appears a few hours later than that of unbagged fruit, and the rate of expansion is half slower than that of unbagged fruit, so there is little fruit cracking.

2. Requirements for environmental conditions

(1) Temperature.

The temperature requirements of pears are different because of their different species, varieties and origins, and distribution ranges are varied. Sand pear originates from the Yangtze River Basin of China, and the required temperature is high, mostly distributed in the north latitude of $22°\sim 32°$, generally can withstand the low temperature above $-23℃$.

The cold tolerance of different organs of pear is different, and the floral organ and young fruit are the least cold tolerant. The phenomenon of frozen flower buds often occurs in Jiangsu and Zhejiang provinces because of the sudden cooling after the warm weather in early spring. Pear is a cross-pollinated fruit tree, which needs insects to propagate. Bees begin to fly and move at $8℃$, while other insects are above $15℃$. The pollen germination of pear requires more than $10℃$, and $18\sim25℃$ is the most suitable. At $16℃$, it takes about 44 hours for Japanese pear to be fertilized. If the temperature increases, the process of fertilizing can be shortened, and on the contrary, it will be prolonged. Therefore, if the weather at the flowering stage is sunny, the temperature is higher, and the fertilizing is generally good, so it is expected to increase the yield in the same year. If it rains continuously, or excessive temperature change will lead to poor pollination and fertilization, and serious drop of flowers and fruits will inevitably lead to yield reduction.

Temperature also has an effect on the quality of pear. If Yali pear, which originated in cold and dry areas, is cultivated in wet area high temperature, the fruit becomes smaller and the flavor becomes lighter. The temperature of Changshiro in the fruiting period (July to August) is positively correlated with quality.

(2) Moisture content.

The growth of pear needs enough water, so if the water supply is insufficient, the branch growth and fruit growth will be inhibited. However, too much water and too much humidity are not suitable, because the root growth needs a certain amount of oxygen, when the oxygen content in the soil is less than 5%, the root growth is poor, and when it falls below 2%, it can inhibit the root growth. When the soil gap is full of water, the root system breathes without oxygen, which will cause the plant to die.

The requirement of water and the degree of moisture tolerance of pear vary according to the different species and varieties. Sand pear has strong moisture resistance and is mostly distributed in areas with precipitation above 1 000mm. Precipitation and humidity have great influence on the color of fruit skin. If the fruits form in rainy and humid climate, the cuticular layer of the pericarp often breaks, generally the fruit point is large, the surface is rough, and lacks the inherent smooth and color of this variety. This is especially obvious on the green varieties such as the 20th-century pear, Jushui pear, etc. At the same time, during the growth of new shoots and the young fruits from April to June, if the rain is too much and the humidity is too high, the disease will inevitably be serious.

(3) Sunshine and wind.

Pear is a light-loving fruit tree, so if the light is not enough, it often grows too much, and its growth is futile, which affects the differentiation of flower bud and fruit development. If the light is seriously insufficient, the growth will gradually weaken and even die.

Wind has a great influence on the pear. The strong wind not only affects the insect pollination, blows off the fruits, but also causes the branches and leaves to be mechanically damaged or even lodging. Wind can also significantly increase the transpiration and reduce the moisture in leaves, thus affecting the normal progress of photosynthesis. Generally speaking, when there is no wind, the moisture content of leaves is the highest and the assimilation amount is the largest. With the increase of wind force, the water content decreases gradually, and the assimilation amount decreases

Chapter V Application of the Integrated Water-Fertilizer Technology to Fruits

accordingly. However, if the orchard is in a windless state for a long time, the air can not be convective, the carbon dioxide content is bound to be too high, which deteriorate the environment, and the assimilation amount will decrease. Therefore, the breeze (wind speed 0.5~1 m/s) is the best for the growth. The establishment of wind-proof forest can improve the microclimate of orchard, reduce wind damage and enhance assimilation.

(4) Soil and topography.

Pears can adapt to all kinds of soil, so whether sandy soil, loam or clay can be proper for cultivation. However, due to the weak physiological drought tolerance, the sandy loam with deep soil layer, loose and fertile soil, good water permeability and water preservation is the most suitable. Pear has a wide range of adaptability to soil acidity and alkalinity, and the soil with pH value of 5~8.5 can be proper, but the pH value of 5.8~7 is the best. Pear tree has strong salt and alkali tolerance, and it can grow normally on the field where the soil salt content is not more than 0.2%.

Pear is not strict on terrain, regardless of mountains, plains, bench lands, all these can be proper to cultivate pear trees. The soil moisture condition of plain and bench land is better and the management is more convenient, but in the area of high temperature and more rains, it is often due to the excessive amount of rain water and the high groundwater level, which affects the tree potential and fruit quality, and the fungal disease is easy to spread. As for the mountain and hilly orchards, its drainage, light and ventilation conditions are good, and it is easy to control the tree growth, diseases can also be reduced. So it is easy to obtain high yield and high quality fruit, but it easily lead to soil and water erosion. At the same time, with the increase of altitude, the temperature gradually decreases, and pear trees are often susceptible to frost damage at flowering stage.

II. The Integrated Water-Fertilizer Management of Pear Trees

1. Fertilization period

In the yearly cycle growth of pear trees, the growth of roots and shoots, flowering, fruiting development and flower bud differentiation have a certain order and mutual restriction. Among them, the transition period from storage nutrient growth to new leaf assimilation nutrient supply and the reproductive period of fruit development and flower bud differentiation are the key periods of cultivation and management. These characteristics should be paid attention to in determining the fertilizing period.

(1) Base fertilizer is applied beforethe leaves falling after picking the fruits. At this time, the soil temperature is high, and the tree is active, which is beneficial for the root healing and growth. The growth of new roots with autumn fertilization is about three times higher than that of spring fertilization. At the same time, the application of base fertilizer in autumn has a significant effect on restoring tree vigor, strengthening assimilation and enhancing tree body nutrition preserve, so it is beneficial to improve fruit setting rate, yield and quality. If you can't apply base fertilizer in autumn because of labor and other reasons, after picking the fruits you should apply nitrogen fertilizer in time to maintain a strong tree. The temperature and evaporation of in southern China are still high in autumn, so it should be irrigated in time after applying base fertilizer in autumn.

(2) The absorption of nutrient elements by topdressing is different at different growth stages. According to the study on nutrient absorption of the 20th-century pear by Sato, the absorption of nitrogen is the most in the new shoots growth stage and the young fruit expansion stage, but the absorption is very few in the physiological fruit falling stage; the absorption increases in the second fruit expansion stage, but it decreases after harvest. The absorption condition of potassium is basically the same as that of nitrogen, but the absorption of potassium is much higher than that of nitrogen in the second fruit expansion stage, while the absorption of phosphorus is less than that of nitrogen and potassium, and it is comparatively even in every growth period. The application of base fertilizer alone can not timely meet the needs of different nutrients in different periods, so topdressing must be applied reasonably according to the characteristics of fertilizer requirement of pear trees. Apply topdressing for 3 times in the whole year. When the weather is dry, it should be combined with irrigation.

①Pre-anthesis fertilizer: apply quick-acting nitrogen fertilizer before flowering and after germination. If human feces and urine, rotten cake fertilizer are used, they should be applied about half a month in advance. Pre-anthesis fertilizer has a certain effect on increasing fruit setting rate, promoting branch and leaf growth, and increasing leaf-fruit ratio. Weak trees and young trees can be applied with single nitrogen fertilizer in order to promote the growth of branches and leaves, accounting for

Chapter V Application of the Integrated Water-Fertilizer Technology to Fruits

about 20% of the yearly application rate. It is generally not suitable to apply nitrogen fertilizer alone in the fruiting tree for the first time and the adult prosperous tree.

②Fertilizer for strengthening the fruits: it is carried out after the peak growth period of the new shoots and before the second expansion of the fruit. The main fertilizer is quick-acting nitrogen fertilizer, combined with phosphorus and potassium fertilizer, and the amount of nitrogen fertilizer accounts for about 20% of the yearly application rate. If fertilizer is applied too early, it will promote the growth of branches and shoots, decrease the sugar content, and affect the quality.

③Pre-harvest fertilizer: it is carried out before fruit picking. The amount of quick-acting nitrogen fertilizer accounts for 20% of the whole year's amount. If the base fertilizer can not be applied in time after fruit picking, the quick-acting nitrogen fertilizer should be properly applied to restore the tree potential and prevent leaves falling early.

2. Fertilizer methods

The root system of pear tree is strong and deeply distributed. Base fertilizer should be applied by annular ditch, rows of ditches, holes, and it should be deeply and fertilized layeredly. The width of the trench is about 0.8 m and the depth is about 0.6 m. Dig trenches in turn, once every 2 years, gradually plow the orchard deep to fertilize, guiding the root system to expand deeply. After 4~5 years, it can be plowed deep by stages and in groups to renew the root system and activate the soil.

According to the type and nature of fertilizer, the method of topdressing should be applied with radioactive ditch, annular trenches or holes, 10~15cm in depth, and cover the soil in time after application. If the soil moisture is insufficient, the fertilization should be combined with irrigation. Ammonia water should be first diluted with water, otherwise the fertilizer effect is not easy to reach or even have destructive effect.

External topdressing has been widely used at present. With the development of mechanization, pipeline irrigation will be widely used in the future. Especially, from April to May, the effect of topdressing on pear treesis more obvious from nutrients storage period to nutrients assimilation period in the same year. The common concentrations are as follows, urea 0.3%~0.5%, human urine 5%~10%, superphosphate 2%~3%, boron 0.2%~0.5%, heavy iron sulfate 0.5%, zinc 0.3%~0.5%, etc.

3. Amount of fertilizer applied

Many experiments have proved that amount of fertilizer should be applied properly, and the necessary elements should not be scarce or excessive, otherwise it will not only cause fertilizer damage to the tree, but also pollute the soil environment. Therefore, both at home and abroad great importance is attached to the amount of fertilizer research.

The fertilizer requirement of fruit trees is greatly affected by variety, tree age and stock. For example, compared with Chang Shiro pear, the 20th - century pear needed more fertilizer, especially phosphorus. In the United States, Bartlett pear (Bali pear) was grafted on different stocks to study the absorption of nutrients by different stocks. The results showed that the absorption of nitrogen by the stock of European pear is more than that of quint and Hawthorn, and the absorption of phosphorus was the highest in the mountain ash and the least in Snow pear. According to foreign research, the ratio of nitrogen, phosphorus and potassium content between vegetative organs (root, stem, leaf) and fruit is different. The ratio of nitrogen, phosphorus and potassium in vegetative organs is 10 : (3.5~5) : (6.4~6.9), and the ratio in the fruits was 10 : (4.5~4.8) : (21~25). The contents of nitrogen and phosphorus are almost the same, while the content of potassium in fruit is more than 3 times higher than that in vegetative organs. Therefore, there should be a difference between adult tree and young tree.

The basis for determining the amount of fertilizer is on the correct nutrition diagnosis. At present, the commonly used diagnostic methods are morphological (tree appearance) diagnosis, leaf analysis, soil analysis and so on, as well as the fertilizer comprehensive diagnosis method (DRIS method), which has been popular in recent ten years. Generally, the amount of the fertilizer is determined according to the size of the tree, the number of flower buds and the yield, and they could be increased or decreased according to the regular and stale index of high yield and stable yield.

In terms of the types of fertilizer, organic fertilizer should be combined with chemical fertilizer. In order to improve the physical and chemical properties of soil, generally the organic fertilizer should account for 1/3~1/2 of the nutritional elements needed for fertilizer application. As to the chemical fertilizer, it should be gradually changed from the use of single chemical fertilizer to the use of compound fertilizer, and formula fertil-

Chapter V Application of the Integrated Water-Fertilizer Technology to Fruits

izer should be applied according to the results of nutritional diagnosis.

4. Irrigation

Pear tree is more moisture tolerant, butif the soil is too wet, the ventilation is poor, the physiological function of root will decrease, the growth of tree body will worsen, especially when the temperature is high and when pear tree grows vigorously, the wet damage is more serious to the pear trees. We often see the plain pear orchards with low terrain and high groundwater level and the inpermeable mountain pear orchards. In the season of more precipitation or concentrated precipitation, the leaves are yellow and green, and the growth is poor. Once in the dry season, the yellow leaves will fall early. Therefore, to construct the orchestras in the southern part, we advocate to dig trenches change soil on the mountainous area, and also dig deep trench and make high border on the flat ground. If the subsoil is sticky and heavy, we must dig trenches, deeply plow the soil, and open up the impermeable hard soil layer. The surrounding trenches should be deep, and the trench can be connected with the river, so that it can drain the surface water and subsurface water to prevent waterlogging. The climatic characteristics of most areas in the Yangtze River Basin and its south are as follows: continuous rain in spring and summer, high temperature, drought and less precipitation in summer and autumn. In short, we should pay attention to drainage in spring and summer, irrigation in summer, autumn and even winter, and in some special drought spring, irrigation is needed. Different tree species, growth period and site conditions are different in water demand and water preservation, so the times of irrigation should be treated differently. Furrow irrigation, drip irrigation, sprinkler irrigation, micro-sprinkler irrigation and hole irrigation are the best methods of irrigation, but the use of high sprinkler irrigation in highly closely planted pear orchards can easily cause excessive microclimate humidity. And fungal disease. If irrigating, the soil should be watered thoroughly. The soil water content in the main layer of the root distribution should reach 70%~80% of the maximum field water holding capacity.

Section III Application of the Integrated Water-Fertilizer Technology to Peach Tree

I. The Requirements of Peach Trees for Environmental Conditions

1. Temperature

The adaptation range of peach is wide, and it can be cultivated in the area where the yearly average temperature is 8~17℃. The suitable annual average temperature for northern cultivar group is 8~14℃. The average temperature during the growth period is 19~22℃, and the flowering stage needs more than 10℃. The suitable temperature for fruit growth and development is 20~25℃. According to American data, when the average monthly temperature is below 18.3℃, the fruit quality is poor, and when the temperature reaches 24.9℃, the yield is high and the quality is good. The temperature of peach areas in China is generally above 24℃ from June to August, so it is beneficial to the growth and development of fruit.

Peach trees need a certain amount of low temperature in winter dormancy period in order to sprout normally and bloom and bear fruits. If the temperature is too high in winter, the dormancy can not be completed smoothly, resulting in late germination in the next spring, irregular flowering, poor pollinating and poor yield. The cumulative hours of cold capacity in dormancy period are 500~1 200 hours when the daily average temperature is less than 7.2℃. Therefore, the high temperature in winter in China limits the development of peach trees.

Different varieties of peach have different resistance to low temperature, and the general varieties can tolerate low temperature of −25℃ to −22℃. The flower buds and young trees of some varieties froze at −18℃ to −15℃. At −25℃, the trees will be injured; at −30℃ to −28℃ the whole plant will be seriously injured, and the whole plant was even killed. After flower bud germination and during the buds' color changing period, the proper freezing temperature is −6.6 to −1.1℃, in the flowering stage it is −2℃ to −1℃, and in the young fruit stage it is −1.1℃.

Chapter V Application of the Integrated Water-Fertilizer Technology to Fruits

2. Moisture

During theyearly growth of peach, proper amount of water is needed. The results showed that the suitable soil water holding capacity of peach tree is 60% ~ 80%. When the field water holding capacity is 20% ~ 40%, the peach could grow normally and when it decreases to 10% ~ 15%, the leaves will wither, which seriously affects the growth and development of peach trees.

Peach trees are resistant to drought andbut cannot be tolerant of waterlogging. Poor drainage and high groundwater level will lead to early root decay, thinning leaves, lightening leaf color, decreasing growth, and then falling leaves, falling fruit, flowing glue and even death. Therefore, attention should be paid to drainage and waterlogging in rainy season.

3. Soil

Peach roots breathe vigorously and require a lot of oxygen. According to the results, when the oxygen content in soil is 10% ~ 15%, the root growth is normal, when the oxygen content decreases to 7% ~ 10%, the root becomes brown and no new root occurs. When it is 2 %, the fine roots begin to die, the new shoots stop growing and even the leaves will fall, so peach trees should be planted in sandy loam or soil with loose soil quality and good drainage. The clay soil's ventilation is poor, so it is easy to suffer from glue disease, neck rot and so on. Peach grows normally at pH 4. 5 ~ 7. 5, and in alkaline soil it is prone to yellow leaf disease due to iron deficiency. Different stocks have different alkali resistance, among which mountain peach is more alkali resistant, so in the soil with slightly high pH value in the north, mountain peach stock is often used. Peach is more tolerant of salt. When the salt content in the soil is 0. 13%, it has no adverse effect on growth. When the salt content reaches 0. 28%, the growth will be poor or die.

4. Light

Peach is native to the continental climate of strong light in northwest China. When the lightis insufficient, the assimilation of the tree decreases significantly; the root development is poor; the branches and leaves are long; the flower bud differentiation is less; the quality is poor, and the falling flowers and fruits are serious. The twigs are easy to die, the fruiting parts are moved up, and the lower parts of the crown are bald. Therefore, peach garden should be selected in a place with good ventilation and light, suitable planting density, reasonable tree shape and moderate branch number.

II. The Integrated Water-Fertilizer Management of Peach Trees

1. Fertilizer application

(1) The law of fertilizing requirement of peach trees.

There are three main periods for the expansion and growth of peach fruits. In the first growing period, from falling flower to hard core, the cells begin to divide rapidly, and there are some differences among varieties. In the second growth period, from the beginning of hardening core to the completion, the number of cells increases rapidly, and the time required for early maturing and late maturing varieties is quite different. The third growth period is from the completion of hardening core to fruit maturity, and the cells grow rapidly at this stage, and the fruits increase significantly 15 days before harvest. From the first stage of fruit growth to the first half of the second stage, the growth of new shoots is growing rapidly, and when the fruit matures gradually, the growth of new shoots is gradually slow and stopped.

Rational fertilization is very important. The peach trees demand for nitrogen and potassium is large, but the demand for phosphorus is less. For each fresh fruit of 1kg, it needs 1.52g of nitrogen, 0.79g of phosphorus and 3.64g of potassium.

The absorption of nutrients by peach trees varies with the growth period in the year. From the beginning of hardening core, the absorption of nitrogen, phosphorus and potassium increases rapidly, and reaches to the peak about 20 days before harvest. During this period, the absorption of phosphorus and potassium increases rapidly, especially potassium.

When micro fertilizeris used as base fertilizer, the dosage of borax per mu is 0.25kg, zinc sulfate 2kg, manganese sulfate 1~2kg, ferrous sulfate 5~10kg (It should be combine with organic fertilizer; the ratio of organic fertilizer to iron fertilizer was 5 : 1). Micro fertilizer can be sprayed on the leaf surface, the concentration of spraying can be controlled at 0.1%~0.5% according to the aging degree of the leaf. The concentration should be sparse when the leaf is tender, and it can be thicker when the leaf is aging.

(2) Base fertilizer.

The application period of base fertilizer can be divided into springfertilization and

Chapter V Application of the Integrated Water-Fertilizer Technology to Fruits

autumn fertilization. It is best to apply base fertilizer in autumn, from the first ten days of September to the middle of October. At this time, the new shoots have stopped growing, but the roots are still active, especially in September, the root growth is at the peak in autumn. The injured roots caused by the base fertilization could resume their growth after about 20 days. After the fertilizer is absorbed by the root system in autumn, it provides a large number of nutritive raw materials for photosynthesis of the leaves, improves the level of tree nutrition storage, and lays a material foundation for the growth and fruiting of the following year. If no base fertilizer is applied in autumn, the next spring must be applied in time after thawing; if the fertilization is too late, the root damage caused by fertilizer will have adverse effects on blooming, fruit setting and spring shoot growth.

The methods ofbase fertilization include scattering application, annular ditch application, strip ditch application, radial ditch application and so on. For the young peach trees, combined with the soil improvement, most of them are fertilized by annular method. The depth and width of the trench are 40~60cm and 50~60cm respectively. The soil of adult peach trees has been improved deeply because of the combination of annular fertilization, so the base fertilization can be applied alternately by strip ditches, radial ditch, large hole and ground bedding method. If the organic matter such as fertilizer or weeds on the ground reaches 10~20cm, and insists on bedding more than 10cm every year, the soil can be changed by big holes every three years. In this way, a lot of labor can be saved, and the effect is also good. The method of changing soil from big holes is as follows: 2~4 holes with 40~50cm square are dug on the outer edge of the crown; the topsoil and raw soil are separated. The topsoil containing a large amount of organic matter is filled into the hole, and the raw soil is filled on the surface. For every 3 years, this is applied and each time changes the position of the holes. For the peach garden without bedding of organic fertilizers, when the deep ditch fertilization is applied, the excavated raw soil and topsoil should be separated. The topsoil should be evenly mixed with the saprophytic organic fertilizer and then filled into the lower layer; the raw soil should be placed on the surface layer, and 30kg of high quality organic fertilizer should be applied per plant, and secondary quality fertilizer should be applied more. Irrigate water in time after fertilization.

(3) Topdressing.

The number and amount of topdressing are related to varieties, soil fertility, tree potential and yield. For example if it is early maturing variety, or the soil is fertile, the tree is strong, the fruiting is not excessive, the fertilizer amount and the frequency should be less, on the contrary, it is more. The method of topdressing soil is as follows: straight ditch or radial ditch method. The ditch is 20 centimeters deep, after the fertilizer is applied, the soil on both sides of the ditch will be refilled into the ditch. Peach garden mainly has the following topdressing: pre-flowering fertilizer, fruit setting fertilizer, fruit expansion fertilizer.

(4) Leaf spray fertilizer.

In order to supplement the urgent need of nutrients in time, external topdressing is also often used, especially for peach trees cultivated on saline-alkali soil, and the effect of topdressing outside roots is better. The sprayed fertilizer enters the leaf through the stomata of the leaf. The intensity and speed of absorption are related to leaf age, fertilizer composition and fertilizer concentration. The physiological function of young leaves was exuberant, and the area of stomata is larger than that of adult leaves, so the absorption of young leaves is faster than that of old leaves. The higher the concentration of fertilizer solution is, the greater the absorption capacity of leaves is; the back of leaves has more stomata, larger cell gap, so the absorption capacity are stronger than the reverse side.

There are many fertilizers sprayed on peach trees, such as urea, potassium dihydrogen phosphate, diammonium phosphate, potassium nitrate and so on. The spraying concentration was 0.3%~0.5% in the growing season and 1% 10 days before the leaves fall.

In theyearly growth of peach trees, the concentration should be increased gradually from anthesis to the falling leaf, preferably before 10:00 and after 16:00. Don't spray fertilizer on windy weather and at hot noon, because spraying fertilizer at this time not only has poor fertilizer efficiency, but also easily causes fertilizer damage because of the rapid evaporation of water.

2. Irrigation and drainage

Peach trees need a certain water supply at all stages of the growing season. According to the condition of soil drought in thegrowing season, irrigation should be

Chapter Ⅴ Application of the Integrated Water-Fertilizer Technology to Fruits

carried out after topdressing, so as to promote the decomposition of fertilizer and the timely absorption by the trees. Soil should be loosened in time after irrigation. Only under the suitable water supply conditions, can the tree carry out physiological activities normally. In the yearly growth cycle, the water demand in spring is the most. There are more wind and less rain in spring in the northern region, so if the water supply is insufficient, the root growth of the tree will be slow, then the growth of the new shoot will be poor, and the fruit setting rate will be low, and finally the fruit is small. Even during dormancy, all kinds of branches of peach trees are still going on transpiration, and the lack of water at this time will cause the tree weak in the next spring, and even the tender twigs will shoot. Peach trees need a certain amount of water, its water resistance is far less than apples, pears and other fruit trees. Too much water supply or soaking for a long time will also affect the growth and development of peach trees, and the soil water accumulation will cause the death of the whole plant. Therefore, suitable irrigation and timely drainage are important conditions to ensure high yield, stable yield and high quality of peach trees.

(1) Irrigation.

Peach tree irrigation methods include border irrigation, furrow irrigation, hole irrigation, drip irrigation, sprinkler irrigation and so on. There are various irrigation methods because of the different water sources and ecological conditions. At present, most peach gardens use the border irrigation method of connecting the parts under the crown. Although this irrigation method is fast and labor-saving, the ditch occupies a lot of land, the irrigation amount is uneven, and it is easy to cause the disease to spread. Sprinkler irrigation is the best choice for peach orchards, which has the advantages of even irrigation and controlled amount, which is beneficial to spray fertilizer and medicine. The general principle is to save water and ensure that water permeates into the soil layer with the most roots in time, so that the soil can maintain a certain amount of water.

The period of irrigation can be divided into the following stages: stage before germination, core hardening stage, second rapid fruit growth period and stage before freezing and after leaves falling. ①Irrigation before germination: In order to ensure the smooth progress of germination and flowering and fruit setting, it is necessary to irrigate and fill up with enough water. The depth of water is 80cm. ②Irrigation in core hardening stage: Although the fruit grows slowly in this stage, the embryo is in the rapid growth period, and the peach tree is sensitive to water, which is the critical period of water demand for peach tree. Excessive moisture will lead to excessive growth of branches and leaves, and thus affect fruit setting; lack of water will cause fruit drop, then affect yield. Irrigation in this period should be shallow irrigation, especially for trees in their initial fruit stage. ③Irrigation in the second rapid growth period of fruit: It is applied about two weeks before harvest. When the fruit is growing and expanding rapidly, sufficient water can obviously increase yield. ④Irrigation before freezing: After the peach leaves fall and before the soil freezes, the land should be irrigated. It is to ensure that the soil has sufficient water, in order to benefit peach trees to survive the winter safely. But it should not be too late, for too much water at the root area and the freeze-thaw alternation in the day and night would lead to stem rot.

(2) Drainage.

Peach trees are afraid of waterlogging. Once the peach garden is flooded and the air in the soil is squeezed away, the peach trees will breathe without oxygen, producing harmful substances such as ethanol and methane, resulting in root poisoning and causing peach trees to die. In general, floods for a day and a night in the rainy season will cause them to die. Therefore, it is an important work to drain to avoid waterlogging in rainy season to cultivate peach trees.

Section Ⅳ Application of the Integrated Water-Fertilizer Technology to Grapes

Ⅰ. The Requirements of Grapes for Environmental Conditions

1. Temperature

Temperature is the most important meteorological factor affecting grape growth andfruiting. When the temperature reaches 7~10℃ in spring (the ground temperature is about 10℃), the root system of grape begins to act; the growth is the fastest at 25~30℃, and the growth is inhibited above 35℃. At 10~12℃, grape roots began to sprout. The suitable temperature for shoot growth, flowering, fruiting and flower bud

Chapter V   Application of the Integrated Water-Fertilizer Technology to Fruits

differentiation is 25~30℃. If there is low temperature during flowering stage (<15℃), grapes can not bloom and can not be normally pollinated. The suitable temperature of table grape and dried grape was 28~32℃, while that of wine grape was 17~24℃. As to different grapes with different maturity stage, only after achieving a certain effective accumulated temperature, can the fruits fully mature.

The harm of low temperature to grapes is a common problem in grape cultivation in the world. There are great differences in cold resistance among different species and varieties, and there are also considerable differences among different tissues and organs. In general, American grapes are more resistant to cold than Eurasian species. The root system is the weakest organ with cold resistance, and the roots of most grapes will be frozen to death at about −5℃. But the mountain grape is resistant to −15.5℃. In order to reduce root freezing damage, mountain grape and Beida (Beta) were used as cold resistant stocks, so that grapes could be cultivated in cold areas by burying soil to prevent cold, which was of great economic value. It is generally believed that the average annual minimum temperature is −15 to −14℃, the grapes can overpass the winter without burying in the soil. But in the temperature below −15℃, grapes can overwinter safely only if they are covered with varying degrees of soil. The winter bud of grape has weaker cold resistance, followed by mature one-year-old branches, and perennial branches, and the trunk is the most cold resistant. The budding eyes of Eurasian grapes are resistant to low temperature of −20 to −18℃ in winter. However, if the maturity of the branches is poor, the buds will be frozen at −15 to −10℃, and when the low temperature of −18℃, 3 or 5 days not only make the eyes of the buds frozen, but also make the branches frozen. The damage caused by low temperature in winter in northern China is often associated with drought and water shortage.

Lack of water leads to a decrease in cold tolerance of grapes. The shoots and young leaves in spring begin to be frozen at −1℃, and the inflorescences are frozen at 0℃.

2. Light

Sunlight is the only energy source for photosynthesis of grapes, and it is the driving force of energy and material cycle of grapes. 90% of grape yield and quality come from photosynthesis. Grapeslike bright and long sunshine. In the growing season of grapes, sufficient light makes the flower bud differentiate well, the leaves grow green, thick, the new shoots are strong, and the fruit is colored well, especially the

European grapes, which are particularly sensitive to light, can be colored normally only under the condition of direct sunlight. The demand for light in grapes is not that the stronger the light, the better. The high temperature at noon in summer and strong light will make the fruit surface temperature reach more than 50℃, and Day Fever often occurs. When the light conditions of the leaves are the best at noon, maybe midday depression will occur which limits the grow.

In China, the utilization rate of solar energy in vineyards is only about 0.5%. Modern science has been pursuing the utilization of solar energy, increasing the conversion rate and tapping the potential of increasing production in order to achieve high yield and high quality.

3. Moisture

The amount of natural rainfall and the seasonal distribution of rainfall strongly affect the growth and development, yield and quality of grapes. It is generally believed that at least 250~350 mm rainfall is required during the growth period. In spring, bud eye germinates new shoot, and if the rainfall is abundant, it is beneficial to the differentiation of inflorescence primordium and the growth of new shoots. Grape blooming requires sunny, warm and relatively dry weather. Wet weather or continuous rain and low temperature will hinder normal flowering and fertilizing, cause young fruit to fall off. Too much or continuous rain water at the maturity stage will lead to the decrease of glucose, the breeding of diseases and the cracking of fruit, which will seriously affect the quality of grapes. When it is rainy at the later stage of grape growth, the new shoots are not mature enough, so it is easy to be frozen when overwintering.

4. Soil

Soil factors that affect the growth and development of grapes include soil ventilation, soil moisture, soil nutrients and soil acidity and alkalinity. Grapes are more adaptable to soil, from sandy soil to clay, but the best soils are loose sand or sandy soil and mountain root soil with a large amount of coarse sand and gravel. The layer of heavy clay soil is thick, and the ability of fertilizer and water conservation is strong, but it is easy to cause the vigorous growth of grapes, and then affect the fruiting and quality. In addition, grape can be planted in slightly acidic soil (pH value is not less than 5) and slightly alkaline soil (pH value is not more than 8.5). When the salt content of saline-alkali land is reduced to less than 0.2%, the grapes can also

Chapter V Application of the Integrated Water-Fertilizer Technology to Fruits

be cultivated well. Different grape varieties can produce high quality fruit only in the soil suitable for their own conditions.

II. Integrated Water-Fertilizer Management of Grapes

1. Fertilizer application

(1) Types and functions of fertilizers.

The essential mineral nutrients for grape plants are mainly obtained from fertilizers. Fertilizers can be divided into two categories: organic fertilizer and inorganic fertilizer. Organic fertilizers often used in production, such as barnyard manure, stable fertilizer, poultry manure, cake fertilizer, human manure, crop straw, etc. They are generally applied to soil after fully decomposed, and most of them are used as base fertilizer. Organic fertilizer not only contains macro-elements, but also contains a variety of trace elements, so it is called complete fertilizer. Most organic fertilizers can be absorbed and utilized by grape roots through microbial action, so they are also called fertilizers with delayed effect. Organic fertilizer can not only supply the nutritional elements needed for the growth and development of grape plants, but also regulate soil permeability, improve soil ability of fertilizer and water conservation, increase soil fertility and create material basic conditions for microbial activities. They play an important role in improving soil structure.

The content of soil organic matter in China is generally low, so it is very important to increase the content of soil organic matter in order to produce high quality grapes.

Inorganic fertilizer, also known as chemical fertilizer, can be used as a supplementto organic fertilizer. Inorganic fertilizer contains single nutrient elements, but it is high in purity and it is soluble in water. Most inorganic fertilizers can be directly absorbed by plant roots, also known as fast-acting fertilizers. Inorganic fertilizer has quick effect after application, and is mostly used as topdressing fertilizer in grape growth period.

(2) Fertilizer application rate.

According to the characteristics of grape's demand, soil conditions, site conditions and fertilizer utilization rate, a reasonable amount of fertilizer is determined in order to fully meet the needs of trees for various nutritional elements. Leaf analysis is a scientific method to determine the amount of grape fertilizer. When it is found that a certain nutritional component is in deficiency, it is necessary to supplement it according to the degree. In practice, the amount of fertilizer is determined by experience and test results.

(3) Autumn application of base fertilizer.

The base fertilizer is usually organic fertilizer with delayed effect, and they can be mixed with a certain amount of chemical fertilizer, such as calcium superphosphate. And it can be applied according to 60%~70% of the whole years fertilizer amount. The time is generally carried out in autumn from harvest to soil freezing, which coincides with the second growth peak of grape roots, with strong absorption capacity and easy healing ability if the roots are injured. In addition, the leaf function has not declined and the photosynthetic capacity is enhanced, which is beneficial to the accumulation of storage nutrition and the cold resistance of grapes. When applied, it should be a little deeper and farther away from the root distribution layer in order to induce the roots to grow deeply and widely and expand the absorption range of the root system.

(4) Topdressing.

The topdressing fertilizer is mainly applied with quick-acting fertilizer, which is applied according to 30%~40% of the annual total fertilizer amount. In production, the period and proportion of grape fertilization shall be determined reasonably according to the load and soil condition. As for the productive orchestra, the fertilizer such as potassium, nitrogen, phosphorus, magnesium, iron, boron, etc. is required, especially the amount of nitrogen fertilizer and potassium fertilizer. As for the fertilization period, nitrogen fertilizer and boron fertilizer are used in the early stage of fruit growth, phosphorus fertilizer and magnesium fertilizer are in the middle stage, potassium fertilizer and iron fertilizer are in the sugar accumulating stage. Grapes need to be fertilized three times a year. ①The topdressing before anthesis is mainly nitrogen fertilizer, supplemented by the combination of nitrogen and phosphorus. The main purpose is to promote the growth of branches and inflorescence differentiation. ② After the flowers fall, young fruits begin to grow, and this is a period needing more fertilizer. The topdressing after anthesis should be quick-acting nitrogen fertilizer in time, combined with appropriate amount of phosphorus and potassium fertilizer, in order to promote the growth of new shoots, ensure the expansion of young fruits and reduce fruits drop. ③The maturity-accelerating fertilizer should be carried out at the beginning of the coloring of berries, and at this time, phosphorus and potassium fertilizer should be the main fertilizer, to

Chapter V Application of the Integrated Water-Fertilizer Technology to Fruits

mainly promote the enrichment of branches, fruit ripening and coloring.

2. Irrigation and drainage

Grape has strong waterlogging resistance and drought resistance, so higher yield and higher quality fruit can be obtained by timely irrigation in dry season and water demand period, and timely drainage in flood orchard. Generally speaking, vineyards of adult age should be watered 5~7 times at the germination stage, before and after flowering stage, fruit expansion stage and after harvest stage, while attention should be paid to controlling water at the flowering stage and maturity stage to prevent flowers and fruit falling, and prevent reducing fruit quality.

(1) Irrigation.

Irrigation before the flowering stage should applied from the sap is flowing, or sprouting to the pre-flowering. It is a period of spring drought and the soil needs to be watered. Water should be filled fully before germination to promote germination and new shoots.

At the stage of fruit expansion, irrigation is carried out from physiological fruit drop to fruit coloring. At this stage, the young fruits grow vigorously, and the air temperature increases continuously, and the leaf water evaporation is large. The leaf is most sensitive to water and nutrients, which is the critical period for grape to need water and fertilizer. Watering should be applied with fruit-accelerating fertilizer, and then decide how much to water according to the weather and soil conditions. In general, when it is dry and rainy, it can be watered every half month to promote fruit expansion.

Before freezing, irrigation of water after applying base fertilizer in autumn is beneficial to root healing and soil subsidence, and overwintering water is irrigated again before freezing, so as to prevent branches from being drained in order to facilitate safe overwintering.

(2) Water control.

Water control should be from the beginning to the end of flowering stage, about 10 to 15 days. At flowering stage, the pollination will be influenced and the size of grapes is affected by rain. Watering at flowering stage leads to vain branches and leaves, and excessive nutrient consumption, which will affect germination, flowers and buds fall, resulting in yield reduction. Proper water control at flowering stage can promote pollination and fertilization and improve fruit setting rate.

When berries are colored and mature, too much water or more rainfall affects coloring andreduces quality; it is easy for the trees to have anthracnose, white rot and so on; some varieties may also have fruit cracking. Therefore, water control in coloring stage can increase sugar content, accelerate coloring and ripening, prevent fruit cracking and improve fruit quality.

(3) Drainage.

When the soil moisture is too much in grape growing season, it is easy to cause branch to grow in vain, reduce fruit sugar content. And when it is serious, the roots lack oxygen, which inhibits respiration, results in plant physiological drought and even causes death. Therefore, we should pay attention to timely drainage in waterlogging and rainy seasons.

Chapter VI Application of the Integrated Water-Fertilizer Technology to Vegetables

China is a big agricultural country, and a large amount of vegetable supply is needed every day. In order to improve the yield of vegetables in China, the integrated water-fertilizer technology is applied in the field of vegetables. The integrated water-fertilizer technology depends on pressure system (or a natural fall of the terrain), mixes the soluble solid or the liquid fertilizer with irrigation water, according to the nutrition contents in the soil and the vegetables' requirement, and then soaks the root of the vegetable in time and quantitatively by means of pipes and sprinkles, so that the growth requirement of the vegetables is met. The technology has so many advantages such as equalizing water and fertilizer, saving water and fertilizer, saving labor, reducing diseases, controlling temperature and humidity, increasing yield, improving quality and achievements.

Section Ⅰ Application of the Integrated Water-Fertilizer Technology to Tomato

Tomatooriginated from the primitive forest of the tropical plateau in South America, and there it is a perennial herb and in a frost area it is an annual one. Another name is Yangshizi (foreign tomato), and the old name is June tomato and Xibaosanyuan (happily succeed in three important exams). In Peru and Mexico, it was originally called "wolf peach". The fruit is rich in nutrition and has special flavor. Its effects include reducing weight, relieving fatigue, stimulating appetite, improving digestion of protein, and reducing flatulence, etc. As a plant of succulent berry, the water content of every 100 g of fresh fruit is about 94 g, the carbohydrate is 2.5~3.8 g, and the protein is 0.6~1.2 g, vitamin C is 20~30 mg, and carotene, mineral salt, organic acid and so on. Fig. 6-1 shows the application of dripping irrigation under the mulching film to tomatoes.

Fig. 6-1 The application of dripping irrigation under the mulching film to tomatoes

I. Water and Fertilizer Management of Tomato

(1) Water management. Tomatoes used for food or processing in China are basically annual. The goal of water management is to keep the root soil in a humid state from planting to the end of harvest. Generally, the 0~40cm soillayer is kept in a humid state. It can be judged by a simple finger method. Dig up the soil under the drippers with a small shovel, when the soil can be kneaded into a lump or rubbed into a mud strip, it shows that the soil is full of water. On the contrary, it indicates that the soil is dry. Usually every drip irrigation takes 1~2 hours, depending on the size of the drip flow. The microspray belt is 5~10 minutes each time. Avoid excessive irrigation and loss of nutrients. Tomato has a long growth period and needs large water consumption. In the growth of tomato, it is necessary to design the irrigation quantity reasonably, and irrigating many times and small amount of each irrigation should be the best one until the root soil is moist. When irrigating and planting it is necessary to observe whether the water pipes in the field are in good condition or not. When the leakage phenomenon occurs, it should be dealt with in time, by replacing with new water pipes. Water amount in planting is 15~20m^3 per mu, drip irrigation or furrow irrigation; water

Chapter VI Application of the Integrated Water-Fertilizer Technology to Vegetables

amount in seedling stage (7 days after planting) is $10\sim12m^3$ per mu. And then from the hardening of the seedling to the first panicle fruit expansion, drip once or not, depending on the actual condition, with the irrigation quota $10m^3$ per mu. After the first fruit expands to 5cm, it drips once for $5\sim7$ days and $2\sim3$ times is ok. Each irrigation quota is $10\sim12m^3$ per mu; when it is in the full productive stage, drip once for $4\sim5$ days, and each irrigation quota is $12\sim15m^3$ per mu. From the fix planting period to fertility, it is about 160 days, the times of dripping water is $20\sim22$, and the total irrigation amount is $260\sim300m^3$ per mu. After tomato is fixed planted, due to the dry soil, it is necessary to apply drip irrigation in time, and the amount of irrigation should be determined according to the soil moisture and seedling growth. During the seedling growth period, drip irrigation should be carried out 3 times a day, each irrigation time being about 1 minute and the irrigation amount being about 500 liters per hm^2 with drippers of 16 mL per minute. In the rapid growth stage, the water consumption will gradually increase. It needs to be irrigated $6\sim8$ times a day and the single irrigation time is 2 minutes, and the amount of irrigation is about 520 liters per hm^2. The water consumption of tomatoes in the process of growth should be determined according to the weather conditions.

(2) Nutrient management. The root system of tomato is developed, the distribution is deep andwide, and the ability of regeneration is strong, so the ability of absorbing nutrients is strong. Tomato needs a large amount of fertilizer, it is more tolerant of fertilizer, and it demands large amount of calcium and magnesium. Tomato is a vegetable with continuous growth and fruiting, and the growth and fruiting period is long. In addition to applying base fertilizer, sufficient topdressing fertilizer is also required. As for the fertilization of seedling bed, generally rotten organic fertilizers are applied with $3\sim5$ tons per mu, superphosphate 30kg per mu, potassium sulfate 10kg per mu, urea 5kg per mu. They are mixed and applied into the 25cm plowing layer, then evenly sowing the seeds. After spouting, if the nutrients are insufficient and the growth rate of the seedlings is poor, the thin human feces and urine can be applied with water. Or $0.1\%\sim0.2\%$ urea solution and 0.2% potassium dihydrogen phosphate solution are sprayed.

The base fertilizer in the field is generally $(5\sim10)\times10^4$ kg manure or compost per hectare, $(4\sim6)\times10^2$ kg superphosphate, 150kg potassium sulfate, which would be evenly applied into the soil before transplanting. Topdressing is mainly by drip fertilization, and the fertilizer should be dissolved in the container before putting into the tank. Dripping fertilizer and dripping water should be applied alternately, that is, dripping fertilizer once, then dripping water again. The first topdressing is applied within one week after planting with the decomposed feces and urine of about 10×10^3 kg per hectare or urea fertilizer of 80~120kg per hectare, which is called "seedling fertilizer"; the second topdressing is applied at the beginning of the expansion of the first cluster of fruits, at the rate of 15×10^3 kg feces and urine or 100~120kg urea per hectare. The third topdressing is carried out after harvest of the first and second cluster of fruits, and when the third and fourth cluster fruits are expanding quickly. The type and quantity of fertilizer are the same as those of the second topdressing, and potassium sulfate could be applied according to 40~60kg per hectare. During fruit harvest, 0.2%~0.4% potassium dihydrogen phosphate and 0.2%~0.5% urea solution can also be sprayed on the leaf.

II. Laying and Sowing of Drip Irrigation Belt

Comprehensive maintenance and debugging on the seeding machine before sowing are needed, and install the drip irrigation pipe device for safe use.

(1) Seed treatment. In order to control early diseases, such as damping-off, 70% mancozeb should be used to mix seeds before sowing, and 0.2% of seed quantity should be used, and then germination test should be carried out.

(2) Herbicides spraying. The dosage of the herbicides is 1 500 g/hm² with Harness herbicide, 2 250 g/hm² with Dual herbicide, and the spraying agent must be even and consistent, to the ends to the sides. Rake diagonally after spraying.

(3) Machine preparation. Before sowing, the sowing machine is repaired and debugged in an all-round way, and the drip irrigation capillary device is installed for safe use.

(4) Sowing method. Strip sowing under the film. Sowing, laying plastic mulching film and laying drip irrigation belt can be completed at one time. One is the dry sowing; the sowing rate is 1 200~1 500 g/hm²; the sowing depth is 1.5~9cm, and the plant spacing is 25cm. Two mulches with eight rows or three mulches with 12 rows; the row spacing is 30cm on the mulch, and the spacing is 60cm at the joint parts. Drip irrigation capillary is laid under the mulching film, two rows of tomato with one

Chapter VI Application of the Integrated Water-Fertilizer Technology to Vegetables

capillary tube. The pipe laying, seeds sowing and film covering are finished at one time. To transplant tomatoes, dripping water should begin from the night before planting, keep the soil moist during planting; plant seedlings near the drip holes, and seal the soil after planting seedlings. Another is seedling raising and transplanting. Seedling cultivation is carried out in greenhouse; sowing date is 40~60 days earlier than that of direct sowing in the same area; and harvest is 15 days earlier. Seedling cultivation can resist the harm of low temperature and frost in early spring and increase the yield. Due to the cutting off of the main root during transplanting, a large number of lateral roots germinated. The number of roots is greatly increased, and the absorption capacity of the roots is enhanced, thus the yield could be increased. The phenomenon of falling flowers in processed tomato is common, such as insufficient water, poor root development, low temperature, fast plant growing and so on. The main measures to prevent falling flowers are as follows: one is to promote seedling strength; the other is to spray flowers with 4-Chlorophenoxyacetic acid (tomato ling) with the concentration of $(30\sim50)\times10^{-6}$ mg/L. When more than 50% of the plants have 2 or 3 flowers in the first panicle, they begin to be sprayed once two weeks and sprayed for 3 times in a row.

(5) Sowing. Strip sowing is usually in the middle of April, when the 5cm ground temperature is stably getting to 12℃, we should start sowing, lay branch pipes after sowing, and connect capillary, ready for seedling water. Sowing must meet the quality requirements, with the depth of 1.5cm, uniform sowing, and there are no broken ridges. The amount of soil on the edge of the film should be strict, and the soil should be pressed horizontally on the film with the distance of 6 meters to prevent wind damage. On the plot applied with dry sowing and wet germinating, after the seedling out, we should release the seedling and seal the hole in time. Some seedlings should be pulled out when it has 2 main leaves and be fixed when it has 4~6 main leaves. After the seedling is fixed, the soil should be sealed to protect the root, and the seedling should be replenished in time when there is no seedling. If there is seedling deficiency, you can leave two plants next to it. Make up for the seedling immediately after dripping.

(6) Dripping water. First is the seedling water which is for the seedling to grow out. After sowing, seedling water begin to drip when the ground temperature under the film reaches 12℃, and the water must be dripped within 5 days after sowing, and the

amount of water dripping was 230m³/hm². Seedling transplanting is prepared. After dripping water, transplant the seedling after frost period, with the row spacing on the film being 30cm, 60cm at the joint parts, plant spacing 30cm. The water for the seedling survival would be applied after seedling planting. Second is the dripping water during growth period. The first water is dripped from 15 days after spouting, and then, according to the comprehensive factors such as weather, soil water content and crop growth, it is determined that the drip of water is applied once for 1~5 days, the amount of water being 180m³/hm², 15 times in the whole growth period and dripping is stopped 7 days before picking. The total amount of water is about 3 000m³/hm².

(7) Releasing the seedling, sealing the hole and fixing the seedling. When dry sowing and wet spouting, the seedling releasing should be carried out in time after seedling emergence, and then fix the seedling, with the plant spacing being 30cm. If there lacks seedling, you can leave two plants next to the seedling, and replenish the seedling immediately after the first water.

(8) Chemical fertilization. All fertilizers are applied with water droplets during the whole growth period. 45kg/hm² special fertilizer is applied with seedling water dripping, and then 60kg/hm² special fertilizer is applied with water every time. Stop dripping fertilizer 5~7 days before the tomato harvest. Calcium nitrate is applied once with water at the stage of first panicle berry expansion, and the rate is 45kg/hm². It can effectively prevent the tomato blossom end rot, so as to improve the quality of tomato.

(9) Regulation and control. The combination of water regulation and chemical control should be carried out to reduce the dosage of chemicals as much as possible to ensure that the products meet the national standards. Chemical agent, 15% paclobutrazol, dripped with water when there are 4~6 leaves, dosage being 45~75 g/hm².

Section II Application of the Integrated Water-Fertilizer Technology to Pepper

Pepper is also called sea pepper, spicy pepper, spicy horn and so on. Pepper plants of the familysolanaceae are perennial or annual. Pepper has high nutritional

Chapter VI Application of the Integrated Water-Fertilizer Technology to Vegetables

value, vitamin C content is among the highest in vegetables, and spicy, which is the favorite pickles and condiment of the people in the northwest such as Gansu and Shanxi, in the southwest such as Sichuan, Guizhou and Yunnan, in central China such as Hunan and Jiangxi. Almost during every meal it is a must-be. The fruit contains capsaicin ($C_{18}H_{97}NO_3$) and chili lycopene, which can promote appetite and help digestion. Dried chili and chili powder produced in China are exported to Singapore, the Philippines, Japan, the United States and other places.

I. The Water Demand Characteristics of Chili Peppers

The water demand of pepper plants is not large, but because of the shallow root and less root quantity, its response to soil water is very sensitive, and the soil water condition is closely related to flowering and fruiting. Pepper is neither tolerant of drought nor waterlogging. Only when the soil is wet can it achieve high yield, but much water will make the plants wither. Generally the water requirement of sweet pepper varieties with large fruit is stricter than that of small fruit type. The water demand of pepper's seedling stage is less, mainly controlled by temperature and ventilation to reduce dampness. In order to meet the needs of plant growth and development after transplanting, water should be increased at the initial flowering stage, and sufficient water should be supplied at fruiting stage. If the soil moisture is insufficient, it is easy to cause the falling of flowers and fruits, fruit expansion problem, fruit surface wrinkles, less luster, and shape bending. When irrigation, accumulated water should not flood in the borders. If the soil moisture is too much, flooding for a few hours, the plant will wither and die. In addition, the requirements for air humidity are also strict. The air relative humidity at flowering and fruiting stage is 60%~80%. Excessive humidity is easy to cause diseases, and too dry is disadvantageous for being pollinated and fruit setting.

Pepper is a kind of vegetable with little water demand, but it is not tolerant of drought and waterlogging with strict water requirement. The water consumption at seedling stage was the least, and when the pepper grew to about the size of 3cm, the dripping water should be less, mainly promoting roots and restraining seedlings properly. After entering the initial fruit stage, the soil moisture is controlled at 70%~80% of the field water by increasing the drip water and irrigation times; and it reaches the peak at the full fruiting stage, and the soil moisture is controlled at 75%~85% of the field water holding capacity.

As for the planting water, the irrigation quota is 15m³ per mu; from planting to fruit period (July to early August) drip once 4~6 days, and the irrigation quota is 6~8m³ per mu; in the initial fruiting period (middle and late August), drip once for 5 days, and the irrigation quota is 8~10m³ per mu; In the productive fruiting period (September), drip once for 5 days, and the irrigation quota is 10~15m³ per mu; for the plant preservation (October - November), drip once in the first ten days of October, and the irrigation quota is 8~15m³ per mu. From planting to the market, the growth period of the pepper is about 130 days, dripping 20~30 times, with the total irrigation amount being 190~230m³ per mu.

II. The Characteristics of Fertilizer Requirement of Pepper

Pepper is a vegetable type needing more fertilizer. To produce 1 000kg fresh pepper, we need Nitrogen 3.5~5.5kg, phosphorus pentoxide 0.7~1.4kg, potassium oxide 5.5~7.2kg, calcium oxide 2~5kg, magnesium oxide 0.7~3.2kg.

The amount of nitrogen, phosphorus, potassium and other nutrients absorbed by pepper at different growth stages is also different. From seedling to budding, due to the small roots and leaves of the plant, the accumulation of dry matter is slow, so the nutrients needed are also less, accounting for about 5% of the total absorption. From budding to early flowering, the growth of plant is accelerated, the vegetative body expands rapidly, so the accumulation of dry matter increases gradually, and the absorption of nutrients increases, accounting for 11% of the total absorption. From the first flowering to the full flowering and fruiting, the vegetative growth and reproductive growth of pepper are vigorous, and it is also the period of the highest absorption of nutrients and nitrogen, accounting for 34% of the total. From the full flowering to the maturing stage, the vegetative growth of the plant is weak. At this time, the demand for phosphorus and potassium is the most, accounting for about 50% of the total absorption. After the fruits are harvested, to promote the growth and development of branches and leaves, a large amount of nitrogen fertilizer is needed at this time.

The absorption of nitrogen by pepper increases steadily with the development of growth, and the absorption of nitrate nitrogen is balanced with the yield of fruit until the end of harvest. Although the absorption of phosphorus increases with the growth, the range of absorption is small. Calcium absorption also increases with the development of growth period. If the supply of calcium during fruit development can not be kept up with, it is prone to have blossom end rot. The absorption of potassium and magnesium is also less at the early stage of growth and increases gradually from the fruiting, and the absorption is the most at the full fruit stage. Potassium deficiency is easy to cause leaves falling, and magnesium deficiency will cause leaves yellowing.

Generally speaking, it is necessary to absorb pure nitrogen 3~5.2kg, phosphorus 0.6~1.1kg, potassium 5~6.5kg, calcium 1.5~2kg, magnesium 0.5~0.7kg to produce every 1 000kg pepper fruit.

III. Integrated Water-Fertilizer Cultivation Techniques of Pepper

1. Variety selection

The environmental characteristics of winter cultivation in greenhouse are low temperature, weak light, poor ventilation and high humidity. Therefore, high yield varieties with tolerance of low temperature, low light and strong disease resistance should be selected for overwintering pepper cultivation in greenhouse, such as medium pepper 108, Guoxi 105, Hongta series, Haifeng series and so on.

2. Stubble arrangements

Peppercultivation in greenhouses in North China generally chooses overwintering stubble type, seedling in early July, planting in late August, harvesting in late October and uprooting in late May of the following year.

3. Planting

(1) Fertilizing the field. The cultivation period of the overwintering stubble pepper is long, so enough base fertilizer must be applied. Generally fertilizers applied per 667 m^2 include high quality saprophytic manure 7 500~10 000kg, calcium superphosphate 75~100kg, potassium sulfate 20~30kg, ammonium bicarbonate 50~100kg, cake fertilizer 50~100kg. The base fertilizer should be applied with the methods of general ground fertilization and concentrated trench fertilization, 2/3 base fertilizer with general ground application, 1/3 base fertilizer with ditch fertilization. Deeply plow the field twice, fully mix the manure and the soil, and then water the ditches.

(2) The planting density. The planting density of the pepper is large, and the planting of the double-row single plant can be adopted, the large-row spacing is 70~

80cm, the small-row spacing is 35~45cm, and the plant distance is 30cm. Cultivation should be on the ridges, and the ridges should not be too high, the general relative height being 20~25cm.

(3) Planting method. Choose fine weather and plant best from 9: 00 a. m. to 3: 00 p. m. Dig holes according to the distance of plants, and each hole is filled with 65% metalaxyl wettable powder 1 000 times liquid and 72% streptomycin sulfate soluble powder 4 000 times solution, so as to prevent the infection of epidemic diseases, root rot diseases and scab bacteria in soil. When fixing the seedling, the seedling lump should be level with or slightly higher than ridge surface. The direction of cotyledon is perpendicular to the row, and the water is fully applied through drip irrigation after planting.

4. Control of major diseases and insect pests

(1) Epidemic disease. Before planting the soil disinfection is carried out with 4% epidemic disease granule 5kg per 667 m^2. In the fruiting stage, 4% epidemic disease granules are applied twice for early prevention, and the dosage is 2 g per hole. After the disease occurs, liquid spray could be used to prevent and cure it with 25% metalaxyl wettable powder 800 times liquid, 75% chlorothalonil wettable powder 600 times solution, or 20% thiobacillus copper suspension 400~600 times.

(2) Powdery mildew. In the early stage, 50% sulfur suspension liquid 200~300 times could be used to control the disease, spray once every 15 days, twice or three times, and the control effect is obvious.

(3) Grey mold. At the beginning of the disease, spray 50% procymidone wettable powder 1 500~2 000 times liquid or 40% pyrimidine wettable powder 800~1 200 times liquid to control the disease. Adding 0. 1% procymidone wettable powder or 50% isobacillus urea wettable powder to tomato solution, and then dip the flowers or coat the fruit stalks to prevent flowers to fall and treat gray mold at the same time.

(4) Viral disease. 1. 5% alkanol · copper sulfate emulsion 500 times solution, or 10% mixed fatty acid water agent 200 times solution, or 20% guanidine · copper acetate wettable powder 500 times solution can be used for prevention and control.

(5) Aphids, white fly. Spray control can be selected with 25% thiamethoxam granule 1 500~2 500 times solution or 2. 5% cyhalothrin EC 4 000~5 000 times liquid.

Chapter Ⅵ   Application of the Integrated Water-Fertilizer Technology to Vegetables

Section Ⅲ   Application of the Integrated Water-Fertilizer Technology to Cucumber

Ⅰ. The Biological Basis of Cucumber Cultivation

Cucumber, also known asHugua melon and Wanggua melon, is an annual climbing herbs as cucurbitaceae, which originated in the rainforest area southern of the Himalayas. Zhang Qian brought back to the Central Plains of China when he went to the western regions in the Han Dynasty. After more than 2000 years of cultivation, the North China cucumber with slender fruit, thin skin and many lumps was formed. In addition, cucumbers were introduced into southern China through Vietnam, and a South China cucumber with short and thick fruit and rare lumps was formed.

Cucumber is a globally popular vegetable, and its cultivation area is second only to tomato, cabbage and onion, ranking at the fourth. Asia has the largest cultivation area, accounting for about 50% of the world's total area, followed by Europe, North America and Central America. The cultivation area of cucumber in China is about 241 000 hm^2, accounting for 28% of the cultivated area in the world, ranking first in all countries. Cucumber has strong adaptability and can be cultivated in many forms and stubble times. It is one of the main vegetables cultivated in northern China.

1. Botany characteristics of cucumber

(1) Root. Cucumber is a shallow root crop, the root systembeing mainly distributed in the soil layer of 25cm, most in the 5cm soil layer. The lateral root is mainly in the radius 30cm range; the root quantity is less; the absorption capacity is weak; the root corkification is early, the regeneration ability is poor, and it is not easy for the new roots to come out after the root cutting. In production, nutrition bowl, nutrition earthwork and other root protection measures should be used to raise seedlings. In addition, cucumber roots like air, so the soil is required to be loose and breathable.

(2) Stem. Cucumber is a trailingand climbing stem, its stem hollow, five-angled, with setae, and the internodes begin to elongate after 5~6 nodes. The height of stem, the length of Internode and the number of branches are affected by varieties, en-

vironmental conditions and cultivation techniques.

(3) Leaves. The leaves of cucumber are simple alternate leaves, large and thin, with pentagonal shape and large transpiration. The surface of the leaf is prickly and stomatal, the thorns are dense on the front of the leaf, sparse on the back, and the stomata are few and small on the front and large on the back, so emphasis should be placed on the back of the leaf when spraying.

(4) Flower. Cucumbers are usually Monoecious flowers. The flowering order is from bottom to top. The node height of the first female flower on the main vine directly affects the time of picking the melon, and this is the important standard to determine the variety. For early cultivation and harvest, the first female flower should be selected with low node.

(5) Fruits and seeds. Cucumber fruit is a pseudocarp; the epidermis is developed from the outer skin of the receptacle, with or without lumps; the cortex is developed from the receptacle cortex and ovary wall. It is of parthenocarpy characteristics, so if the environmental conditions are uncomfortable and the management is not appropriate, there will be abnormal fruits. Cucumber seeds are lanceolate, flat, yellow and white. Each melon has 150~300 seeds. The seeds are born on the lateral membrane placental, and the 1 000-grain weight is 22~42 g.

2. Growth cycle of cucumber

The growth cycle of cucumber can be divided into germination stage, seedling stage, budding stage and fruiting stage.

(1) Germination period. It takes 5 to 6 days from seed germination to the appearance of the first leaf. The germination period mainly depends on the nutrients stored in the seeds. Full and big seeds should be selected in the production, and higher humidity, temperature and sufficient light should be provided.

(2) Seedling stage. From theappearance of the first true leaf to 5~6 leaves, the stems and vines begin to elongate, taking 20 to 30 days. At this stage, seedlings grow upright and differentiate a large number of leaf buds and flower buds. When cucumber seedlings are 1~2 leaves, even the true leaves are fully expanded, the flower buds begin to come out. And at the beginning of differentiation it has both pistils and stamens. However, in the process of flower buds' development, some pistils degenerate and develop into stamens, and some stamens degenerate into pistils, which is affected

by internal and external factors. The number and proportion of pistils and stamens are plastic. At present, it is considered that the main internal factors of cucumber type differentiation are the ratio of carbon to nitrogen (C/N) in plants, metabolic levels and plant hormones. Too much nitrogen compounds in the plant can promote the differentiation of stamens, and much the carbon compounds can promote the differentiation of pistils. The lower metabolic level of the plant is beneficial to the differentiation of pistils, and if the growth of stem and leaf is exuberant and the metabolic level is high, it is beneficial to the differentiation of stamens. Gibberellin can promote stamens differentiation; auxin and ethylene can promote pistils differentiation. Low temperature, especially low night temperature, short time of sunshine, sufficient phosphorus and potassium nutrition, suitable water, ethephon, abscisic acid, carbon dioxide, carbon monoxide and so on, can promote the differentiation of pistils. However, long time of sunshine, excessive nitrogen fertilizer, malnutrition and the use of gibberellin are all beneficial to stamens. Suitable environmental conditions should be created in production to promote the differentiation of pistils in seedlings.

(3) shooting stage. It is from 5~6 leaves to the first pistil blooming. In this stage, the plant change to trailing growth, the plant growth is exuberant, and change from vegetative growth to reproductive growth.

(4) Flowering and fruiting period. It is from the first pistil blooming to seedlings. The length of fruiting period is closely related to the cultivation season and cultivation environment. Spring cucumber is generally 50~60 days, and cucumber overwintering in greenhouse can reach 6~8 months.

II. Cucumber Variety Selection

1. Cucumber variety types

Cucumber cultivation has a long history, wide distribution and rich variety types. According to its distribution region and biological traits, it can be divided into the following types.

(1) South Asia type. It distributes all over South Asia. The stem and leaf of the plant are thick and easy to branch. The fruit is large, and the weight of single melon is 1~5kg. The fruit is short cylindrical or long cylindrical, with light skin color. The lump is sparse; the thorn is black or white; the skin is thick and the taste is light. It likes

dampness and heat, and the it strictly requires short sunshine. There are many local varieties, such as Sikkim cucumber, Chinese Banna cucumber and Yunnan Zhaotong cucumber.

(2) South China type. It is distributed in the south of the Yangtze River in China and all over Japan. The stems and leaves are luxuriant, wet-tolerant and heat-tolerant, and they are short-sunshine plants. The fruit is small, and the lump is sparse with many black thorns. The young fruit is green, green and white, yellow and white, with light taste; the mature fruit is yellow-brown, with net texture. Representative varieties are Kunming early cucumber, Guangzhou Erqing, Shanghai Yanghang, Wuhan qingyudan (green fish gallbladder), Chongqing Dabai and Japanese Qingchang and Xiangmobanbai (Sagami Hanjiro) and so on.

(3) North China type. This type distributes in the north of the Yellow River Basin in China and North Korea, Japan and so on. The plant grows with medium level, likes humid soil and sunny weather, and is insensitive to the length of sunshine. Young fruits are rod-shaped, green, with dense lumps and thorns; mature fruits are yellow-white with no net texture. The representative varieties are Shandong Xintai Mici, Beijing Dacigua, Jinyan system cucumber, Jinza system cucumber and Jinyou system cucumber and so on.

(4) The open field type in Europe and the United States. It distributes all over Europe and North America. Lush stems and leaves, cylindrical fruit, medium size, thin lump, white thorns, light taste; mature fruit is yellowish brown. There are Eastern Europe, Northern Europe, Northern America and other species groups.

(5) The greenhouse type in northern Europe. It distributes in the United Kingdom, the Netherlands. Lush stems and leaves, tolerant to low temperature and weak light, smooth fruit surface, light green, fruit length up to 50cm. There are British greenhouse cucumbers, Dutch greenhouse cucumbers and so on.

(6) Mini-cucumber. It distributes in Asia and Europe and the United States. The plant is short, and the branch is strong with many flowers and fruits. It is mainly for salinized processing. Representative varieties are Yangzhou milk cucumber and so on.

There are two main ecological types of cucumber cultivated in China: North China type and South China type. According to the cultivation season, North China cucumber

Chapter Ⅵ Application of the Integrated Water-Fertilizer Technology to Vegetables

can be divided into spring cucumber type, spring & summer cucumber type and autumn cucumber type. Spring cucumber types are cold tolerant and early maturing; spring & summer cucumber types have strong growth potential and adaptability with heat resistance and disease resistance, most of them being medium; autumn cucumbers are mainly medium and late mature varieties, with thick dark green leaves and strong adaptability. In recent years, with the development of cucumber industry, there have been some varieties suitable for winter production, such as Xintai Mici, Jinchun No. 3, Jinyou No. 2, Jinyou No. 3 and so on.

2. Main excellent cucumber varieties

(1) Jinyou No. 2. It is an early-maturing hybrid generation bred by Tianjin Cucumber Research Institute, which was approved by Tianjin Crop Variety Approval Committee in 1998. The plant's growth potential is strong, fruiting on the main vine, and the fruits are dense, almost each node having fruits and many Huitou fruits (Branch fruits). The fruits are long-rod shaped, about 34cm long, with short handle, dark green lustrous color, medium lumps and white thorns. The fruit pulp is dark green, crispy in taste and excellent in quality. It is tolerant to low temperature and low light and it is high resistant to downy mildew, powdery mildew and blight. From sowing to harvest, it is 60 days, and the harvest time is 80~100 days. The weight of a single fruit is about 200g. It is suitable for winter&spring stubble and early spring in the greenhouse in northern China.

(2) Jinyou No. 30. It is a hybrid generation bred by Tianjin Cucumber Research Institute. The plant growth potential is strong, fruiting on the main vine. The fruit strip is straight, about 35cm long. The single fruit weighs about 220g. It is green and luster, with prominent lumps and dense white thorns. The pulp is light green, crispy and sweet on the taste. It is resistant to low temperature and low light, and it is resistant to blight, downy mildew and powdery mildew. Suitable for overwintering stubble and winter & spring stubble cultivation in greenhouses in North China, Northeast China, Northwest China and East China.

(3) Jinyou No. 3. The early-maturing hybrid generation bred by the Tianjin Cucumber Research Institute. The plant growth potential is strong, tolerant of low temperature and weak light; the plant type is compact, the leaf color is dark green, and the leaf is of medium size. It fruits on the main vine; the branching is weak; the fruits are

dense; and the huitou fruits are enough, so under good cultivation, the fruits can be repeated in a plurality of times. The fruit is straight; the handle is short; the length is 30~33cm; the color is dark and green; the lump is remarkable; and white thorns are thin. One single fruit weighs about 200g. This variety is one of the high-yield varieties of the overwintering stubble in the greenhouse with strong disease resistance and high yield.

(4) Xintai Mici. The local variety of Xintai City, Shandong Province, was approved by Shandong Crop Variety Approval Committee in 1987. The plant growth potential is stronger; the stem is thick; the internode is short; it fruits on the main vine; the first pistil is born in the 4~5th section; one section may have many fruits, and the huitou fruits are more. The fruit is 25~35cm long; the transverse diameter is 3~4cm; the handle is long, dark green; the lumps and thorns are dense with white thorns and unobvious edges. The taste is crisp and tender, and the quality is good. It is resistant to low temperature and weak light; resistant to blight but irresistant to downy mildew and powdery mildew. It is early mature, and single one weighs about 250g. And is suitable for the cultivation in the greenhouse.

(5) Zhongnong 202. It is the latest hybrid generation with extremely early maturity produced by the Institute of Vegetables and Flowers of the Chinese Academy of Agricultural Sciences. The plant growth potential is strong; the growth rate is fast, and it fruits on the main vine. The fruit is long – rod – shaped; the handle is short; the strip is straight; the length is about 35cm, and the transverse diameter is about 4cm. The color is dark green, lustrous, without edges on the surface, and the lumps are small, with medium density and white thorns. The pulp is thick; the cavity is small; the quality is crispy; the taste is slightly sweet, so the market is good. The first pistil is born on the 2~3 knots, and the weight of single fruit is about 250g. This variety is planted only in conservation areas.

(6) Zhongnong No. 9. It is a new hybrid generation of medium and early maturity and less thorns produced by the Institute of Vegetables and Flowers of the Chinese Academy of Agricultural Sciences. The plant's growth potential is strong; the first pistil begins from the 3~5 nodes on the main vine; the pistils interval is 2~4 nodes; in the early period it fruits on the main vine, and in the middle and late periods it fruits on the lateral vines; most of the pistils nodes are double melons. The shape is like a short

tube, 15~20cm long, with dark green color, with luster, without pattern, with short handle, sparse lumps, white thorns and no edge. And its quality is medium, resistant to downy mildew, blight, black star disease and other diseases. Suitable for spring stubble in the greenhouse and autumn delay cultivation.

(7) Zhongnong No. 12. It is a new hybrid generation of medium and early maturity bred by the Institute of Vegetables and Flowers of the Chinese Academy of Agricultural Sciences. The plant growth rate is fast, fruiting on the main vine with dense fruits. It is long-rod-shaped, about 30cm long, with dark green color and luster, with small lumps and white thorns. It tastes crisp and sweet, with good quality. It is resistant to downy mildew, powdery mildew, mosaic virus, moderate resistant to black star disease, bacterial keratosis, blight. It is early mature, and the weight of a single fruits is 150g. Suitable for early spring conservation area, early spring open field and autumn delay cultivation.

(8) Zhongnong No. 19. It is a fruit type of pistil hybrid generation variety. The fruit is short and tube-shaped, 15~20cm long, with bright green color, no pattern, smooth surface. And it tastes crisp and sweet. It is strong resistant to low temperature, weak light, resistant to blight, black star disease, downy mildew and powdery mildew, etc. It has a strong ability to fruit continuously, and the weight of single melon is about 100g. Suitable for overwintering stubble and early maturing cultivation in greenhouse.

III. Seed Treatment

Seed treatment before sowing is one of the important technical links in vegetable seedling. The main purpose and function is to improve the value of seeds, sterilize, promote germination, promote growth and so on. There are many methods of treating seeds, which can be divided into the following according to the purpose of treatment.

1. Seed selection

To remove impurities and immature seeds is to improve the seed purity, thus it can improve the value of seeds. The simple and feasible methods are wind selection, water selection, screening selection and manual selection.

2. Seed disinfection

Many vegetable diseases are transmitted by seeds, most of which are parasitic on the surface of seeds. Seed disinfection can kill pathogenic bacteria and avoid the occurrence of diseases. There are many methods of seed disinfection, and there are five commonly used methods.

(1) Soaking seeds with warm water. The temperature used in soaking seeds with warm water is about 55℃, and the water amount is 5~6 times of that of seeds. First soak the seeds with warm water for 15 minutes, and then soak the seeds in 55~60℃ hot water, and stir continuously during the period, and keep the water temperature for 10~15 minutes, then reduce the water temperature to 30℃ and continue to soak the seeds for 4 hours.

(2) Soaking seeds with hot water. This method is generally used for seeds which are difficult to absorb water. The water temperature is 70~75℃, and the water quantity should not exceed 5 times of the seed. And the seeds should be fully dried. Use two containers to pour hot water back and forth when socking. The first few movements should be fast, so that the hot gas releases and provides oxygen. This action goes on until the water temperature drops to 55℃, then change it to constantly stir. The seeds are dumped for 7~8 minutes and stirred with 55℃ warm water for 10~15 minutes. Stop stirring when the water temperature drops to 30℃, and then soak the seeds for 3~4 hours.

(3) Soaking seeds with drug solution. The commonly-used agents for seed disinfection are 1% potassium permanganate solution, 10% trisodium phosphate solution, 1% copper sulfate solution, 40% formaldehyde 100 times solution, etc. The seeds are generally soaked for 5~10 minutes in the solution and then washed repeatedly with clean water until the seeds are tasteless. Soaking seeds with 10% trisodium phosphate solution for 20~30 minutes or 40% formaldehyde 100~200 times solution for 15~20 minutes and then washing them with clean water can also prevent cucumber virus disease.

(4) Stirring the seeds with medicine. If you sow with dry seeds, you can mix the agent evenly with the seeds so that the agent adheres to the surface of the seeds, and then sow again. The dosage of the agent is generally 0.2%~0.3% of the weight of the seeds. The commonly used agents are 70% fenaminosulf soluble powder, 50% carbendazim wettable powder, 40% Fumei seedvax wettable powder, 25% metalaxyl wettable powder. Stirring the seeds with 50% carbendazim wettable powder by 0.4 % of the seed quantity, can prevent cucumber fungal disease.

(5) Dry heat treatment. Dry heat treatment can passivate the virus and improve seed vigor by placing fully dried seeds (water content less than 4%) at high temperature above 75℃. Suitable for heat-resistant vegetable seeds, such as seeds of the melons and solanaceous vegetables. Treatment under the temperature of 70℃ for 2 days could completely deprive cucumber green mosaic virus and make it die.

3. Budding treatment

Budding is a technical measure taken to promote seed germination after disinfection and soaking. In the process of germination, the temperature, humidity, ventilation and light conditions needed for seed germination should be satisfied, so as to promote the rapid decomposition and transport of seed nutrients and supply the growth needs of seed embryos. The temperature should be low at the early stage and be gradually increased at the later stage, and then decreased when the seeds' white is coming out, so that the radicle can thrive. The humidity is suitable when the seed coat is not slippery and not white. The suitable temperature of budding is 25~28℃, and after 24 hours they will meet the requirements of sowing. In the process of budding, attention should be paid to turning the seeds every once in a while to make the seeds heat evenly, so as to ensure the water demand oxygen demand of the seeds. During the budding period, the sprouts can also be exercised by placing the seeds that have just shown its white buds with a cloth at −2 to −1℃ for 12 hours, and then at 18~22℃. They are repeated for 3 days, which can significantly improve the tolerance of cold and enhance the adaptability of seedlings to low temperature. In production, if the weather condition is not suitable for sowing after seed soaking, the bud exercise method can also be used to control the budding time of seeds.

4. Sowing

The sowingamount per 667 m^2 is 120~220kg. Before sowing, the bottom water is poured sufficiently before sowing. After the water is seeped down, the seedling bed is planned according to 8~10 square cm, and then one seed is sown in the center of each square. When the seedling is raised in containers, one seed is sown in each nutrition bowl, and the seed should be covered with screened fine soil after sowing. The thickness of covering soil is generally 1~2cm, for too thick is not conducive for the bud to come out, too thin is easy to lead to "wearing cap" budding. In order to better the temperature, aeration and budding of the seedling bed or nutrition bowl, it is best to form a small pile of soil centered on seed and about 5cm in diameter. And then spread 5~7mm thick screening fine soil evenly across the entire border, which is conducive to the preservation of soil moisture and the prevention of cracks. Finally, the plastic film is used to seal the field and cover it with straw mattress at night.

IV. Seedling Techniques

1. Seedling raising by routine way

During 7 to 10 days before sowing, the greenhouse is sterilized by mixed fumigation with 250g sulfur powder and 500g sawdust for 12 hours per 100m^3. At the 50cm to the south of the column in the greenhouse, make a furrow with a width of 1~1.5 meters, and a ground hot line can be laid if conditions permit. The nutritious soil was prepared by mixing 60% field soil with 40% rotten organic fertilizer, adding 50% carbendazim wettable powder 180g per cubic meter, 58% methyl cream manganese zinc wettable powder 50g. After screening, this soil is covered with agricultural film for 24 hours. The nutrition bowl with diameter of 10cm and height of 10cm is used and filled with the nutritious soil to 8cm high. Water it and after seepage, one germinated seed is sown in each nutrition bowl, the soil covered is about 1cm thick, and the plastic film is covered to preserve soil moisture. After emergence, in addition to rainy days and nights, try to make the seedlings ventilated and exposed to the sunlight. In the seedling stage the temperature is high and evaporation is large, so the seedling bed must be provided with enough water. If water is in shortage, it should be replenished in time. In order to promote the formation of pistils, ethephon treatment is used, generally spray 40% ethephon water agent 4 000 times in the evening at the stage of 2 true leaves. Seedlings can be planted from 2 leaves 1 heart to 3 leaves 1 heart.

2. Seedling raising by grafting

Cucumber is the main vegetable planted in greenhouse; it is difficult for the rotation, and the blight caused by continuous cropping is becoming more and more serious. Seedling raising by grafting is an effective measure to control blight. Moreover, the root system of stock is developed, drought tolerant, cold tolerant, and the absorption of water and fertilizer are strong, which is beneficial to high yield and good quality. Cucumber stock is mainly pumpkin, concluding mainly black seed pumpkin, Nanzhen No. 1, Xintuzuo, Zhuangshi, Gongrong and other varieties.

Chapter VI Application of the Integrated Water-Fertilizer Technology to Vegetables

(1) Grafting method. The main grafting methods of cucumber are wedge graphing, approach grafting, inserting grafting, cutting root grafting, core grafting, two-stage grafting and so on.

①Approach grafting. Usually, the scion cucumber is sown 2~5 days earlier than the stock pumpkin. 10~12 days after sowing, the first true leaf is exposed to semi-spreading, the cotyledon of the stock pumpkin is fully spread and the first true leaf exposes, it is time for grafting. If the grafting is too early, the seedling is too small for convenient operation. And the survival rate is low when the grafting is too late. Before grafting, the substrate of stock seedling and scion seedling should be sprayed wet, and covered with wet cloth after being dug out of the seedling tray. During grafting, the scion is slanted upward at the 1~1.5cm lower of cotyledon at an angle of 15°~20°, with a depth of 3/5~2/3 of the diameter of hypocotyl; Remove the growth point and true leaf of the stock, cut at the 0.5~1cm under the cotyledon, and the cutting angle is 20°~30°and the depth is 1~2 of the diameter of hypocotyl, and the incision length of stock and scion is 0.6~0.8cm respectively. Finally, the cuttings of the stock and scion are inserted into each other and fixed with grafting clips or bound with plastic strips. When the stock-scion complex is planted into the nutrition bowl, the distance between the roots and stems is 1~2cm, which is beneficial to the root removal after survival, as shown in Fig. 6-2.

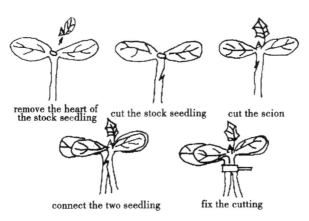

Fig. 6-2 **Approach grafting technique of cucumber**

Itis easy to manage the seedling by approach grafting, with high survival rate, neat growth and easy operation. However, the grafting speed is slow, the interface needs to be fixed, the stem is needed to be cut off and the root removal is needed after survival, and the interface position is low, so it is easy to be polluted by the soil and often adventitious roots occur, and the interface site is easy to be separated when they are transported and managed.

②Inserting grafting. The suitable period for grafting is when the cotyledons of the scion are fully spread, and the cotyledons of the stock are open spread and the first true leaf is just spread. According to the seedling season and environment, pumpkin stock is sown 2~5 days earlier than cucumber, and cucumber will be grafted 7~8 days after sowing. When grafting, spray wet the substrate of the scion and stock seedling in the bowl (plate). The growth point of stock seedling is eliminated, then insert from the center of the top with a bamboo stick at an angle of 40° to 0.5cm deep, and it is proper when the pressure of the stick tip in the stock is felt by the finger. Then the scion seedling is cut into wedges at 0.5cm under the cotyledon. Pull out the bamboo stick on the stock and insert the scion into the small hole of the stock so that the two are closely connected, and the extension direction of the stock and scion cotyledons is in the shape of a cross, which is easy for them to accept sunlight, as shown in Fig. 6-3.

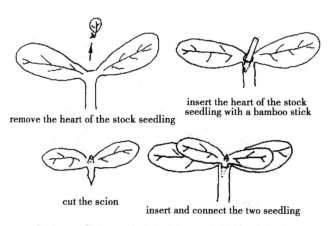

Fig. 6-3   Sketch map of inserting grafting of cucumber

Chapter VI Application of the Integrated Water-Fertilizer Technology to Vegetables

Using the Inserting grafting method, the stock seedling doesn't need to be moved, but also reduces the procedure of planting and grafting clip, and does not need to cut off the stem and remove the root. The grafting speed is fast, the operation is convenient, and the labor is saved. The grafting site is close to the cotyledon node; the cell division is exuberant; the vascular bundles are concentrated; the healing speed is fast; the interface is firm; the stock, the scion are not easy to break, so the survival rate is high; the interface position is high; it is not easy to be polluted and infected again; and the disease prevention effect is good. However, the skill of grafting, the age of grafting seedlings and the management of survival period are strict. When the technology is not skilled, the survival rate of grafting is low and the growth is poor in the later stage.

③Cutting root grafting. At the proper length of the stock hypocotyl, cut off and graft to make the root grow into a complete plant. This method often use Xintuzuo as the stock. The black seed pumpkin hypocotyl is too short, cotyledon is too large, so the application is less. According to the temperature conditions at the time of grafting, the stock is sown 2~3 days earlier than the scion, or the stock and scion are sown at the same time, when the first true leaf of the stock is 0.5~1cm and the first true leaf of the scion is 0.2~0.5cm, the grafting can begin. One or two days before grafting, proper cooling the temperature and controlling the water can promote hypocotyl hardening. Fully water the seedling bed on the day of grafting, so the plants can absorb enough water. It is best to spray fungicides with low concentration once. During grafting, cut off the hypocotyl at the 5~6cm under the cotyledon node of the stock (the closer it is to the root, the better the rooting ability of the hypocotyl is), and the scion is cut off at the 2~3cm under the cotyledon node. The two are placed in moist containers and covered with wet cloth to prevent wilting. Pay attention that the amount of the cut stocks and scions should not be too large, it is best to graft immediately after cutting. After grafting, the stock scion complex is inserted into the seedling bowl (plate) with substrate, the planting depth is 2~3cm, and then it is covered in the shed to avoid sunlight. The cutting root grafting method is simple and labor-saving; the stock and scion are not attached to soil at all; and the grafting efficiency is high. Seedlings take root again, and the number of lateral roots is large. And the plant growth vigor is strong, which can improve the yield. But the management requirement after grafting is strict, which requires a lot of work compared with the common grafting.

④Wedge grafting (as shown in Fig. 6-4). The suitable grafting period is when the 2 leaves of the scion are fully spread and the first true leaf of the stock appears. The stock is sown 3 days earlier than the scion. During grafting, the core leave of the stock are removed, then cut vertically in the center or side of the hypocotyl. The cut length is 1~1.5cm, and then the scion hypocotyl is cut into a wedge, and the length of the cutting surface corresponds with the cutting length of the stock. Finally, the scion is inserted into the stock and fixed with grafting clip or wrapped in plastic bag. It is difficult to manage cucumber after wedge grafting, the survival rate is low, so it is less used in production.

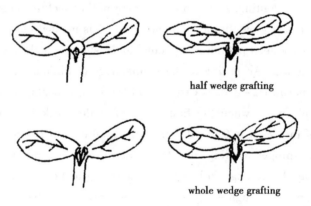

Fig. 6-4 Wedge grafting of cucumber

(2) Management after grafting. Within 3 days after grafting, the seedling bedshould not be ventilated, and no light is seen. The seedling bed temperature is kept at 25~28℃ during the day, 18~20℃ at night, and the relative humidity of air is kept 90%~95%. 3 days later, according to the seedling condition, a small amount of ventilation is carried out for a short time with the standard of no seedling wilting. In the future, the ventilation will be gradually increased. After one week, the cutting is recovered, so the straw mattress can be gradually removed and the ventilation will be increased. The seedling bed temperature should be kept at 22~26℃ during the day and 13~16℃ at night. If the seedling bed temperature is lower than 13℃, the straw mat-

Chapter VI Application of the Integrated Water-Fertilizer Technology to Vegetables

tress should be used to cover the shed. In the seedling period, according to the seedling condition, water the field for once or twice. After the cutting was recovered by using the approach grafting method, the root of the scion should be cut off in time.

In order to improve the stress resistance of cucumber and cultivate strong seedlings, large temperature difference management should be carried out. After the grafting seedlings survived, the temperature should be kept at 25~30℃ during the day. It should not be not ventilated when it is not more than 35℃. 15~18℃ is kept before the midnight, 11~13℃ after the midnight, about 10℃ in the morning before opening the straw mattress, and sometimes reduced to 5~8℃ in a short period of time, the ground temperature is kept above 13℃. Water does not need to be overcontrolled, and suitable water, sufficient light and large temperature difference between day and night are used to prevent seedlings from growing fast. The seedling age of the winter & spring stubble cucumber should not be too old, and 3~4 leaves with 1 heart, and 10~13cm plant should be proper. The calendar seedling age should be about 35 days and should not exceed 40 days.

3. Strong seedling standards

The strong seedling standard are as follows, the cotyledons of the seedling are good; the stems are strong; the leaves are strong and green; it has no disease-pests infection; the leaves are 4~5 true leaves; the plant is about 15cm high; the root system is developed; and the seedling age is 25~30 days.

V. Management of Integrated Water-Fertilizer

After planting to fruit setting, there is no need of topdressing, which can be sprayed with 0.2% potassium dihydrogen phosphate + 0.2% urea or 0.3% ternary compound fertilizer. The first irrigation is applied at the time of the planting, and the water consumption is 15~20m^3 per 667m^2. When the plant has 8~10 leaves and the first fruit is about 10cm long, the second irrigation and fertilizer are carried out, with urea being 10kg per 667m^2, potassium sulfate being 10~12kg per 667m^2, and water consumption being 15m^3 per 667m^2. Before winter, apply topdressing every 15 to 20 days. In addition to topdressing with water, water should be controlled from planting to deep winter season. If the plant showed lack of water, small water could be applied. After the late of February, the water and fertilizer amount of the cucumber is increased, so the watering times and amount are appropriately increased, and the water is irrigated every 7~10 days. After entering the full productive period, the number of irrigation times can be appropriately reduced, and the chemical fertilizer can be applied for every 10~15 days in combination with watering, with 8~10kg of urea per 667m^2 and 12~15kg of potassium sulfate per 667m^2. During the whole growth period, the cucumber can be watered for 12~15 times and be applied with topdressing for 8~10 times. In the late stage, the application of the urea solution or the potassium dihydrogen phosphate solution with 0.2~0.3 percent or the potassium dihydrogen phosphate solution can be used for foliage topdressing to strengthen the seedling and prevent the early senescence of the seedling.

VI. Prevention and Control of the Main Diseases and Insect Pests

(1) Powdery mildew. In the early stage of the disease, theleaf is found to have a round spot, and it can rapidly spread, and when it is seriously ill, the whole leaf is full of white powder. The leaves in the late stage of the disease are yellow and dry. The bacteria passed over the winter on the diseased body and spread through the air. When the growth of the plant is poor, the sunlight is insufficient or drought, the disease is rapidly developed. To prevent and control it, 1 500 times of liquid with 25% trialone wettable powder is used to spray.

(2) Blight. The root and stem are mainly damaged, the injured root system is brown decayed, and the stem cortex is sometimes longitudinally split. When it is wet, pink mold is produced; the vascular with the disease does not change to brown, and the whole plant wilts and dies. Soil germs invade from root, and it is serious in the land with continuous cultivation, excessive nitrogen fertilizer or poor drainage. Controlling methods are as follows: rotation of the fields, selection of disease-free seeds, seedling raising with disease-free soil, reasonable watering and fertilization, removal of diseased plants and sprinkling lime in and around the diseased hole, grafting with black seed pumpkin stock. In the early stage of the disease, 50% carbendazim wettable powder 800 times liquid or 70% methyl thiosulfurin wettable powder 1 000 times liquid could be used for root control, 0.25kg per plant.

(3) Carbon worm disease. The leaf is infected to have yellow-brown round disease spot; the edge is obvious; the outer edge is yellowish, easy to break. When the tem-

Chapter Ⅵ Application of the Integrated Water-Fertilizer Technology to Vegetables

perature and humidity is high, pink pastes are on the diseased spot, and the stem and fruit damage are not serious. The bacteria pass over winters in the soil with the sick plants, and the seeds carry bacteria, causing the incidence of cotyledons. Germs spread through rain water (or dripping water in the greenhouse).Rainy weather, storm and high temperature and humidity are easy for the disease to occur. Controlling methods are as follows: use disease-free seeds or disinfected seeds, and raise seedlings with disease-free soil. Pay attention to ventilation and prevent drainage, using high border and plastic film. 2.5% uracil wettable powder 1 000 times liquid can be sprayed.

(4) Pests. Cucumber pests mainly include aphids and whitefly, which mainly harm leaves, thus the injured leaves will curl. When it is serious, the whole leaf will curl and even the whole plant wilts and dies. The old leaves do not curl after infected, but will die of early dry. Controlling methods are: aphids are controlled by spraying 50% anti-aphid wettable powder 2 000 times liquid. Whiteflies are treated with 25% thiazinone wettable powder 2 500 times liquid.

Section Ⅳ Application of the Integrated Water-Fertilizer Technology to Eggplant

Ⅰ. The Water Demand of Eggplant

Eggplant grows well under the condition of high temperature and high humidity, and the demand for water is large. However, the water requirements at different growth and development stages are different. In the early stage of seedling, under the suitable conditions such as light intensity and temperature, the seedling bed is full of water, which can promote the strong growth of seedlings and the smooth differentiation of flower buds. And it can improve the quality of flowers. Therefore, when raising the seedlings, the soil with strong water content should be selected as bed soil, and the bottom water should be applied at the same time in order to reduce the number of watering after sowing and stabilize the temperature of seedling bed. At flowering and fruit setting stage, because the eggplant is transiting from vegetative growth to reproductive growth, in order to maintain the balance, we should avoid much nourishing. In water management, control is the main job, that is if it is not drought, there is no need of watering. In the fruiting period, the water demand is less before the first eggplant "opened its eye" (fast growing), but after the rapid growth of the first eggplant, the water demand increases gradually, and the water requirement is the largest before the harvest. Eggplant fruiting rate and yield are negatively correlated with rainfall and air humidity at that time. The relative humidity of air is proper when it is 70%~80%, and over 80% for a long time will easily cause diseases. The relative water content of soil should be well kept at 60%~80%, generally not less than 55%, otherwise there will be rigid seedlings and rigid fruits. In production, we should try our best to meet the water demand of eggplant, otherwise it will affect its growth. Lack of water will lead to less, small and rough fruits with poor quality. Eggplant is not tolerant to soils with poor ventilation and wet moist, so it is necessary to prevent the soil from being too wet; otherwise, it is easy for root rotting.

II. The Fertilizer Demand of Eggplant

Vegetables like eggplant are a kind of plants with developed roots, and it has long picking period with high yield and large nutrient absorption. Eggplant has not strict requirements for soil, but the sandy soil is the best, which has rich organic matter, deep soil layer, strong capability of water and fertilizer and good ventilation and drainage, and the pH value of soil is 6.8~7.3.

1. Demand for different nutrient elements of eggplant

(1) The demand for nitrogen fertilizer. Eggplant is mainly harvest for tender fruits, and the effect of nitrogen on yield is especially obvious. When nitrogen fertilizer is sufficient, the plant grows vigorously, the flower bud develops well, and the fruiting rate is high. Eggplant needs nitrogen fertilizer supply from planting to pulling the seedlings. The ability of eggplant to tolerate high concentration nitrogen is stronger than that of tomato after planting, and it is not easy for the plant of eggplants to grow futile because of too much nitrogen.

(2) The demand for phosphorus. Phosphorus has a direct effect on the differentiation and development of the flower bud. Sufficient phosphorus fertilizer at seedling stage is not only beneficial to root development, but also can be favorable for good flower bud differentiation. If the phosphorus is insufficient, it will cause the flower bud to grow

slowly or not grow, or to form the flower that can not bear fruits. After blooming and fruiting, the total demand for nitrogen and potassium increase, but the absorption of phosphorus begin to decrease. Too much phosphorus would harden the pericarp and affect the fruit quality.

(3) The demand for potassium. Although potassium had no significant effect on flower bud's growth, lack or insufficient potassium or excessive potassium could delay the flower bud differentiation. Before the middle growth stage of eggplant, the demand for nitrogen and potassium of eggplant is basically the same, and the demand for potassium increases obviously at the productive fruiting stage. Potassium deficiency in the whole period of eggplant will affect the yield, so the application of potassium should be paid attention to in the whole growth.

(4) The demand for magnesium and calcium. The demand for magnesium in eggplant begins to increase obviously after fruiting. Magnesium deficiency could make the flower bud grow slowly or not grow, or form unfruiting flowers. The color near the main vein of leaves turns yellow from green and the leaves fall early, which affect the yield. If the soil is too wet or contains too much nitrogen, potassium and calcium, it will cause magnesium deficiency, which shows that the fruit or leaf reticular vein browns and produces rusts. Eggplant is not as sensitive to calcium deficiency as tomato is.

2. The absorption of nutrients by eggplant at different growth stages

The absorption of nutrients in eggplant seedling stageis not large, but it is very sensitive to the abundance or shortage of nutrients. The nutrient supply affects the growth of the seedlings and flower bud differentiation. From seedling stage to flowering and fruiting stage, the absorption of nutrients increases gradually, and after the harvest of fruit, the need for nutrition is the greatest, that is, the absorption of nitrogen and potassium increases sharply, and the absorption of phosphorus, calcium and magnesium also increases, but not as obvious as potassium and nitrogen. The absorption characteristics of eggplant to various nutrients are also different. Nitrogen is very important to eggplant at all growth stages, so any period of nitrogen deficiency will have a negative effect on the flowering and fruiting. The absorption of potassium is similar to that of nitrogen till the middle of growth. To the picking period, the absorption of potassium is significantly higher. In the full productive period, the absorption of nitrogen and potassium increases, so application of nitrogen, phosphorus and potassium can promote each other.

Generally, to produce 1 000kg eggplant, we need 3.65kg of nitrogen, 0.85kg of phosphorus, 5.75kg of potassium, 1.8kg of calcium and 0.4kg of magnesium. The base fertilizer of eggplant should be organic fertilizer, and nitrogen, phosphorus and potassium fertilizer should be applied properly at the same time. In the former period of the fruiting, attention should be paid to the use of nitrogen and phosphorus fertilizer, and in the latter period of the fruiting, the combined application of nitrogen, phosphorus and potassium, and the proper application of magnesium, calcium and other micro fertilizers.

III. Integrated Water-Fertilizer Cultivation Techniques of Eggplant

1. Variety selection

The environmental characteristics inthe greenhouse in winter and spring are low temperature, weak light, poor ventilation and high humidity. Therefore, such varieties should be selected as compact plant type, strong cold tolerance and disease resistance, strong continuous fruiting ability, and dense and tender pulp. The main cultivation varieties are Qingxuan long eggplant, Shenqie 1, Qiqie 1, Qizaqie 2, green eggplant, Yuanza 2, Tianjin Kuaiyuan eggplant and so on.

2. Stubble arrangements

In North China, overwintering stubble cultivation is generally selected, raising seedlings from late August to early September, planting in the first and middle ten days of October, the calendar seedling age being 50~55 days; harvest in the first and middle ten days of December, and pulling the seedling in late June of the following year.

3. Planting

(1) Plowing and fertilizing the ground. Eggplant should avoid continuous cropping, and after harvest of the previous crop, deep plough and expose to the sun. The base fertilizer is generally applied decomposed organic fertilizer 5 000 kg, urea 46kg, superphosphate 100kg, potassium sulfate 30kg per $667m^2$. 2/3 of the base fertilizer is scattered and 1/3 is applied in the planting ditch.

(2) Planting method and density. The planting time should be decided at the beginning of a sunny day after a cloudy day. When planting, first dig trench according to

Chapter VI Application of the Integrated Water-Fertilizer Technology to Vegetables

50cm narrow row, 70cm wide row alternately, the trench depth is 5~6cm, and then lay the seedlings with the distance of 45cm between the plants, cover soil and water it. When the soil is in dry-wet condition, intertill and loosen the soil. Make the height of ridge 4cm higher than that of the field, so that the seedling line forms a ridge platform of about 10cm, and then cover the ridges with a wide plastic film of 80~90cm between the two ridges, the seedling is made out of the film at the openings, and the irrigation under the film is carried out. More than 2 500 seedlings are planted per 667m^2.

4. Control of major diseases and insect pests

Eggplant diseases in greenhouse are mainly greensickness, cotton blight, brown grain disease and so on. The main pests are cotton red spider and tea yellow mites.

(1) Greensickness (Verticillium wilt). The seeds could be soaked with 0.2% of 50% carbendazim wettable powder for 1 hour before sowing. Spray 50% carbendazim wettable powder 600~700 times at seedling stage and before planting. After planting and at the beginning of the disease, water the field with 15% mixed copper, zinc, manganese and magnesium 500 times liquid, or water the roots with 2% pyrimidine nucleosides antibacterial agent 200 times liquid. 150~250 g per plant, once every 10 days for twice in succession.

(2) Cotton blight. Also called Diaodan (eggplants drop), rotten eggplant, Shuilan (rotten eggplant by water), is one of the important diseases of eggplant. At the beginning of the disease, spray 75% chlorothalonil wettable powder 500 times liquid, 64% oxo frost manganese zinc wettable powder 400 times liquid, or 90% ethyl aluminum manganese zinc wettable powder 600 times solution, once every 7~10 days. It can also be controlled with chlorothalonil smoke detergent or dust agent.

(3) Brown grain disease. Spraying 75% chlorothalonil wettable powder 600 times liquid, or 40% metalaxyl wettable powder 600~700 times liquid, or 64% oxo frost manganese zinc wettable powder 500 times liquid. The three liquids can be used alternately, once for about 10 days depending on the weather and disease condition. Prevention and treatment are carried out for 3~4 times in succession.

(4) Pests. To kill the red spiders, spray with 25% acaricidal wettable powder 1 000 times liquid, spraying 2~3 times in succession. To treat the tea yellow mites, spray with 73% propargite 2 000~3 000 times liquid, spraying once every 7~10 days for 3 times in succession.

Section V Application of the Integrated Water-Fertilizer Technology to Summer Squash (Zucchini)

I. The Water Demand of Zucchini

Zucchini, also known as Jiaogua, is native to arid tropical areas. The root system of zucchini is developed, the number of lateral roots is large, the absorption ability is strong, and the drought tolerance is strong. The root system of zucchini is mainly distributed in the deep plowing soil, but generally the common cultivated layer is shallow, with limited ability of water storage and fertilizer storage, so it is often easy to dry and lack fertilizers; and the stems and leaves of zucchini are luxuriant; the leaves are large and many; the transpiration is strong, and the water consumption is large. Therefore, it is necessary to strengthen irrigation and maintain suitable water content of the soil in order to obtain high yield.

Zucchini requires comparatively dry air conditions. In greenhouse, the air humidity must be adjusted by reducing water evaporation and increasing the ventilation on the ground, and the suitable air relative humidity is 45%~55%. Virus disease is easy to occur under high temperature and drought; powdery mildew is easy to occur under high temperature and high humidity. To plant zucchini in greenhouse, we should be more careful to control temperature and humidity to prevent the occurrence and spread of virus disease, powdery mildew and other diseases.

II. The Fertilizer Demand of Zucchini

The root system of zucchini is developed, the depth of themain root of sowing seed being more than 2 meters, the depth of the main root of seedling transplanting is more than 1 meter, and the lateral expansion range of the root is about 2 meters. The branching ability of lateral roots is also very strong, and most of the lateral roots are distributed in the deep plowing layer of 30cm. Zucchini is tolerant to drought, barren soil; soil requirements are not strict, so clay, loam, sand loam are all suitable for cultivation. However, because of its developed roots, it is appropriate to select deep soil layer, with fertile and loose sandy soil, so as to maintain strong growth potential and absorption capacity of roots at low temperature, and to harvest earlier and prolong the fruiting period. The suitable soil pH value is 5.5~6.8.

Zucchini needs different fertilizer types and nutrient ratios at different growth stages. In the early stage, the plant grows slowly and absorbs less nutrients. From seedling to fruiting, sufficient nitrogen should be supplied to promote plant growth and lay the foundation for fruit growth. Zucchini has a good response to manure and compost, so more organic fertilizer should be applied as base fertilizer in production. The application of high quality organic fertilizer and ternary compound fertilizer and proper control of the amount of nitrogen fertilizer are beneficial to balance vegetative growth and reproductive growth and increase yield. If the amount of nitrogen fertilizer is too large, it is easy to cause the stem and leaf to grow, resulting in the occurrence of falling flowers and fruits and diseases.

Zucchini has strong ability to absorb fertilizer, and potassium is the most, followed by nitrogen, calcium and phosphorus. For the production of 1 000kg zucchini fruit, it is necessary to absorb pure nitrogen 3.92~5.47kg, phosphorus 2.13~2.22kg, potassium 4.09~7.29kg, calcium 3.2kg, magnesium 0.6kg. The amount of nitrogen, phosphorus and calcium required by zucchini is higher than that of cucumber, except the requirement of potassium.

III. Integrated Water-Fertilizer Cultivation Techniques of Zucchini

1. Variety selection

The environmental characteristics inthe greenhouse in winter are low temperature, weak sunlight, poor ventilation and high humidity. So Zucchini in overwintering greenhouse should be varieties with disease resistance, low temperature and low light resistance, strong resistance to stress and good quality and high yield, such as the Zaoqing First Generation, Zhonghu No.3, Dongyu Zucchini, cold jade zucchini (Hanyu) and so on.

2. Stubble arrangements

In North China, overwintering stubble cultivation in greenhouse is generallyapplied, sowing and raising seedlings in the first and middle ten day of October, planting from late October to early November, picking in late November, and removing seedlings in late May of the following year.

3. Seedling raising techniques

(1) Seed treatment.

①Bathing the seeds in the sun. Before sowing, seed selection is carried out, and new seeds with luster, full grain, no disease spot, no insect injury and no mildew are selected. The selected seeds are bathed in the dustpan under the sunshine for 1 to 2 days in order to improve germination rate, germination potential, and speed the germination and seedlings. At the same time, it also has the effect of sterilization and disinfection. If old seeds are used for sowing, it is particularly important to sunbathe the seeds before sowing.

②Seed disinfection. Soak the seeds with 10% trisodium phosphate solution for 20 minutes, or with 50% carbendazim wettable powder 500 times for 30 minutes, then rinse with clean water and soak the seeds with warm water.

③Soaking the seeds for germination. 400~500g seed is required per $667m^2$. Put warm water of 50~55℃ in the container, and put the seeds in the water. Then continuously stir the seeds, and stop stirring when the water temperature drops to 30℃ and soak for 3~4 hours. Then seeds are taken out of the water, spread and dry for 10 minute, wrap them with clean wet cloth, then place it at 28~30℃ to stimulate for bud. 1~2 days later, when about 70% of the seeds germinate, it is ok for sowing.

(2) Sowing.

A flat bedfor seedling with a width of 1.2m and a depth of 10cm is built in the greenhouse. Make the nutrition soil with 60% fertile field soil and 40% decomposed farm fertilizer, which are mixed and screened. Each cubic meter of the nutrition soil should be added with the following fertilizer, that is 15kg of saprophytic chicken manure, 2kg of superphosphate, 10kg of grass and wood ash, or 3kg of ternary compound fertilizer, 80g of 50% carbendazim wettable powder. And mix them sufficiently evenly. Put the prepared nutritious soil in the nutrition bowl or paper bag, and the nutrition bowl is arranged closely on the seedling bed. You can also buy commodity matrix. Pour the bottom water before sowing, sow the seeds in the nutrition bowls, and cover with dry and fine soil for 1.5~2cm thick after sowing. Finally the seedling bed is covered with plastic film or with arch film.

(3) Seedling management.

Chapter Ⅵ Application of the Integrated Water-Fertilizer Technology to Vegetables

①Temperature management. The temperature of seedling stage is kept at 25~28℃ during day and 15~20℃ at night, and the temperature in cloudy day is lower.

②Light management. When thesunlight is strong, the sun-shading net shall be covered and the water should be sprayed for cooling. The seedlings shall be provided with sufficient light conditions as far as possible after the seedlings are excavated out of the soil, so as to prevent the insufficient sunlight from causing it to grow in vain.

③Water management. The bed soil is kept wet during the seedling, and then it is appropriately watered in the future according to the soil moisture condition.

④Exercising the seedlings. One week before planting, we should not water it but enhance ventilation to exercise the seedling with low temperature. This is beneficial to shorten the seedling survival stage. When the young seedlings have 3 leaves 1 heart, and plant height is 12~15cm, they can be planted.

4. Prevention and control of main diseases and pests

The main diseases of zucchini in greenhouse aredamping-off disease, powdery mildew, gray mold, blight disease and so on. The main pests are aphid, whitefly, red spider, spotted fly and so on.

(1) Damping-off disease. In the early stage of the disease, spray 64% oxocream manganese zinc wettable powder 500~600 times liquid, or 72.2% downy mildew water agent 5 000 times solution, or 15% oxamol water agent 450 times liquid.

(2) Powdery mildew. At the beginning of the disease, spray 25% triadimefon wettable powder 2 500 times solution, 40% fluorosilazole EC 8 000~10 000 times solution, or 15% pyrimidine EC 2 000~3 000 times solution to control the disease. It can also be sprayed with 4% pyrimidine nucleoside antibiotic water agent 600~800 times.

(3) Gray mold. At the beginning of the disease, spray 40% methylpyrimidine suspension 1 200 times liquid, or 65% methyl thiocarbamate wettable powder 1 000~1 500 times solution, or 50% procymidone wettable powder 1 000~2 000 times solution.

(4) Blight disease. At the beginning, spray 58% methyl cream manganese zinc wettable powder 750~1 500 times liquid, 90% aluminum triethylphosphonate wettable powder 500 times liquid, or 20% thiobacillus copper suspension 400~600 times liquid. Or spray 1 million units of new phytomycin wettable powder 2 000~3 000 times liq-

uid.

(5) Aphids. The aphids are sprayed with 10% imidacloprid wettable powder 2 000~3 000 times liquid, or 3% acetamiprid EC 1 500~2 000 times liquid, or 0.3% azadirachtin EC 1 000~1 500 times liquid.

(6) White fly, red spider, spotted fly. In the initial stage of damage, spay 10% bifenthrin EC 4 000~8 000 times liquid, 1.8% avermectin EC 3 000~4 000 times, or 20% fenvalerate EC 1 500~2 000 times liquid, or 0.3% azadirachtin EC 1 000~1 500 times liquid.

Section VI Application of the Integrated Water-Fertilizer Technology to Watermelon

Watermelon belongs to cucurbitaceae, native to tropical desert areas of South Africa. It is a heat-resistant crop, loving sunlight, tolerant to high-temperature and dry. Fruit belongs to gourd, sweet and juicy, rich in vitamins, minerals and sugar. It is an important summer fruit, widely cultivated in China. The planting area of watermelon in China accounts for more than 55% of the total area of the world, and the total output accounts for more than 70% of the total output of the world. Watermelon occupies an important position in the world horticulture industry, whose production scale ranks fifth, after grapes, bananas, oranges and apples.

Watermelon grows fast and needs to supply nutrients in time, but in watermelon planting, farmers mostly use traditional method of flood irrigation, so blind and excessive fertilization not only result in a large amount of waste of fertilizer and water, but also it is easy to pollute and acidify the soil, reduce the fertilizer utilization rate, and seriously affect the yield and quality of watermelon. Therefore, integrated irrigation techniques such as drip irrigation, infiltration irrigation, sprinkler irrigation and others should be used to realize high quality and high yield of watermelon. In addition, watermelons are mostly planted in light or sandy soils, which tend to have poor capacity of conserving fertilizer and water. The application of integrated water-fertilizer technology with a small amount but many times can solve this problem. The research on drip irrigation and micro-spraying of watermelon in China began in the 1970s. In recent years,

Chapter VI Application of the Integrated Water-Fertilizer Technology to Vegetables

many watermelon growers have adopted the integrated technology of drip irrigation and fertilizer to manage water and fertilizer, and the effect has been remarkable. More and more attention has been paid to its application.

I. The Laying of Drip Irrigation Pipes

For watermelon drip irrigation, when laying network, the water supply pipe is generally three-stage, that is, main pipe, branch pipe and drip irrigation capillary, and the flow rate of the capillary dripper is 2.8 liters per hour, and the distance between drippers is 30cm. The water inlet is connected with the outlet of the pump, and the water supply pipe is equipped with a four-way joint with switch at the corresponding place of the melon line, which is connected directly to the water supply pipe, and each side is connected with a dripping pipe, using a plastic film with the width of 90cm. One dripping capillary is laid under each film, the distance between the two adjacent capillaries is 2.6 m, and the amount is 390 meters per hm^2. Once the dripping pipe is installed, use bamboo every 60cm to make a bow, and the distance is 0.5cm between the top of the bow and the dripping pipe filled with water, which is beneficial to the fact that the film is not close to the drippers and the sediment does not block the outlets of the drippers after covering the film. Finally cover with plastic film, and add a small arch shed to protect against cold in spring.

II. Water Management

The suitable irrigation mode of watermelon isthe spraying belt under the film, and there is also drip irrigation under the film and so on. Usually for one line of watermelon installs a spraying belt, the holes face up, and covers the field with film. The sand soil is loose and the requirement for water is not high, but the water on clay soil should be small to prevent surface runoff. The diameter of the spraying belt is related to the length of the spraying belt, and it is better when the evenness of effluent of the whole belt is 90%. If the spacing is 40~50cm, the flow rate can be 1.5~3.0L per hour. And for the sandy soil, large flow drippers are selected, and for the clay smaller ones. Watermelon irrigation applies the principle of "abundant water in the middle and controllable water at the two ends". The irrigation amount can be controlled by irrigation time, and the length of irrigation time should be determined with the weather, plant growth and other factors. In the whole growth period of watermelon, the dripping times is 9~10, and the amount of water is 350~400m^3 per hm^2. After sowing, the dripping water is 20m^3 per hm^2. Seedling water should be sufficient, soak the sowing field to ensure that the soil moisture is connected to the soil moisture, and the amount is 45m^3 per mu. After the seedlings came out, harden the seedlings according to soil moisture. When the main vine grows to 30~40cm, drip once and the rate is 40m^3 per mu. From flowering to fruiting expansion, a total of 6 times are dripped, once every 5 to 7 days, each dripping amount is 40m^3 per mu, during which the water requirement at fruiting stage is much bigger about 45~50m^3 per mu, and in the fruit expansion period, the water is kept at 50m^3 per mu. In the mature period, drip once, but we should reduce the amount of irrigation in order to ensure the quality and flavor of watermelon. According to the growth of melon vine, it should be kept at 35~45m^3 per hm^2. The rate of irrigation should not be over the ridges, and the dripping should stop 7 days before fruit harvest.

III. Fertilizer Management

The growth stage of watermelonconsists of germination stage, seedling stage, vine extension stage and fruiting stage, and the demand and absorption ratio of nitrogen, phosphorus and potassium are different at different growth stages. The absorption at germination stage is the least, accounting for only 0.01% of the total fertilizer absorption; in the seedling stage, it accounts for about 0.54%; during the vine extension stage, it increases, accounting for 14.67%, and in the fruiting period, the fertilizer absorption is the largest, accounting for about 85% of the total. During the growth period of watermelon, the absorption of the potassium is the highest, that of the nitrogen is the second, and that of the phosphorus is the least. The proportion of nitrogen, phosphorus and potassium was 3.3 : 1 : 4.3 (nitrogen : phosphorus : potassium). Potassium fertilizer should be sufficient to increase the sugar content of watermelons. According to research, to produce 1 000kg fruit, N 4.6kg, P_2O_5 3.4kg, K_2O 3.4kg are needed. During the whole growth period, base fertilizer, seed fertilizer and topdressing fertilizer are mainly applied, and the topdressing fertilizer is mainly for the seedling, vine extension and fruiting. A total of 7 times of fertilization is applied in the whole growth period, and the amount of fertilizer is 35kg per mu. At seedling stage and flowering stage, drip the nutrition fertilizer with water, 75kg per hm^2. After the melon is set,

Chapter VI Application of the Integrated Water-Fertilizer Technology to Vegetables

the nutrition fertilizer is applied 3 times, 60kg per hm^2 each time, and no fertilizer is applied at mature stage. Fertilization should be "enough, fine, skillful", that is, the bottom fertilizer should be sufficient, the seed fertilizer should be fine, and the top-dressing should be skillful.

(1) Seedling fertilizer. The fertilization in the resettling stage, that is, 5-leaf stage, can mainly promotes growth; the vines extend rapidly, assimilation area expands, and lays a material foundation for bud differentiation. In the seedling stage, the range of root system is still small, so a small amount of quick-acting fertilizer should be selected to promote root development and shoot growth. Topdressing for the young seedlings: urea 50g, compound fertilizer 50g, water 50kg; or urea 100g, water 50kg; or compound fertilizer 125g, water 50kg; human feces and urine can be mixed with twice as much water. Topdressing at the middle seedling stage: urea 100g, compound fertilizer 50g, water 50kg; or compound fertilizer 100g, urea 75g, water 50kg. Wheat gluten fertilizer should be soaked in water for half a month in advance; the concentration of fertilizer should not exceed 0.8%, and the human feces and urine can be added with 1/3 water for spraying the seedlings.

(2) Vine extending fertilizer. According to the climate and plant growth, vine extending fertilizer should be applied to promote the rapid growth of stems and leaves, but not to grow in vain. The extension fertilizer is applied before and after the plant's "swinging the dragon head". In the middle of the two seedlings, a topdressing ditch with a depth of 10cm, a width of 10cm and a length of 40cm is opened, and about 500g of high quality fertilizer is applied to each plant, such as decomposed cake fertilizer 100g or decomposed farm fertilizer 500g.

(3) Fruiting fertilizer. It could be applied to the fruit when most of the plants in the field have fruits with the egg size (7 days after the flowers fall), in order to promote the fruit expansion and maintain the growth potential. The furrow is opened at one side of the plant 30~40cm away from the root, the amount of fertilization per mu combined with watering are as follows, compound fertilizer 7.5~10kg, potassium sulfate 5.0~7.5kg, or compound fertilizer 10~15kg. Or spray on the leaf surface with 0.5% urea and 0.1% potassium dihydrogen phosphate. 5~7 days before harvest, stop drip irrigation and fertilization to ensure the normal transformation of pigment and sugar, which is good for the melon not to be rotten during storage and transportation.

Section VII Application of the Integrated Water-Fertilizer Technology to Chinese Cabbage

Chinese cabbage is native to the north of China, commonly known as Dabaicai. It is introduced to the south and cultivated in all parts of the north and south. In the Yellow River Basin it can be cultivated in three seasons: spring, summer and autumn stubbles in one year, and in the northeast it can also be cultivated in spring and autumn. In the Qinghai-Tibet plateau and the northern region of Daxinganling, it can be cultivated only once a year, while in the southern China it can be cultivated annually. At present, it is mainly cultivated in autumn in most of our country. There are also be cultivated with facilities for the cabbage oversummerring.

I. Chinese Cabbage Irrigation Type

Chinese cabbage is mostly cultivated in open field, and the application ofintegrated water-fertilizer technology is less. The most suitable method is micro-sprinkler irrigation. Micro-sprinkler irrigation can be divided into mobile sprinkler irrigation, semi-fixed sprinkler irrigation and fixed sprinkler irrigation. In areas with plenty of water sources (furrow water storage), ship-type sprinkler is used. In some farms, drip irrigation pipe was used, the distance between the drippers is 20~30cm, the flow rate is 1.0~2.5 liters per hour, and thin-wall irrigation belt is used.

The water spray diameter of the micro sprinkler is generally 6 meters. In order to maintain the evenness of irrigation, use the circular overlap method of the spray area, setting the installation distance at 2.5 meters, so that the water spray area of the two adjacent sprinklers can overlap each other.

II. Water Management of Chinese Cabbage

1. The law of water demand of Chinese cabbage

The leaves of the Chinese cabbage are many, the cutin on the leaves is thin, and the amount of water transpiration is very large. The water demand of Chinese cabbage is different at different growth stages. At seedling stage, the soil water content is 65%~

Chapter VI Application of the Integrated Water-Fertilizer Technology to Vegetables

80% (soil wetting). The rosette stage is the fastest growing period for leaves, but the water demand is less, and the general soil water content is 15%~18%. The heading period is the period in which Chinese cabbage needs the most water, so it is necessary to keep the water content 19%~21%, and when the water is insufficient, it needs irrigation.

2. Water management of Chinese cabbage

The water demand of Chinese cabbage at germination stage and seedling stage is less, but there should be enough water for seed germination; the root system is weak and shallow at the seedling stage, and when it is dry drought, it should be watered in time to keep the ground moist, in order to benefit the seedlings to absorb water and control excessive surface temperature to burn roots. In the rosette period, much water is needed. We should master the dampness to promote and control the growth of rosette leaves. During the heading period, the water demand is the most, so water should be irrigated at the right time. After the heading, water should be controlled to facilitate storage.

(1) Experientialmethod. Soil water content can be judged by experience in practice. For example, as to the loam and sandy loam, clench by hand to form a soil ball, and then press it. If it is not easy to break up, it indicates that soil moisture is about 50% of the maximum water capacity, so it generally needs not to be irrigated; if it is pinched and the soil ball can not be formed, and it is easy to break up, this proves that the moisture content is low, so irrigation should be applied in time. In the drought period in summer and autumn, the irrigation period can also be determined according to the weather conditions, generally if high temperature and drought are more than 15 days, irrigation need to be applied, and in autumn and winter the drought period can last more than 20 days before irrigation.

(2) Tension meter method. Chinese cabbage is a shallow-root crop, and most roots are distributed in 3cm soil layer. When the moisture is measured by tension meter, the meter can be buried 20cm deep in the soil layer of vegetable garden. When the soil moisture is kept at 60%~80% of the field capacity, that is to say, the soil tension is at 10~20cm, it is beneficial to the growth of cabbage. More than 20cm indicates that the soil becomes dry and needs irrigation till the tension meter is back to zero. When drip irrigation is used, the tension meter is buried directly below the dripper.

(3) Watering in time. The root water should be filled in time after planting, and

the soil is intertilled for 1 and 2 times, then the soil is properly irrigated according to the weather conditions to keep the soil moist. The irrigation time is 3~4 hours, the soil wetting layer is 15cm, and the sprinkler irrigation time is usually chosen in the morning or afternoon. At this time, the ground temperature can rise rapidly after irrigation. The time and interval of water spraying can be determined according to the different growth period and water demand. Chinese cabbage can be sprayed several times from the resettling stage to rosette stage, but the water can be controlled for several days at the end of rosette. When Chinese cabbage enters the heading stage, it needs the most water. Therefore, as soon as the hardening of the seedling is finished, it is necessary to spray once, and the sprinkler time is 3~4 hours. And then spray for the second time again after 2~3 days, and then spray water once for every 5~6 days to make the soil wet. And the amount of water to be irrigated in the earlier stage is less than that of the later stage, which can ensure high yield.

III. Nutrient Management of Chinese cabbage

1. The law of fertilizer requirement of Chinese cabbage

Chinese cabbage is growing rapidly, the yield is very high, and the demand for nutrients is higher, as shown in Table 6-1. To produce 1 000kg, 1.3~2.5kg of nitrogen, 0.6~1.2kg of phosphorus pentoxide and 2.2~3.7kg of potassium oxide should be absorbed by Chinese cabbage. The approximate proportion of the three main elements is 2.5 : 1 : 3. It can be seen that the absorption of potassium is the most, followed by nitrogen, phosphorus is the less.

Table 6-1 Absorption Amount of N, P, K with Different Yield

Yield (kg/mu)	Nutrition Absorption Amount (kg/mu)		
	N	P	K
5 000	12.1	1.6	8.6
6 000	14.5	2.3	9.3
8 000	19.3	2.9	10.4
10 000	22.5	3.4	13.9

The nutrient requirements of Chinese cabbageare significantly different in each period. In the general seedling stage (about 31 days from sowing), the nutrient

Chapter VI Application of the Integrated Water-Fertilizer Technology to Vegetables

absorption is less, the nitrogen accounting for 5.1%~7.8%, phosphorus accounting for 3.2%~5.3%, potassium accounting for 3.6%~7.0%. When entering the rosette stage (about 31~50 days from sowing), the growth of Chinese cabbage accelerates and the nutrient absorption increases rapidly, the nitrogen accounting for 27.5%~40.1%, phosphorus accounting for 29.1%~45.0%, and potassium accounting for 34.6%~54.0% of the total. At the beginning and middle of the heading period (about 50~69 days from sowing), it is the fastest growing period with most nutrient absorption, the nitrogen accounting for 30%~52%, phosphorus accounting for 32%~51%, potassium accounting for 44%~51% of the total. From the late heading stage to the harvest period (about 69~88 days from sowing), the nutrient absorption decreases obviously, the nitrogen accounting for 16%~24%, the phosphorus accounting for 15%~20%, and the potassium accounting for less than 10% of the total. It can be seen that the period when Chinese cabbage needs the most fertilizer is the rosette stage and the early stage of heading, and it is also the crucial period of yield formation and high quality management. Special attention should be paid to fertilization.

2. Nutrient management of Chinese cabbage

Chinese cabbage is prohibited to rotatewith cruciferae vegetables, and the former crops are the best as watermelon, cucumber, beans, Chinese onions, garlic and rice. The growth period of Chinese cabbage in autumn and winter is long, the yield is high, and the amount of fertilizer required is large. In the process of growth, it experiences seedling stage and rosette stage, then enters the heading period. Fertilization methods are different in different growth stages. At germination stage, the growth amount is small and the speed is fast, and the nutrients needed are mainly supplied by seeds and do not need to be fertilized. At seedling stage, in order to ensure the rapid growing of the seedlings, it is necessary to apply seedling fertilizer, urea or ammonium phosphate 30~60kg hm^2. The growth rate of rosette stage is faster, the growth amount is gradually increased, and the absorption of nutrients and water is increased. So in rosette stage, fully fertilizing and watering is the key to ensure the robust growth and high yield. If drip irrigation is adopted, we should observe the principle of "watering in need". In rosette period nitrogen fertilizer of 100~150kg hm^2 should be applied, and the growth in the heading stage is the largest. In this period, 175~300kg per hm^2 can be fertilized, together with 100kg potassium sulfate and calcium superphosphate respectively. We should keep the soil moist all the time, maintaining dripping irrigation for 1 hour or sprinkling irrigation for 30 minutes every day. The general fertilizer absorption characteristics of Chinese cabbage are as follows: less nutrients are absorbed at seedling stage, and the absorption of nitrogen, phosphorus and potassium is less than 1% of the total absorption; and in the rosette stage, it is obviously increased, which accounts for 30% of the total; in the heading period, it absorbs the most nutrients, accounting for about 70% of the total. During the growth period, the absorption of potassium is the largest, followed by nitrogen and calcium, and the absorption of phosphorus and magnesium is smaller. To produce 1 000kg cabbage, it needs about 1.82kg N, 0.36kg P_2O_5, 2.82kg K_2O, 1.61kg CaO, and 0.21kg MgO, and the ratio is 5 : 1 : 7.8 : 4.5 : 0.6.

Section VIII Application of the Integrated Water-Fertilizer Technology to Lettuce

I. The Water Demand of Lettuce

Lettuce is a genusLactuca of Compositae, herbal plant for 1~2 years, which is divided into two categories: leaf use and stem use. Lettuce, also known as spring lettuce and shengcai, began to be developed in Beijing and some coastal cities in the late 1980s. Leaf lettuce has many leaves, with large leaf area and large transpiration, so it consumes more water and demands more water. The growth period of lettuce is about 65 days; the water demand is about 215m^3 per 667m^2, and the average daily water demand is 3.3m^3. Leaf lettuce has different needs for water at different growth stages. When seeds germinate, it is necessary to keep the soil moist in the seedling bed in order to facilitate seed germination. In the young seedling period, the watering should be properly controlled, the soil remaining dry and wet. Too much soil moisture will make the seedlings easy to grow, and lack of moisture will make the seedling aging. At the resettling stage, proper hardening of the seedlings should be applied to promote root growth. In the heading stage, enough water should be supplied. Lack of water will lead to loose heading or no heading. At the same time, lactucin in the plant increases and the bitter taste is aggravated. In the later stage of heading, there should not be too much watering to prevent the occurrence of breakage and diseases such as soft rot disease and sclerotinia.

II. The Fertilizer Requirement of Lettuce

The absorption capacity of lettuce's roots is weak, and the demand for oxygen is high. The root growth in the sandy or clay cultivation is poor. Therefore, loam or sandy loam with high organic content and good permeability should be selected for lettuce cultivation. Lettuce likes slightly acidic soil, the suitable soil pH value is about 6. The pH value higher than 7 or lower than 5 is not conducive to the growth and development of lettuce.

The growth period of leaf lettuce is short; the edible part is leaf. The demand for nitrogen is large; the whole growth period requires sufficient nitrogen. At the same time, phosphorus and potassium should be applied together, and potassium should be fully supplied at the heading stage. At the beginning of lettuce growth, the fertilizer demand is less. With the increase of growth, the demand for nitrogen, phosphorus and potassium increase gradually, especially at the heading stage. During the whole growth period of lettuce, the demand for potassium is the highest, followed by nitrogen, and phosphorus is the least. To produce 1 000kg leaf lettuce, we need pure nitrogen 2.5kg, phosphorus 1.2kg, potassium 4.5kg, calcium 0.66kg, magnesium 0.3kg.

Nitrogen deficiency at seedling stage could inhibit leaf differentiation and decrease the number of leaves, and nitrogen deficiency at rosette andheading stage has the greatest effect on yield. The lack of phosphorus in seedling stage not only leads to small number of leaves, but also causes a short plant and a decrease in yield. Potassium deficiency affects the heading of lettuce, with loose bulb, light leaf, decreased quality and yield. Leaf lettuce also needs calcium, magnesium, boron and other trace elements. Calcium deficiency often causes edge dry of the leaf, commonly known as dry heartburn, resulting in the bulb decay. Magnesium deficiency leads to the loss of green leaves. Trace elements in production can be supplemented by leaf fertilization, usually in the rosette stage, but the effect of spraying in the later stage is poor.

III. Integrated Water-Fertilizer Cultivation Techniques of Lettuce

1. Variety selection

Inchoosing the autumn and winter stubble lettuce in greenhouse, such lettuce with high quality, disease resistance and strong adaptability should be selected as "five lakes" lettuce, Great Lake 659, Santa Lina, American PS and so on.

2. Stubble arrangements

In North China, the stubble lettuce in greenhouse is generally cultivated in autumn and winter, and most of them are planted by seedlings. The general seedling age is 30-35 days, and the harvest can be obtained 60~65 days after planting. In the middle and last ten days of September, the seeds are sowed for seedlings, and the lettuce can be planted in late October. The following year at New Year's Day it will supply the market.

3. Sowing and seedling raising

(1) Preparing the seedling bed. The fertile sandy loam with good fertilizer capacity should be selected for the preparation of seedling bed, and the soil should be fully exposed to the sun after deep plough. Seedling bed is fertilized per $10m^2$ with 10kg high quality decomposed farm fertilizer, 0.3kg ammonium sulfate, 0.5kg superphosphate, 0.2kg potassium chloride. All the fertilizers are fully mixed with soil, and the bed surface is smoothed ready for sowing.

(2) Soaking seeds for germination. The seeds are wrapped in gauze and soaked in water at 20℃ for 3 hours. After taken out of water, the seeds are germinated at 15~20℃ for 2~3 days, and the seeds could be sowed after they showed white bud spot.

(3) The sowing method isbroadcast sowing, and the seedling rate of the greenhouse per $667m^2$ needs $50m^2$ seedling bed, and the seed amount is 30g. In order for the even sowing, proper amount of fine soil particles can be mixed with the seeds when sowing, and the soil is 0.5~1cm thick after sowing. The film can be covered on the seedling bed in a low temperature season so as to remain warmth and moisture. After sowing, the temperature of the seedbed is kept at 20~25℃; after the budding, the temperature of the seedling bed is kept at 18~20℃ in the day time, 8~10℃ at night; and the seedlings can be planted when it has 4 leaves 1 heart.

4. Planting

Before planting, the fertilizerused per 667 m^2 is 4 000kg organic fertilizer, 30~

Chapter Ⅵ Application of the Integrated Water-Fertilizer Technology to Vegetables

40kg ternary compound fertilizer. And the plough depth is 25cm. Then make the border, the width being about 1m, and most of them are cultivated in a flat border. The cultivation density depends on the variety, and generally the spacing of the plants and rows is 30cm ×40cm. For early maturing varieties, the plant's expansion degree is small, so it can be planted a little closer, the plant and row spacing being 30cm × 30cm. The middle and late maturing varieties have a large extension, and the plant and row spacing is 35cm ×40cm. The planting should be fixed with soil lump, the planting depth should be flat with the ground, and the watering must be in time after planting to promote the seedling survival.

5. Environmental management after planting

At the seedlingsurvival stage, the heat preservation is strengthened; the closed shed is not ventilated; the daytime temperature is $20\sim25°C$; the night temperature was above $10°C$; the 10cm ground temperature is above $15°C$. In the early resettling stage, the water should be properly controlled and ventilation should be increased, and the temperature should be kept at $15°C$ during the day and $10°C$ at night. In the early heading stage, increase the water and fertilizer, timely intertill and weed. And if it is drought, water in time to maintain the balance of soil water content, otherwise it is easy to for the fruit to crack and affect the quality. At the later stage of heading, increase ventilation, keep the soil moist, and water in time to prevent cracking.

6. Prevention and control of main diseases and insects.

The main diseases of lettuce are softrot disease, downy mildew, sclerotinia and gray mold, and the main pests are aphids, ground tigers and so on.

(1) Soft rot disease. At the beginning of the disease, 72% streptomycin sulfate soluble powder 200~250mg/L solution could be sprayed once every 7~10 days for 2~3 times.

(2) Downy mildew. 25% mancozeb wettable powder or 25% metalaxyl wettable powder 600 times liquid, or 75% chlorothalonil wettable powder or 65% zincin wettable powder 600 times liquid could be sprayed once every 7~10 days for 2~3 times.

(3) Sclerotinia. It can be sprayed with 50% carbendazim wettable powder 600 times liquid, or 50% procymidon wettable powder or 40% carbendazim wettable powder 1 000~1 500 times, once every 7~10 days, twice a day, 2~3 times in succession.

(4) Grey mold. It can be sprayed with 50% procymidon wettable powder 1 000~

1 500 times, or 50% thiophylline wettable powder or 50% carbendazim wettable powder 500~600 times, once every 7~10 days, 2~3 times in succession.

(5) Aphids. It can be controlled by 10% imidacloprid wettable powder 1 500 times liquid.